Nonnegative Matrices, Positive Operators,
AND APPLICATIONS

Jiu Ding
The University of Southern Mississippi, USA

Aihui Zhou
Chinese Academy of Sciences, China

NONNEGATIVE MATRICES,
POSITIVE OPERATORS,
AND APPLICATIONS

World Scientific

NEW JERSEY · LONDON · SINGAPORE · BEIJING · SHANGHAI · HONG KONG · TAIPEI · CHENNAI

Published by

World Scientific Publishing Co. Pte. Ltd.

5 Toh Tuck Link, Singapore 596224

USA office: 27 Warren Street, Suite 401-402, Hackensack, NJ 07601

UK office: 57 Shelton Street, Covent Garden, London WC2H 9HE

British Library Cataloguing-in-Publication Data
A catalogue record for this book is available from the British Library.

Cover image by Victoria Ding.

ISBN-13 978-981-283-917-6
ISBN-10 981-283-917-8

Printed in Singapore.

Dedicated to our respective families

Preface

The theory and methods of nonnegative matrices and positive operators have found more applications in science, engineering, and technology during the past decades. Traditionally, nonnegative matrices have direct applications in Markov chains, game theory, iterative methods for solving linear algebraic equations, and etc. A recent example of nonnegative matrices is the so-called Google matrix, the world's largest one, for the computation of the PageRank used by the Google Web search engine [Bryan and Leise (2006); Langville and Meyer (2006)]. The Google matrix is a stochastic matrix, that is, it is a nonnegative square matrix each row of which sums to be one. Because of its huge size presently in the order of at least 10 billions, seeking faster and more efficient numerical algorithms for the PageRank computation of the Google matrix is a current research topic.

On the other hand, mathematical modeling of complicated physical and social systems often leads to the statistical study of deterministic dynamical systems, which may be the main practical and reliable approach to obtaining a meaningful and reasonable solution to the problem [Albert and Barabasi (2002); Dorogovstev and Mendes (2002); Newman (2003); Ding and Zhou (2009)]. For example, for a given chaotic discrete dynamical system $S : X \to X$ on a state space X, instead of observing the deterministic properties of the individual sequences of the iterates which exhibit sensitive dependence on initial conditions, one can consider the probabilistic distribution of such eventually unpredictable or random-like iterates. Functional analysis plays a key role for the mathematical investigation of the dynamics from the global point of view. In particular, infinite dimensional positive operators are introduced to describe the evolution of the probability density functions associated with the underlying dynamical system.

An extremely important class of positive operators is that of Markov operators which transform probability density functions into themselves. Markov operators appear in the stochastic analysis of deterministic dynamical systems. They are widely used in Markov processes, stochastic differential equations, and applied probability. A subclass of Markov operators consists of Frobenius-Perron operators that describe the evolution of statistical quantities associated with measurable transformations. There are two important and related problems concerning Frobenius-Perron operators. One is about the existence of stationary density functions of the operator, which give absolutely continuous invariant probability measures that describe the statistical properties of the dynamics, such as the probability distribution of the sequences of the iterates for almost all initial points and the rate of the decay of correlations of such sequences. The other problem, which is more important in applications, is about efficient computation of such invariant probability measures to any prescribed precision. The basic idea behind the numerical analysis of Frobenius-Perron operators is to approximate the infinite dimensional Markov operator with finite dimensional linear operators that keep the structure of the approximated operator. Here, positivity is the structure of the original operator. Thus, the resulting finite dimensional Markov operator can be represented by a nonnegative (usually stochastic) matrix with respect to any basis consisting of density functions, so the theory and methods of nonnegative matrices and the theory of positive operators can work together to address the computational challenges of such important positive operators.

This textbook aims at providing the fundamental theory and methods of nonnegative matrices and positive operators, and their modern applications in several applied fields, such as the life science, electrical engineering, and the Web information retrieval technology. Naturally the book is divided into two parts. Part I consists of Chapters 1 to 4 concerning nonnegative matrices and their applications, and Chapters 5 through 11 constitute Part II on positive operators, their finite dimensional approximations via nonnegative matrices, and several representative applications including modern ones. There are excellent textbooks and monographs on nonnegative matrices and positive operators, respectively, such as [Berman and Plemmons (1979); Minc (1988); Bapat and Raghavan (1997)] for the former and [Schaefer (1974); Walters (1982); Aliprantis and Burkinshaw (1985); Krengel (1985); Lasota and Mackey (1994); Boyarsky and Góra (1997); Zaanen (1997); Baladi (2000)] for the latter. But due to ever growing applications of nonnegative matrices in physical sciences, engineering, technology, and social sciences, and in the numerical analysis of positive operators arising from various areas of computational science, we feel a need for writing the present textbook that combines the two topics, which are closely related but often separately presented, into a single unity. Because of the intended large readership, we have kept in mind the consideration of the material coverage and tried to balance the presentation of abstract theories and practical methods to benefit the readers studying or working in different disciplines. For example, we only give an introduction to the theory of abstract positive operators to lessen an unnecessary burden of the reader; on the other hand, we present in more detail the theoretical and numerical analysis of several concrete classes of positive operators, such as Markov operators, Frobenius-Perron operators, and composition operators widely used in applied fields, so that most of the book can be understood by as many readers as possible with various backgrounds. Thus, lots of more general and advanced theorems on abstract positive operators are not included, but they can be found in monographs on Banach lattices and positive operators.

Specifically, the textbook has eleven chapters, each of which has five sections. Chapter 1 provides basic algebraic, spectral, and combinatorial properties of nonnegative matrices, and we also study an important class of matrices whose inverse is a nonnegative matrix. Chapter 2 is the main body of the theory of nonnegative matrices, in which we shall study the classic Perron-Frobenius theory for irreducible nonnegative matrices. In Chapter 3 we focus on stochastic matrices, a special class of nonnegative matrices with spectral radius 1 which are closely related to any field involving probability. Applications of nonnegative matrices to Markov chains, the Web information retrieval, dynamical geometry, and the iterative method for solving linear equations are discussed in Chapter 4. Chapter 5 begins the study of positive operators, and we are devoted to the general properties of abstract positive operators on ordered vector spaces and in particular Banach lattices. The spectral analysis and ergodic theory of positive operators are the main themes of Chapter 6, which are of fundamental importance in applications of positive operators in physical sciences. Markov operators, which play a main role in stochastic processes and the statistical analysis of deterministic dynamical systems, will be introduced in Chapter 7 and the asymptotic properties of their iterates will be investigated for a class of Markov operators, based on a powerful decomposition theorem. A special class of Markov operators, called Frobenius-Perron operators which are used for the existence and computation of absolutely continuous invariant probability measures of nonsingular transformations of measure spaces, will be explored in detail in Chapter 8. We shall present some existence theorems for several classes of one or multi-dimensional transformations. Topological dynamical systems will be briefly studied in Chapter 9 since composition operators defined on the Banach lattice of continuous functions on a compact metric space are used for exploring the asymptotic behavior of continuous transformations. In Chapter 10 we study structure-preserving approximations of Markov operators and in particular Frobenius-Perron operators via finite dimensional Markov operators. The convergence of Ulam's piecewise constant method and the piecewise linear Markov method will be presented for several classes of transformations, and their error estimates will be given for a class of interval mappings. The last chapter provides several representative applications of positive operators to approximating continuous functions, evolutionary partial differential equations, computational molecular dynamics, and wireless communications.

This textbook can be used for students in mathematical sciences as an upper level undergraduate or beginning graduate course on nonnegative matrices or/and positive operators. Besides, the first four chapters can be used for a one semester course on the theory of nonnegative matrices, and some other ones can form a course on positive operators or applied ergodic theory. Materials of the book can also be adopted as a textbook for a specialized course in different areas of physical sciences with a purpose of mathematics enhancement for the interdisciplinary research. For students or researchers in engineering subjects such as electrical engineering, who are interested in applied ergodic theory, this textbook can be used for a graduate special topic course or a mathematical tool book. A good background of linear algebra and functional analysis is sufficient for understanding an essential part of the book, but we have included Appendix A for basic matrix theory and Appendix B for fundamental functional analysis for the reader's convenience. Some of the exercises in each chapter are complements to the main text, so the reader should try as many as possible, or at least take a look and read appropriate references if possible. In writing this textbook and selecting exercise problems, we have consulted many books and research papers on nonnegative matrices, positive operators, and more specialized topics, including those mentioned above as well as ours. We thank these authors and the others for their works' influences on ours.

Some parts of the textbook, such as approximations of Frobenius-Perron operators in Chapter 10, contain our joint research results in the past sixteen years. Jiu Ding owes a deep gratitude to his Ph.D. thesis advisor, Professor Tien-Yien Li, for introducing him into the field of computational ergodic theory. Aihui Zhou is very grateful to his Ph.D. thesis advisor, Professor Qun Lin, for his constant encouragement in the past two decades. Our research collaboration and writing of this book have been supported by the National Natural Science Foundation of China, the National Basic Research Program of China, the Academy of Mathematics and Systems Science and the State Key Laboratory of Scientific and Engineering Computing at the Chinese Academy of Sciences, the Chinese Ministry of Education, K. C. Wong Education Foundation of Hong Kong, and the A. K. Lucas Endowment at the University of Southern Mississippi, for which we express our sincere thanks.

Just at the completion of this work, we are deeply saddened to learn the untimely death of Dr. Xiangzhong (Jerry) Yang at 49 on February 5, whose *China Bridge Foundation* at the University of Connecticut has supported our joint research. This great scientist will be in our memory forever.

Some chapters from the initial draft of the textbook were used for a graduate course, Topics in Algebra, at the University of Southern Mississippi for the spring semester of 2007. Jiu Ding thanks the students Rebecca Eckhoff, Johnathan McEwen, Ryanne McNeese, Margaret Moore, Kedar Nepal, Corwin Stanford, Anita Waltman, Erin Westmoreland, and Nina Ye in the class for making it fun to teach with success. Several topics and materials in the textbook have been addressed in a graduate course, Numerical Analysis, at the Graduate University of the Chinese Academy of Sciences since 2005. Aihui Zhou would like to acknowledge the many students who took the course for giving him an opportunity to test our ideas.

Finally, we thank the editors Lai Fun Kwong and Yubing Zhai of the World Scientific for their editorial assistance and help, which has made writing and publishing this textbook a pleasant experience.

Jiu Ding and Aihui Zhou, February 11, 2009

Department of Mathematics
The University of Southern Mississippi
Hattiesburg, MS 39406, USA

Institute of Computational Mathematics and Scientific/Engineering Computing
Academy of Mathematics and Systems Science
Chinese Academy of Sciences
Beijing 100190, China

Contents

Chapter 1

Elementary Properties of Nonnegative Matrices

The theory of nonnegative matrices is a specialized topic in matrix theory. Nonnegative matrices can be viewed as finite dimensional positivity-preserving linear operators defined on Euclidean spaces and can be used for structure-preserving approximations of infinite dimensional positive operators in the numerical analysis of such important linear operators with many applications. The first four chapters of this textbook will provide a prototype of general positive operators to be studied later. We assume that the reader has taken a first course in linear algebra and is familiar with the fundamental theory of matrices, which and some more advanced concepts are included in Appendix A for reference.

In the first chapter, we present elementary properties and some basic results for nonnegative matrices. In Section 1.1 we give various algebraic properties of nonnegative matrices, and Section 1.2 is devoted to their spectral properties. As a prerequisite for the next chapter, combinatorial properties of nonnegative matrices and several related concepts will be introduced in Section 1.3. An important class of square matrices whose inverses are nonnegative ones will be studied in Section 1.4 because of their wide applications in various areas.

1.1 Algebraic Properties of Nonnegative Matrices

Because of the *total ordering* property of the set $\mathbb{R}$ of all real numbers, that is, exactly one of the relations $x < y$, $y < x$, or $x = y$ holds for any two real numbers x and y, it is convenient to be able to compare n-dimensional (column) vectors in the n-dimensional Euclidean space $\mathbb{R}^n$ component by component and $m \times n$ matrices in the mn-dimensional vector space $\mathbb{R}^{m \times n}$ entry by entry, thus providing a *canonical ordering* relation for real vectors

1

and real matrices. A component-wise *partial ordering* on $\mathbb{R}^n$ is defined by the following relation

$$x \le y \quad \text{if and only if} \quad x_i \le y_i, \quad i = 1, \cdots, n. \tag{1.1}$$

Two vectors $x, y \in \mathbb{R}^n$ are said to be *comparable* under the ordering if $x \le y$ or $y \le x$ which can also be written as $x \ge y$. The following properties of the natural ordering can be immediately verified, the first three of which explain why the relation $\le$ is a partial ordering [Zaanen (1997)].

Proposition 1.1. *The canonical ordering relation $\le$ defined on $\mathbb{R}^n$ by (1.1) satisfies the properties:*
(i) $x \le x$ *for all $x \in \mathbb{R}^n$ (reflexivity);*
(ii) $x \le y$ *and $y \le x$ imply $x = y$ (transitivity);*
(iii) $x \le y$ *and $y \le z$ imply $x \le z$ (anti-symmetry);*
(iv) $x \le y$ *implies $ax \le ay$ for any number $a \ge 0$ and $ax \ge ay$ if $a \le 0$;*
(v) *if $x \le y$, then $x + z \le y + z$ for all vectors $z \in \mathbb{R}^n$.*

If a vector $x \in \mathbb{R}^n$ is such that $x \ge 0$, then x is said to be *nonnegative*, and the set of all nonnegative vectors in $\mathbb{R}^n$ is a cone, called the *positive cone* of $\mathbb{R}^n$ and is denoted by $\mathbb{R}^n_+$. Thus, if $x, y \in \mathbb{R}^n_+$, then $x + y \in \mathbb{R}^n_+$ and $ax \in \mathbb{R}^n_+$ for all nonnegative numbers a. The special vector e of all components 1 is the *order unit* of $\mathbb{R}^n$ which is the *ideal* generated by e (see Section 5.3 for general definitions of such terms). The *absolute value vector* of any complex (column) vector x in the n-dimensional unitary space $\mathbb{C}^n$ is defined to be the nonnegative vector

$$|x| = (|x_1|, \cdots, |x_n|)^T$$

in $\mathbb{R}^n_+$. It is again straightforward to verify the following properties of absolute value vectors of complex ones.

Proposition 1.2. *Let $x, y \in \mathbb{C}^n$. Then*
(i) $|x| \ge 0$, *and $|x| = 0$ if and only if $x = 0$;*
(ii) $|ax| = |a||x|$ *for all complex numbers a;*
(iii) $|x + y| \le |x| + |y|$ *(triangle inequality).*

Note that properties (i)-(iii) in Proposition 1.2 for the absolute value operation $|\ |$ look similar to those for a norm $\|\ \|$ in format, but the difference is that while absolute value vectors are nonnegative vectors, norms are nonnegative numbers.

A norm $\|\ \|$ on $\mathbb{R}^n$ is called a *lattice norm* if for any $x, y \in \mathbb{R}^n$,

$$|x| \le |y| \quad \text{implies} \quad \|x\| \le \|y\|. \tag{1.2}$$

It can be shown that condition (1.2) is equivalent to the equality

$$\| \, |x| \, \| = \|x\|, \ \forall \, x \in \mathbb{R}^n.$$

Thus the p-norm $\| \, \|_p$ on $\mathbb{R}^n$ with $1 \le p \le \infty$ is a lattice norm.

A matrix $A = (a_{ij}) \in \mathbb{R}^{m \times n}$ of all $m \times n$ real matrices is said to be *nonnegative* if $a_{ij} \ge 0$ for all indices i and j, and this is denoted by $A \ge 0$. Similarly A is called a *positive matrix* and is written as $A > 0$ if all $a_{ij} > 0$. The set of all nonnegative matrices in $\mathbb{R}^{m \times n}$ is denoted by $\mathbb{R}_+^{m \times n}$. Given any two real matrices A and B of the same order, we write $A \ge B$ (or equivalently, $B \le A$) if $A - B \ge 0$, and $A > B$ if $A - B > 0$. Thus, $A \ge B$ (or $A > B$) means that $a_{ij} \ge b_{ij}$ (or $a_{ij} > b_{ij}$) for all pairs (i, j) of row and column indices. It is easy to see that If $A \le B$ and $C \ge 0$, then

$$AC \le BC \quad \text{and} \quad CA \le CB$$

whenever the matrix multiplication is well-defined. It follows that when A and B are nonnegative square matrices, the inequality $A \le B$ implies that $A^k \le B^k$ for all nonnegative integers k.

Let $A \in \mathbb{C}^{m \times n}$, the space of all $m \times n$ complex matrices. Then the corresponding *absolute value matrix* is defined to be

$$|A| = (|a_{ij}|) \in \mathbb{R}_+^{m \times n}.$$

Propositions 1.1 and 1.2 also hold for matrices. Moreover, we have

$$|Ax| \le |A| \, |x|, \ \forall \, A \in \mathbb{C}^{m \times n}, \ x \in \mathbb{C}^n$$

and

$$|AB| \le |A| \, |B|, \ \forall \, A \in \mathbb{C}^{m \times k}, \ B \in \mathbb{C}^{k \times n}.$$

In particular, for any $A \in \mathbb{C}^{n \times n}$ and any nonnegative integer k,

$$|A^k| \le |A|^k.$$

The partial ordering relation (1.1) on $\mathbb{R}^n$ induces the concept of *monotone operators* on $\mathbb{R}^n$ in the following definition.

Definition 1.1. A mapping $S : D \subset \mathbb{R}^n \to \mathbb{R}^m$ is said to be *isotone* if $S(x) \le S(y)$ whenever $x \le y$, $x, y \in D$. Similarly, S is said to be *antitone* if $S(x) \ge S(y)$ for all $x, \, y \in D$ such that $x \le y$.

Clearly S is antitone if and only if $-S$ is isotone. The concepts of nonnegative matrices and monotone mappings are closely related as the next proposition shows, which can be proved immediately.

Proposition 1.3. *A matrix $A \in \mathbb{R}^{m \times n}$ is nonnegative if and only if the resulting linear operator $A : \mathbb{R}^n \to \mathbb{R}^m$ is isotone, which is true if and only if $Ax \ge 0$ whenever $x \in \mathbb{R}_+^n$.*

1.2   Spectral Properties of Nonnegative Matrices

We now present several elementary spectral properties of nonnegative matrices. More advanced spectral properties are the ingredients of the classic Perron-Frobenius theory of nonnegative matrices and will be the main part of the next chapter. The *spectral radius* $r(A)$ of any complex matrix $A \in \mathbb{C}^{n \times n}$ is the maximal absolute value of all the eigenvalues of A. The following comparison theorem for the spectral radii of two matrices of the same order is very useful in many occasions.

Theorem 1.1. *Let $A \in \mathbb{R}^{n \times n}$ be nonnegative and let $B \in \mathbb{C}^{n \times n}$. Then the inequality $|B| \le A$ implies the inequality $r(B) \le r(A)$.*

Proof Let $\| \ \|_1$ be the matrix 1-norm on $\mathbb{C}^{n \times n}$. The assumption implies that for every positive integer $k = 1, 2, \cdots$, we have $|B^k| \le |B|^k \le A^k$. Thus, $\|B^k\|_1 \le \||B|^k\|_1 \le \|A^k\|_1$, which guarantees that

$$\|B^k\|_1^{\frac{1}{k}} \le \|A^k\|_1^{\frac{1}{k}}$$

for all k. Since $r(B) = \lim_{k \to \infty} \|B^k\|_1^{1/k}$ and $r(A) = \lim_{k \to \infty} \|A^k\|_1^{1/k}$, letting $k \to \infty$ in the above inequality, we deduce that $r(B) \le r(A)$. □

Corollary 1.1. *Let $A, B \in \mathbb{R}^{n \times n}$. If $0 \le B \le A$, then $r(B) \le r(A)$.*

Corollary 1.2. *Let $A = (a_{ij}) \in \mathbb{R}^{n \times n}$ be nonnegative. If B is any principal sub-matrix of A, then $r(B) \le r(A)$. In particular, $\max_{1 \le i \le n} a_{ii} \le r(A)$.*

Proof Without loss of generality, assume that B is a $k \times k$ leading principal sub-matrix of A. Let an $n \times n$ matrix A' be defined by

$$A' = \begin{bmatrix} B & 0 \\ 0 & 0 \end{bmatrix}.$$

Then $0 \le A' \le A$, so $r(B) = r(A') \le r(A)$ by Corollary 1.1. □

The next result strengthens Banach's lemma in Appendix A.4 for the case of nonnegative matrices. Let I denote the identity matrix as usual.

Theorem 1.2. *Let $A \in \mathbb{R}^{n \times n}$ be such that $A \ge 0$. Then the inverse matrix $(I - A)^{-1}$ exists and is nonnegative if and only if $r(A) < 1$.*

Proof Suppose that $r(A) < 1$. Then Banach's lemma (Theorem A.15) ensures that the inverse matrix $(I - A)^{-1}$ exists and equals $\sum_{k=0}^{\infty} A^k$. Since each term of the series is a nonnegative matrix, $(I - A)^{-1} \ge 0$. Conversely, assume that $I - A$ is invertible and $(I - A)^{-1} \geq 0$. Let λ be any eigenvalue of A and let x be a corresponding eigenvector. Then

$$|\lambda||x| = |\lambda x| = |Ax| \leq |A||x| = A|x|,$$

so $(I - A)|x| \leq (1 - |\lambda|)|x|$. Hence,

$$|x| \leq (1 - |\lambda|)(I - A)^{-1}|x|.$$

Since $x \neq 0$ and $(I - A)^{-1} \geq 0$, the above inequality implies that $|\lambda| < 1$. Therefore, $r(A) < 1$. $\qquad\square$

A lower (or upper) triangular matrix is said to be *strictly lower* (or *upper*) *triangular* if all of its diagonal entries are zero.

Corollary 1.3. *Let $A \in \mathbb{R}^{n \times n}$ be a triangular matrix such that $a_{ii} > 0$ for all $i = 1, \cdots, n$ and $a_{ij} \leq 0$ for all $i \neq j$. Then A^{-1} exists and $A^{-1} \geq 0$.*

Proof Without loss of generality, assume that A is lower triangular. Let $D = \mathrm{diag}(a_{11}, \cdots, a_{nn})$ and $L = D - A$. Then D is invertible with $D^{-1} \geq 0$, and L is strictly lower triangular and nonnegative. Since $r(D^{-1}L) = 0$, by Theorem 1.2, $(I - D^{-1}L)^{-1} \geq 0$. So

$$A^{-1} = (D - L)^{-1} = [D(I - D^{-1}L)]^{-1} = (I - D^{-1}L)^{-1}D^{-1} \geq 0.\square$$

1.3   Combinatorial Properties of Nonnegative Matrices

Many properties of nonnegative matrices, such as irreducibility and primitivity, depend on the structure of the zero entries of the matrix. In this section we use the zero pattern of nonnegative matrices to study their combinatorial properties which have applications to their spectral analysis in the next chapter. For this purpose, it is convenient to relate a special matrix of entries 0 or 1 with the same zero pattern and a directed graph to a given nonnegative matrix to reflect the structure of the nonzero entries. A matrix each of whose entries is either 0 or 1 is called a $(0, 1)$-*matrix*.

Definition 1.2. Let $A \in \mathbb{R}^{m \times n}$ be nonnegative. A $(0, 1)$-matrix $M = (m_{ij}) \in \mathbb{R}^{m \times n}$ is said to be the *indicator matrix* of A provided $m_{ij} = 1$ if and only if $a_{ij} > 0$.

Definition 1.3. A *directed graph* is the pair $G = (V, E)$ of a set V of n elements, called *vertices* of G, and a set E of ordered pairs, called *directed edges* of G. A directed edge $(i, j) \in E$ is said to *join* vertex i to vertex j.

A *subgraph* of a directed graph G is a directed graph whose vertices and directed edges belong to G. A *spanning subgraph* of G is a subgraph of G containing all the vertices of G. A sequence of directed edges

$$(i, i_1), (i_1, i_2), \cdots, (i_{k-2}, i_{k-1}), (i_{k-1}, j)$$

in G is called a *path* connecting vertex i to vertex j. The *length* of the path is the number of the directed edges in the sequence. A path of length k connecting vertex i to itself is called a *cycle* of length k. A cycle is called a *circuit* if each of its vertices appears exactly once as the first vertex of a directed edge. A cycle of length 1 is called a *loop*. A spanning circuit is called a *Hamiltonian circuit*.

Definition 1.4. Let G be a directed graph with n vertices. The *adjacency matrix* $A(G)$ of G is the $(0,1)$-matrix whose (i, j)-entry is 1 if and only if (i, j) is a directed edge of G. A directed graph is said to be *associated with* a nonnegative matrix A, denoted as $G(A)$, if its adjacency matrix is the indicator matrix of A.

Definition 1.5. A directed graph G is said to be *strongly connected* if for any ordered pair of vertices i and j, there is a path connecting i to j.

Let k be a nonnegative integer. We use $a_{ij}^{(k)}$ to denote the (i, j)-entry of the kth power A^k of a matrix $A = (a_{ij}) \in \mathbb{R}^{n \times n}$.

Theorem 1.3. *Let $A = (a_{ij}) \in \mathbb{R}^{n \times n}$ be a $(0,1)$-matrix and let $G(A)$ be its associated directed graph. Then the number of distinct paths of length k connecting vertex i to vertex j in $G(A)$ is equal to $a_{ij}^{(k)}$.*

Proof. Since $a_{ij}^{(2)} = \sum_{s=1}^{n} a_{is} a_{sj}$ and $a_{ij}^{(3)} = \sum_{s=1}^{n} \sum_{t=1}^{n} a_{is} a_{st} a_{tj}$, by induction we have

$$a_{ij}^{(k)} = \sum_{s_1=1}^{n} \sum_{s_2=1}^{n} \cdots \sum_{s_{k-2}=1}^{n} \sum_{s_{k-1}=1}^{n} a_{i,s_1} a_{s_1,s_2} \cdots a_{s_{k-2},s_{k-1}} a_{s_{k-1},j}.$$

On the other hand, since A is a $(0,1)$-matrix, there is a path

$$(i, s_1), (s_1, s_2), \cdots, (s_{k-2}, s_{k-1}), (s_{k-1}, j)$$

connecting i to j in $G(A)$ if and only if $a_{i,s_1} = a_{s_1,s_2} = \cdots = a_{s_{k-2},s_{k-1}} = a_{s_{k-1},j} = 1$, that is, if and only if

$$a_{i,s_1} a_{s_1,s_2} \cdots a_{s_{k-2},s_{k-1}} a_{s_{k-1},j} = 1.$$

Hence the (i, j)-entry of A^k is exactly the number of distinct paths of length k connecting i to j. This proves the theorem. $\square$

An immediate consequence of the above theorem is

Corollary 1.4. *Let $A \in \mathbb{R}^{n \times n}$ be a nonnegative matrix. Then its associated directed graph $G(A)$ has a path of length k connecting vertex i to vertex j if and only if $a_{ij}^{(k)} > 0$.*

We now introduce the concept of irreducibility for an arbitrary square matrix, which plays a key role in the next chapter.

Definition 1.6. An $n \times n$ complex matrix A is said to be *reducible* if there is a permutation matrix P such that

$$P^T A P = \begin{bmatrix} B & 0 \\ C & D \end{bmatrix},$$

where B and D are square matrices. A is *irreducible* if it is not reducible.

Definition 1.6 simply says that $A \in \mathbb{C}^{n \times n}$ is reducible if and only if there is a nonempty and proper subset J of $\{1, \cdots, n\}$ with the property that

$$a_{ij} = 0, \quad \forall\, i \in J,\, j \notin J. \tag{1.3}$$

Before presenting an equivalent condition for a nonnegative matrix to be irreducible in terms of the corresponding directed graph, we give a characterization of irreducibility for complex matrices in terms of the concept of chains. Given $A \in \mathbb{C}^{n \times n}$, a *chain* for indices i and j is a sequence of nonzero entries in A of the form

$$\{a_{ii_1}, a_{i_1 i_2}, \cdots, a_{i_m j}\}.$$

Lemma 1.1. *A matrix $A \in \mathbb{C}^{n \times n}$ is irreducible if and only if there is a chain for any two indices $1 \leq i, j \leq n$.*

Proof Suppose that A is reducible. Let J be an index set with property (1.3). Choose $i \in J$ and $j \notin J$. If $a_{ik} \neq 0$, then $k \in J$, and so there cannot be a chain for i, j. This proves the sufficiency part of the lemma.

Conversely, assume that the matrix A is irreducible. For any given index i, set $J = \{k : \text{a chain exists for } i, k\}$. the index set J is not empty since, otherwise, the ith row of A would be zero, and this would contradict the irreducibility of A. Now suppose that there is no chain for i and some j. Then J is a proper subset of $\{1, \cdots, n\}$. For any $k \in J$ and $l \notin J$, if $a_{kl} \neq 0$, then there is a chain for k, l. Since there is a chain for i, k, so a chain exists for i, l, which means that $l \in J$, a contradiction. Hence, (1.3) is satisfied, which is impossible since A is assumed to be irreducible. $\quad\square$

Lemma 1.1 gives rise to another characterization of irreducibility in terms of strong connectedness.

Theorem 1.4. *Let $A \in \mathbb{R}^{n \times n}$ be nonnegative. Then A is irreducible if and only if its associated directed graph $G(A)$ is strongly connected.*

Proof By Lemma 1.1, A is irreducible if and only if there is a chain for any two indices $1 \leq i, j \leq n$, which is equivalent to the statement that for each i and j there exists a path in $G(A)$ connecting i to j; in other words, if and only if $G(A)$ is strongly connected. □

The following characterization of irreducibility will be useful in the spectral analysis of irreducible nonnegative matrices in Chapter 2.

Theorem 1.5. *A nonnegative $A \in \mathbb{R}^{n \times n}$ is irreducible if and only if*

$$(I + A)^{n-1} > 0.$$

Proof It is easy to see that

$$(I + A)^{n-1} = I + \binom{n-1}{1} A + \cdots + \binom{n-1}{n-2} A^{n-2} + A^{n-1} > 0,$$

where $\binom{k}{l} = \frac{k!}{l!(k-l)!}$ is the number of combinations choosing l from k elements, if and only if for each pair of vertices (i, j) with $i \neq j$, at least one of the terms $A, A^2, \cdots, A^{n-1}$ has a positive (i, j)-entry. But Corollary 1.4 says that this is true if and only if there is a path in $G(A)$ connecting i to j. Thus, $(I + A)^{n-1} > 0$ if and only if $G(A)$ is strongly connected, which is equivalent to the irreducibility of A by Theorem 1.4. □

Next, we introduce the concept of *primitivity*, which is a stronger notion than irreducibility and is due to Frobenius in 1912.

Definition 1.7. A nonnegative matrix $A \in \mathbb{R}^{n \times n}$ is said to be *primitive* if it is irreducible and has only one eigenvalue of maximal modulus.

More generally, if an irreducible nonnegative matrix $A \in \mathbb{R}^{n \times n}$ has exactly s peripheral eigenvalues, then s is called the *index of imprimitivity* of A, or simply the *index* of A. Definition 1.7 says that A is primitive if and only if $s = 1$. If $s > 1$, A is said to be *imprimitive*.

Again, the combinatorial method can be used to determine whether a nonnegative matrix is primitive or not, and to find the index of imprimitivity of an irreducible matrix. For this purpose, we need

Definition 1.8. Let G be a directed graph which is strongly connected. The greatest common divisor (g.c.d.) of the lengths of all cycles in G is called the *index of imprimitivity* of G.

Proposition 1.4. *Let G be a strongly connected directed graph with index of imprimitivity k. Suppose that k_i is the g.c.d. of the lengths of all cycles of G through vertex i. Then $k_i = k$.*

Proof It is enough to show that k_i divides k since k divides k_i. Let C_j be a cycle in G passing through vertex j with length l_j. Since G is strongly connected, it contains a path P_{ij} connecting vertex i to vertex j with length l_{ij} and a path P_{ji} connecting vertex j to vertex i with length l_{ji}. Now, the path consisting of the paths P_{ij} and P_{ji}, and the path consisting of the path P_{ij}, the cycle C_j, and the path P_{ji} are two cycles through the same vertex i, whose lengths are $l_{ij} + l_{ji}$ and $l_{ij} + l_j + l_{ji}$, respectively. Since k_i divides both $l_{ij} + l_{ji}$ and $l_{ij} + l_j + l_{ji}$, it must divide l_j. Hence, k_i divides the length of any cycle in G, so it divides k. Therefore $k_i = k$. $\square$

Using the above proposition and the *Frobenius standard form* (Theorem 2.20 in Section 2.4) of irreducible nonnegative matrices, one can prove the following important relation between the two indices of imprimitivity for a nonnegative matrix and its associated directed graph; see [Minc (1988)].

Theorem 1.6. *The index of imprimitivity of an irreducible nonnegative matrix is equal to the index of imprimitivity of its associated directed graph.*

Since a nonnegative matrix has a nonzero diagonal entry if and only if its associated directed graph has a loop, that is, a cycle of length 1, Theorem 1.6 implies the following simple sufficient condition for primitivity.

Corollary 1.5. *An irreducible nonnegative matrix with a nonzero diagonal entry is primitive.*

We conclude this section with another sufficient condition for primitive matrices due to Lewin [Lewin (1971/72)].

Theorem 1.7. *Let $A \in \mathbb{R}^{n \times n}$ be nonnegative. If A is irreducible and $a_{ij} a_{ij}^{(2)} > 0$ for some i and j, then A is primitive.*

Proof Let $G(A)$ be the associated directed graph of A. By Corollary 1.4, vertex i is connected to vertex j by paths of lengths 1 and 2. Since A is irreducible, $G(A)$ is strongly connected due to Theorem 1.4. Thus, there is a path connecting vertex j to vertex i with length k, say. Then $G(A)$ contains cycles of lengths $k + 1$ and $k + 2$. Since g.c.d.$(k + 1, k + 2) = 1$ it follows, by Theorem 1.6, that A is primitive. $\square$

1.4 *M*-matrices

In this section we study a class of nonsingular matrices whose inverses are nonnegative. We do so by extending Corollary 1.3 from special triangular matrices to more general matrices. This can be done by introducing the class of M-matrices as follows.

Definition 1.9. A matrix $A \in \mathbb{R}^{n \times n}$ is called a *Z-matrix* if $a_{ij} \leq 0$ for all $i, j = 1, \cdots, n, i \neq j$. A Z-matrix A is called an *M-matrix* if A is invertible and $A^{-1} \geq 0$. A symmetric M-matrix is called a *Stieltjes matrix*.

The following theorem gives a characterization of M-matrices in terms of the spectral radius.

Theorem 1.8. *Let $A \in \mathbb{R}^{n \times n}$ be a Z-matrix. Then A is an M-matrix if and only if the diagonal entries of A are positive and $r(I - D^{-1}A) < 1$, where $D = \mathrm{diag}(a_{11}, \cdots, a_{nn})$.*

Proof Suppose that $r(I - D^{-1}A) < 1$. Then, since $I - D^{-1}A \geq 0$, $(D^{-1}A)^{-1} = [I - (I - D^{-1}A)]^{-1} \geq 0$ by Theorem 1.2. So A^{-1} exists, and $A^{-1} \geq 0$ since $D^{-1} \geq 0$. This means that A is an M-matrix.

Conversely, suppose that A is an M-matrix. If $a_{jj} \leq 0$ for some j, then the jth column a_j of A is non-positive for A is a Z-matrix. Since $A^{-1}A = I = [e_1, \cdots, e_n]$ and since $A^{-1} \geq 0$, we have $e_j = A^{-1}a_j \leq 0$, which is a contradiction. Thus, $a_{jj} > 0$ for all $j = 1, \cdots, n$. Hence, $D = \mathrm{diag}(a_{11}, \cdots, a_{nn}) \geq 0$ and D is invertible. It follows that $A^{-1}D \geq 0$. Thus, $r(I - D^{-1}A) < 1$ by Theorem 1.2. $\square$

The second characterization of M-matrices in the theorem below is often taken as the definition of M-matrices in many books.

Theorem 1.9. *Let $A \in \mathbb{R}^{n \times n}$. Then A is an M-matrix if and only if there exist a nonnegative matrix B and a real number $c > r(B)$ such that*

$$A = cI - B. \tag{1.4}$$

Proof Suppose that A is an M-matrix. Then A is a Z-matrix. Denote $c = \max_{i=1}^{n} a_{ii} > 0$ and define $B = cI - A$. Clearly $B \geq 0$. Since

$$\left(I - \frac{1}{c}B\right)^{-1} = c(cI - B)^{-1} = cA^{-1}$$

exists and is nonnegative, it follows from Theorem 1.2 that $r(c^{-1}B) < 1$. That is, $r(B) < c$. Thus, (1.4) is valid with $B \geq 0$ and $c > r(B)$.

Conversely, let $A = cI - B$ with $B \geq 0$ and $c > r(B)$. Then A is obviously a Z-matrix, and A^{-1} exists for 0 is not an eigenvalue of A. Since $r(c^{-1}B) = c^{-1}r(B) < 1$, Theorem 1.2 ensures that $(I - c^{-1}B)^{-1} \geq 0$. Hence $A^{-1} = c^{-1}\left(I - c^{-1}B\right)^{-1} \geq 0$, and thus A is an M-matrix. $\square$

Corollary 1.6. *A principal sub-matrix of an M-matrix is an M-matrix.*

Proof Let A be an $n \times n$ M-matrix. By Theorem 1.9, $A = cI - B$ with $B \geq 0$ and $c > r(B)$. Suppose that A_k is a $k \times k$ principal sub-matrix of A. Then $A_k = cI_k - B_k$, where I_k is the $k \times k$ identity matrix and B_k is the corresponding principal sub-matrix of B. Since $r(B_k) \leq r(B)$ from Corollary 1.2, Theorem 1.9 again ensures that A_k is an M-matrix. $\square$

The next theorem gives a way to obtain new M-matrices from a given M-matrix, namely, by increasing the diagonal entries and decreasing to some extent the off-diagonal ones.

Theorem 1.10. *Let $A = D - O \in \mathbb{R}^{n \times n}$ be an M-matrix with the diagonal part D and the off-diagonal part $-O$. If $D' \in \mathbb{R}^{n \times n}$ is some nonnegative diagonal matrix and $O' \in \mathbb{R}^{n \times n}$ is a nonnegative matrix with all 0 diagonal entries such that $O' \leq O$, then $A + D' + O'$ is an M-matrix and*

$$(A + D' + O')^{-1} \leq A^{-1}. \tag{1.5}$$

Proof The given conditions imply that

$$0 \leq (D + D')^{-1}(O - O') \leq D^{-1}O,$$

so from Corollary 1.1,

$$r\left[(D + D')^{-1}(O - O')\right] \leq r(D^{-1}O) < 1.$$

By Theorem 1.8, $A + D' + O'$ is an M-matrix. The inequality (1.5) follows by multiplying $A + D' + O' \geq A$ with nonnegative matrices $(A + D' + O')^{-1}$ from the left and A^{-1} from the right. $\square$

Corollary 1.7. *Let $A \in \mathbb{R}^{n \times n}$ be an M-matrix and let $D \in \mathbb{R}^{n \times n}$ be any nonnegative diagonal matrix. Then $A + D$ is an M-matrix, and the inequality $(A + D)^{-1} \leq A^{-1}$ holds.*

Corollary 1.7 implies the following result for Stieltjes matrices.

Theorem 1.11. *A Stieltjes matrix* $A \in \mathbb{R}^{n \times n}$ *is* positive definite, *that is,* $x^T A x > 0$ *for all nonzero vectors* $x \in \mathbb{R}^n$.

Proof A real symmetric matrix is positive definite if and only if all of its eigenvalues are positive numbers [Horn and Johnson (1985)]. If A has some eigenvalue $\lambda \leq 0$, then Corollary 1.7 implies that $A - \lambda I$ is an M-matrix, which contradicts the singularity of $A - \lambda I$.   □

Finally, we give a sufficient condition for a matrix to be an M-matrix, which is easily verified. First, we state a definition.

Definition 1.10. A matrix $A \in \mathbb{C}^{n \times n}$ is *diagonally dominant* if

$$\sum_{j=1, j \neq i}^{n} |a_{ij}| \leq |a_{ii}|, \quad i = 1, \cdots, n, \tag{1.6}$$

and *strictly diagonally dominant* if the strict inequality holds in (1.6) for all i. A is *irreducibly diagonally dominant* if A is irreducible, diagonally dominant, and the strict inequality in (1.6) holds for at least one i.

The following lemma is needed for the proof of the next theorem.

Lemma 1.2. (The diagonal dominance theorem) *Let* $A \in \mathbb{C}^{n \times n}$ *be either strictly or irreducibly diagonally dominant. Then* A *is nonsingular.*

Proof Suppose that A is singular. Then there is $x \neq 0$ such that $Ax = 0$. First assume that A is strictly diagonally dominant. Let

$$|x_k| = \max_{1 \leq i \leq n} |x_i|.$$

Then

$$|a_{kk}| \, |x_k| = \left| \sum_{j \neq k} a_{kj} x_j \right| \leq |x_k| \sum_{j \neq k} |a_{kj}|.$$

Since $|x_k| > 0$, it follows that

$$|a_{kk}| \leq \sum_{j \neq k} |a_{kj}|,$$

a contradiction to the strict diagonal dominance.

Assume now that A is irreducibly diagonally dominant. Let

$$|a_{kk}| > \sum_{j \neq k} |a_{kj}|$$

for some k. Define

$$J = \{l : |x_l| = \max_{1 \leq i \leq n} |x_i|, \ |x_l| > |x_i| \text{ for some } i\}.$$

Then $J \neq \emptyset$ is a proper subset of $\{1, \cdots, n\}$, and for any $l \in J$,

$$|a_{ll}| \leq \sum_{j \neq l} |a_{lj}| \frac{|x_j|}{|x_l|}.$$

It follows that $a_{lj} = 0$ whenever $|x_l| > |x_j|$, or else the diagonal dominance is contradicted; that is,

$$a_{lj} = 0, \ \forall \, l \in J, \ j \notin J.$$

By (1.3), A is reducible. This contradicts the irreducibility assumption. $\square$

Theorem 1.12. *If a Z-matrix $A \in \mathbb{R}^{n \times n}$ is strictly or irreducibly diagonally dominant with all positive diagonal entries, then A is an M-matrix.*

Proof Let $B = I - D^{-1}A$ and $D = \text{diag}(a_{11}, \cdots, a_{nn})$. The diagonal dominance of A implies that

$$\sum_{j=1}^{n} |b_{ij}| \leq 1, \quad i = 1, \cdots, n. \tag{1.7}$$

Thus, $r(B) \leq \|B\|_\infty \leq 1$.

Suppose that A is strictly diagonally dominant. Then the strict inequality holds in (1.7) for all i, so

$$r(B) \leq \|B\|_\infty < 1.$$

Suppose now that A is irreducibly diagonally dominant. If there is an eigenvalue λ of B with $|\lambda| = 1$, then $\lambda I - B$ is singular, so, by (1.7),

$$|\lambda - b_{ii}| \geq 1 - |b_{ii}| \geq \sum_{j \neq i} |b_{ij}|, \ \forall \, i = 1, \cdots, n,$$

and a strict inequality holds for at least one i. Since $\lambda I - B$ is irreducible, Lemma 1.2 implies that $\lambda I - B$ is invertible. This is a contradiction. Thus, $r(B) < 1$ and A is an M-matrix by Theorem 1.8. $\square$

To finish this section, we give an application of the M-matrix theory to a power control problem in wireless communications [Ding *et al.* (2006)].

Example 1.1. In the third generation of direct sequence code-division multiple-access (DS-CDMA) wireless communications, the power and data rate allocation problem can be reduced to solving an algebraic inequality system, which is constructed by service requirements and restrictions.

Suppose that a DS-CDMA communication system supports k multi-class services with n_i active users in the ith service class that is specified by its target signal energy to interference power spectral density ratio (SIR), γ_i, and processing gain, g_i, which is the ratio of the chip rate to the data rate. The received SIR for the ith service class can be expressed as

$$\text{SIR}_i = \frac{p_i g_i}{\sum_{j=1}^{k} n_j p_j - p_i + \sigma_i^2}, \quad \forall\, i, \tag{1.8}$$

where p_i and σ_i^2 are respectively the allocated power and background noise seen by an ith class user. The noise component may include the additive white Gaussian noise and inter-cell interference. The numerator of (1.8) represents the power for the desired use multiplied by the processing gain. The denominator is the total interference, including the power sum from all other users in the system and the noise component.

A satisfactory system operation requires that [Zhao *et al.* (2006)]

$$\text{SIR}_i \geq \gamma_i, \tag{1.9}$$

where γ_i is the target SIR for the ith class user, which is specified by the target bit error rate, modulation, and coding schemes. By manipulating (1.8) and (1.9), we can obtain a set of linear inequality requirements as

$$p_i \left(1 + \frac{g_i}{\gamma_i}\right) \geq \sum_{j=1}^{k} n_j p_j + \sigma_i^2, \qquad 1 \leq i \leq k. \tag{1.10}$$

Let $d_i = 1 + g_i/\gamma_i$ and $b_i = \sigma_i^2$ for $i = 1, \cdots, k$. We define the power vector $p = (p_1, \cdots, p_k)^T$ and the nonnegative noise vector $b = (b_1, \cdots, b_k)^T$. Then we can write the inequality system (1.10) as

$$\left(D - ev^T\right) p \geq b, \tag{1.11}$$

where $D = \text{diag}(d_1, \cdots, d_k)$, $e = (1, \cdots, 1)^T$, and $v = (n_1, \cdots, n_k)^T$. The optimal power allocation problem is to find the power vector solution p^* such that any other solution p of (1.11) satisfies $p \geq p^*$.

The capacity problem is to determine the set of $(n_1, n_2, \cdots, n_k)$ for which (1.11) is solvable. This subset of $\mathbb{R}^k$ specifies the maximal allowable number of users that can be supported by the system. It turns out that the maximal capacity set is determined by an optimal solution of the inequality system (1.11) whose solution is based on the following perturbation result for M-matrices from [Ding *et al.* (2006)]:

Lemma 1.3. *Let $A \in \mathbb{R}^{n \times n}$ be an M-matrix, and let u and v be two n-dimensional nonnegative vectors. Then the Z-matrix $A - uv^T$ is an M-matrix if and only if $v^T A^{-1} u < 1$.*

Based on the above result for M-matrices, one can show that the system (1.11) has a positive solution if and only if

$$\sum_{i=1}^{k} \frac{n_i}{1 + \frac{g_i}{\gamma_i}} < 1.$$

Furthermore, if the above inequality is satisfied, then the inverse matrix $(D - ev^T)^{-1}$ of the Z-matrix $D - ev^T$ exists and is nonnegative. In this case, namely when $D - ev^T$ is an M-matrix, the optimal power allocation is given by the following expression

$$p^* = \left(D - ev^T\right)^{-1} b = \left(I + \frac{1}{\eta} D^{-1} e e^T\right) D^{-1} b$$

$$= \frac{1}{\eta} \begin{bmatrix} (\frac{\gamma_1 n_1}{\gamma_1 + g_1} + \eta)\frac{\gamma_1}{\gamma_1 + g_1} & \frac{\gamma_1 \gamma_2 n_2}{(\gamma_1 + \gamma_2) + g_1 + g_2} & \cdots & \frac{\gamma_1 \gamma_k n_k}{(\gamma_1 + \gamma_k) + g_1 + g_k} \\ \frac{\gamma_1 \gamma_2 n_1}{(\gamma_1 + \gamma_2) + g_1 + g_2} & (\frac{\gamma_2 n_2}{\gamma_2 + g_2} + \eta)\frac{\gamma_2}{\gamma_2 + g_2} & \cdots & \frac{\gamma_2 \gamma_k n_k}{(\gamma_2 + \gamma_k) + g_2 + g_k} \\ \vdots & \vdots & \vdots & \vdots \\ \frac{\gamma_1 \gamma_k n_1}{(\gamma_1 + \gamma_k) + g_1 + g_k} & \frac{\gamma_2 \gamma_k n_2}{(\gamma_2 + \gamma_k) + g_2 + g_k} & \cdots & (\frac{\gamma_k n_k}{\gamma_k + g_k} + \eta)\frac{\gamma_k}{\gamma_k + g_k} \end{bmatrix} \begin{bmatrix} \sigma_1^2 \\ \sigma_2^2 \\ \vdots \\ \sigma_k^2 \end{bmatrix},$$

where

$$\eta \equiv 1 - \sum_{i=1}^{k} \frac{n_i}{1 + \frac{g_i}{\gamma_i}} > 0.$$

1.5 Notes and Remarks

1. Almost all the materials presented in this chapter are standard ones and can be found in textbooks which study the theory of nonnegative matrices exclusively, such as [Berman and Plemmons (1979)], [Minc (1988)], [Berman *et al.* (1989)], and [Bapat and Raghavan (1997)], or other textbooks or monographs that contain the topics of nonnegative matrices and Markov chains, such as [Chung (1967)], [Ortega and Rheinboldt (1970)], [Horn and Johnson (1985)], and [Varga (2000)]. The classic monograph [Schaefer (1974)] on positive operators covers nonnegative matrices as its

first chapter, but its presentation is more advanced with its approach mainly based on functional analysis.

2. The theory of M-matrices has many applications [Bapat and Ragha-van (1997); Varga (2000); Berman *et al.* (1989); Ding *et al.* (2006)]. In some books, the definition of an M-matrix is weaker than ours in Defini-tion 1.9 in the sense that there is no requirement for the non-singularity of the M-matrix, for example, Definition 6.4.1 of the book [Minc (1988)]. Theorem 1.5.2 in the book [Bapat and Raghavan (1997)] lists ten equivalent statements for a nonsingular Z-matrix to be an M-matrix, some of which are put in the exercise set of this chapter.

3. The concept of M-matrices can be extended to that of *extended M-matrices with respect to a cone* $\mathcal{C} \subset \mathbb{R}^n$, as defined in the book [Berman *et al.* (1989)] whose Chapter 5 is devoted to this class of matrices. When $\mathcal{C} = \mathbb{R}^n_+$, the definition of extended M-matrices is equivalent to that of M-matrices, Definition 1.9.

4. Some mathematical analysis behind the application of M-matrices in Example 1.1 is contained in the paper [Ding *et al.* (2006)], in which several general perturbation results concerning M-matrices are obtained. In particular, Lemma 1.3 is a corollary of the following theorem:

Theorem 1.13. *Suppose that A is an $n \times n$ M-matrix and U, V are two $n \times j$ nonnegative matrices. Then the Z-matrix $A - UV^T$ is an M-matrix if and only if the $j \times j$ matrix $I - V^T A^{-1} U$ is an M-matrix or equivalently,*

$$\det(I_i - V_i^T A^{-1} U_i) > 0, \ \forall \, i = 1, \cdots, j,$$

where I_i is the $i \times i$ identity matrix, and U_i and V_i are the $n \times i$ sub-matrices of U and V by deleting their last $j - i$ columns, respectively.

The reader is suggested to provide a proof of the above perturbation result in Exercise 1.26.

Exercises

1.1 Show that if $A > 0$ and $x \geq 0$, $x \neq 0$, then $Ax > 0$.
1.2 Show that if $|A| \leq |B|$, then $\|A\|_2 \leq \|B\|_2$.
1.3 Show that $\|A\|_2 = \| \, |A| \, \|_2$.

1.4 Suppose that $A \in \mathbb{R}^{n \times n}$ is nonnegative. Show that
  (i) if the row sums of A are constant, then $r(A) = \|A\|_\infty$;
  (ii) if the column sums of A are constant, then $r(A) = \|A\|_1$.
1.5 Prove that if $A \in \mathbb{C}^{n \times n}$ is an irreducible matrix, then it cannot have zero rows or columns.
1.6 Find a permutation matrix P such that

$$P^T A P = \begin{bmatrix} 8 & 3 & 1 & 0 \\ 0 & 7 & 4 & 1 \\ 2 & 3 & 5 & 1 \\ 4 & 7 & 1 & 3 \end{bmatrix} \quad \text{for} \quad A = \begin{bmatrix} 3 & 1 & 7 & 4 \\ 1 & 5 & 3 & 2 \\ 1 & 4 & 7 & 0 \\ 0 & 1 & 4 & 8 \end{bmatrix}.$$

1.7 Let $A \in \mathbb{C}^{n \times n}$. Show that A is reducible if and only if there is an $n \times n$ permutation matrix P such that

$$P^T A P = \begin{bmatrix} B & C \\ 0 & D \end{bmatrix},$$

where B and D are square matrices.
1.8 Let A and B be $n \times n$ nonnegative matrices. Which of the following statements are correct?
  (i) If A is irreducible, then A^T is also irreducible.
  (ii) If A is irreducible, then A^k is also irreducible for any positive integer k. How about the converse statement?
  (iii) If A and B are irreducible, then their sum $A + B$ is also irreducible. How about their product AB?
  (iv) AB is irreducible if and only if BA is irreducible.
  (v) A is irreducible if and only if $I + A$ is irreducible.
  (vi) If A is reducible, then A^2 is also reducible.
  (vii) If A is irreducible, then A^2 is also irreducible.
1.9 Show that if $A \in \mathbb{R}^{n \times n}$ is nonnegative and irreducible with a positive trace, then $A^k > 0$ for all k large enough.
1.10 Show that the $n \times n$ matrix $[e_n, e_1, \cdots, e_{n-1}]$ is irreducible.
1.11 Show that an $n \times n$ matrix $A \geq 0$ is reducible if and only if there is $\{j_1, \cdots, j_k\} \subset \{1, \cdots, n\}$ with $1 \leq k < n$ such that

$$\text{span} \{A e_{j_1}, \cdots, A e_{j_k}\} \subset \text{span} \{e_{j_1}, \cdots, e_{j_k}\}.$$

1.12 Let $A \in \mathbb{R}_+^{n \times n}$ be irreducible and let m be the degree of its minimal polynomial. Show that for any (i, j) with $i \neq j$, there is an integer $0 < k < m$ such that the (i, j)-entry $a_{ij}^{(k)}$ of A^k is positive.
1.13 Provide a proof of Proposition 1.3.

1.14 Assume $A \geq 0$ and $A \neq 0$. Show that if A has a positive eigenvector or a positive left eigenvector, then $r(A) > 0$.

1.15 Show that a norm $\| \; \|$ on $\mathbb{R}^n$ is a lattice norm if and only if $\| \, |x| \, \| = \|x\|$ for all vectors $x \in \mathbb{R}^n$.

1.16 Show that a Z-matrix A is an M-matrix if and only if any Z-matrix $B \geq A$ is nonsingular.

1.17 Show that a Z-matrix is an M-matrix if and only if all of its real eigenvalues are positive.

1.18 Show that the determinant of an M-matrix is positive.

1.19 Show that a Z-matrix is an M-matrix if and only if the determinants of its principal sub-matrices are all positive.

1.20 Show that an $n \times n$ Z-matrix is an M-matrix if and only if there exists $x \in \mathbb{R}^n_+$ such that $Ax > 0$.

1.21 Show that a Z-matrix A is an M-matrix if and only if there is a diagonal matrix D with positive diagonal entries such that the matrix $AD + DA^T$ is positive definite.

1.22 Find the range of the values of c such that the matrix

$$\begin{bmatrix} 2 & -3 & 1 \\ -2 & c & -4 \\ -3 & -1 & 1 \end{bmatrix}$$

becomes an M-matrix.

1.23 Let $A \in \mathbb{R}^{n \times n}$ be a tridiagonal matrix such that the sup-diagonal and sub-diagonal entries of A are negative, and all row sums are positive. Show that A is an M-matrix.

1.24 Find the spectrum $\sigma(A)$ for the $n \times n$ permutation matrix $A = [e_n, e_1, \cdots, e_{n-1}]$

1.25 Let $A \in \mathbb{R}^{n \times n}$ be an M-matrix and denote by $\tau(A)$ the minimal value of the real part of all eigenvalues of A. Show that:

(i) $\tau(A)$ is an eigenvalue of A;

(ii) $\tau(A) = 1/r(A^{-1})$;

(iii) $\tau(A) = c - r(B)$ if $A = cI - B$ as in Theorem 1.9.

1.26 Prove Theorem 1.13 in Section 1.5.

Hint: The Sherman-Morrison-Woodbury formula [Ortega and Rheinboldt (1970)], the determinant identity (2.3), Theorem 1.10, and Exercise 1.19 may be useful for the proof.

Chapter 2

The Perron-Frobenius Theory for Nonnegative Matrices

The most fascinating facts about nonnegative matrices were discovered by Perron and Frobenius at the beginning of the twentieth century, and their original results and modern developments are generally referred to as the *Perron-Frobenius theory* for nonnegative matrices. The purpose of this chapter is to present an introduction to this beautiful theory.

In Section 2.1, Perron's theorem on positive matrices will be given first, and Frobenius' theorem for more general irreducible nonnegative matrices will be studied in Section 2.2, in which we also give various estimates for the maximal eigenvalue of a nonnegative matrix and further explore the peripheral spectrum of irreducible nonnegative matrices. Ergodic theorems for nonnegative matrices form the topic of Section 2.3. Several equivalent conditions for the irreducibility or the primitivity of nonnegative matrices are contained in Section 2.4.

2.1   Perron's Theorem for Positive Matrices

We first study the spectral theory for positive square matrices, which is largely due to Perron's pioneering work [Perron (1907)] about one hundred years ago. This theory will be extended in the next section to a class of nonnegative matrices.

It is well-known that the spectral radius of a square matrix may not be its eigenvalue, as the simple 2×2 orthogonal matrix

$$A = \begin{bmatrix} 0 & 1 \\ -1 & 0 \end{bmatrix}$$

shows. However, one of the most attractive spectral properties of *any* nonnegative square matrix is that its spectral radius is always an eigenvalue, which is called the *maximal eigenvalue* of the matrix in the literature. More-over, there is a nonnegative eigenvector, called a *maximal eigenvector* asso-ciated with the maximal eigenvalue. These facts were discovered by Perron in his first major theorem for positive square matrices.

Before we prove Perron's theorem, we state Brouwer's fixed point theo-rem first with a short and elegant proof from Milnor's book [Milnor (1965)] which is based on Hirsch's original idea. A set $C \subset \mathbb{R}^n$ is said to be *convex* if for any $x, y \in C$, the line segment $\{ax + (1 - a)y : 0 \le a \le 1\} \subset C$. The empty set or any set with exactly one element are convex.

Lemma 2.1. (Brouwer's fixed point theorem) *Let* $K \subset \mathbb{R}^n$ *be a closed, bounded, and convex set. If a mapping* $S : K \to K$ *is continuous, then there is* $x^* \in K$ *such that* $S(x^*) = x^*$.

Proof From topology, the set K is homeomorphic to the k-dimensional *closed unit ball* $\mathbb{D}^k \equiv \{x \in \mathbb{R}^k : \|x\|_2 \le 1\}$ for some $1 \le k \le n$, so without loss of generality, we assume that $K = \mathbb{D}^n$. First suppose that S is smooth. If the theorem is not true, then $S(x) \ne x$ for each $x \in \mathbb{D}^n$. Now define a mapping $H : \mathbb{D}^n \to \partial \mathbb{D}^n$ from the closed unit ball onto its boundary as follows: for $x \in \mathbb{D}^n$ let $H(x)$ be the (unique) intersection point with $\partial \mathbb{D}^n$ of the ray from $S(x)$ toward x. Actually

$$H(x) = x + \left(u^T x + \sqrt{1 - \|x\|_2^2 + (u^T x)^2} \right) u,$$

where

$$u = \frac{S(x) - x}{\|S(x) - x\|_2},$$

from which we see that H is also smooth. It is clear that $H(x) = x$ whenever $x \in \partial \mathbb{D}^n$. In other words, $H|_{\partial \mathbb{D}^n} : \partial \mathbb{D}^n \to \partial \mathbb{D}^n$ is the identity mapping. This is impossible from topology since there does not exist any smooth mapping from $\mathbb{D}^n$ into $\partial \mathbb{D}^n$ such that its restriction to $\partial \mathbb{D}^n$ is the identity one. The resulting contradiction proves the theorem for smooth S.

Now assume that $S : \mathbb{D}^n \to \mathbb{D}^n$ is only continuous. If $S(x) \ne x$ for all $x \in \mathbb{D}^n$, then the minimal value ϵ of the continuous function $\|S(x) - x\|_2$ over $\mathbb{D}^n$ is positive. The Weierstrass approximation theorem (see also Lemma 5.3) ensures that there is a mapping P, whose component functions are polynomials of n variables, such that $\|P(x) - S(x)\|_2 < \epsilon/2$ for all $x \in \mathbb{D}^n$. Let $\hat{P}(x) = (1 + \epsilon/2)^{-1} P(x)$. Then $\hat{P}$ maps $\mathbb{D}^n$ into $\mathbb{D}^n$ and for all $x \in \mathbb{D}^n$,

$$\|\hat{P}(x) - S(x)\|_2 \le \|\hat{P}(x) - P(x)\|_2 + \|P(x) - S(x)\|_2 < \frac{\epsilon}{2} + \frac{\epsilon}{2} = \epsilon.$$

Clearly $\hat{P}(x) \neq x$ for all $x \in \mathbb{D}^n$ from the construction of $\hat{P}$, which means that $\hat{P}$ is a smooth mapping from $\mathbb{D}^n$ into itself without a fixed point. This contradicts the first part of the proof. $\square$

Let A be an $n \times n$ nonnegative matrix. Then A, as a linear operator from $\mathbb{R}^n$ into itself, leaves the convex positive cone $\mathbb{R}^n_+$ invariant. Suppose that $Ax = \lambda x$ for some number λ and some nonzero vector $x \geq 0$. Then $e^T x > 0$, which implies that

$$\lambda = \frac{e^T Ax}{e^T x} \geq 0.$$

Actually this x is a fixed point of the nonlinear mapping

$$z \rightarrow \frac{e^T z}{e^T Az} Az, \quad \forall \, z \in \mathbb{R}^n_+.$$

This observation motivates the proof of part (i) of the following classic result via Brouwer's fixed point theorem.

Theorem 2.1. (Perron's theorem) *Let $A \in \mathbb{R}^{n \times n}$ and assume that $A > 0$. Then the following statements are true:*

(i) *The spectral radius $r(A)$ of A is a positive eigenvalue with a positive eigenvector x and a positive left eigenvector y.*

(ii) *The eigenvalue $r(A)$ is geometrically simple.*

(iii) *Any nonnegative eigenvector of A is a positive scalar multiple of x corresponding to $r(A)$.*

Proof (i) Let

$$K = \{z \in \mathbb{R}^n : z \geq 0, \ e^T z = 1\}$$

be the *standard simplex* in $\mathbb{R}^n$. Then K is closed, bounded, and convex. Define the mapping

$$S(z) = \frac{1}{e^T Az} Az, \quad \forall \, z \geq 0, \ z \neq 0.$$

If $z \in K$, then $e^T Az > 0$ and $e^T S(z) = (e^T Az)^{-1} e^T Az = 1$. So $S : K \rightarrow K$ is well-defined. It is easily checked that S is continuous on K. By Lemma 2.1, there is an $x \in K$ such that $S(x) = x$. In other words,

$$Ax = (e^T Ax)x.$$

Since $x \neq 0$ and $A > 0$, we have $Ax > 0$, and so $\lambda \equiv e^T Ax > 0$, which implies that $x = \lambda^{-1} Ax > 0$. Therefore, x is a positive eigenvector of A corresponding to a positive eigenvalue λ.

If we apply the above result to A^T, we conclude that there are $\lambda_1 > 0$ and $y > 0$ such that $y^T A = \lambda_1 y^T$. Obviously $\lambda_1 = \lambda$ since $\lambda_1 y^T x = y^T A x = \lambda y^T x$ and $y^T x > 0$.

We next show that $\lambda = r(A)$. Let $Av = \gamma v$ for some complex number γ and a nonzero vector $v \in \mathbb{C}^n$. Without loss of generality, we can assume that $e^T |v| = 1$. Then

$$A|v| \geq |Av| = |\gamma v| = |\gamma||v|.$$

Multiplying y^T to the both sides of the above inequality, we have

$$\lambda y^T |v| \geq |\gamma| y^T |v|.$$

Since $y > 0$ and $|v| \neq 0$, it is true that $\lambda \geq |\gamma|$. Hence, $\lambda = r(A)$.

(ii) It is enough to show that if $Av = r(A)v$ for some real vector $v \neq 0$, then v is a multiple of x. If v and x are linearly independent, then there exists a real number a such that the vector $x - av$ is nonzero and nonnegative with at least one zero component. Then, by $A(x - av) = r(A)(x - av)$,

$$x - av = \frac{1}{r(A)} A(x - av) > 0,$$

which gives a contradiction. Thus v is a scalar multiple of x.

(iii) Suppose that $Av = \gamma v$ for some complex number γ and some nonzero vector $v \geq 0$. Then

$$\gamma y^T v = y^T A v = r(A) y^T v.$$

Since $y^T v > 0$, we have $\gamma = r(A)$ and by (ii), $v = ax$ for some number a. Since $v \geq 0$, $v \neq 0$, and $x > 0$, we have $a > 0$. This proves that v is a positive eigenvector of A corresponding to eigenvalue $r(A)$. $\square$

Corollary 2.1. *Let $A \in \mathbb{R}^{n \times n}$ be positive. Then there exists a unique vector x such that $Ax = r(A)x$, $x > 0$, and $e^T x = 1$.*

The unique normalized eigenvector x in Corollary 2.1 is called the *Perron vector* of the matrix A, and the Perron vector for A^T is called the *left Perron vector* of A. So $y \in \mathbb{R}^n$ is a left Perron vector of A if and only if $y^T A = r(A)y^T$, $y > 0$, and $y^T e = 1$.

The next theorem indicates that for any positive square matrix A, the spectral radius $r(A)$ is the only eigenvalue of A on the circle centered at the origin with radius $r(A)$.

Theorem 2.2. *Let $A = (a_{ij}) \in \mathbb{R}^{n \times n}$ be positive. Then $|\lambda| < r(A)$ for every eigenvalue $\lambda \neq r(A)$.*

Proof Let $Ax = \lambda x$ be such that $x = (x_1, \cdots, x_n)^T \neq 0$ and $|\lambda| = r(A)$. Then from the inequality

$$r(A)|x| = |\lambda||x| = |Ax| \leq |A||x| = A|x|,$$

we have $u \equiv A|x| - r(A)|x| \geq 0$. If $u \neq 0$, then for $v \equiv A|x| > 0$,

$$0 < Au = A(A|x| - r(A)|x|) = Av - r(A)v.$$

Thus $0 < y^T Av - r(A)y^T v = r(A)y^T v - r(A)y^T v = 0$, where y is the left Perron vector of A. This contradiction shows that $A|x| = r(A)|x|$, and so $|x| = r(A)^{-1}A|x| > 0$. Now,

$$r(A)|x_k| = \left| \sum_{j=1}^{n} a_{kj}x_j \right| \leq \sum_{j=1}^{n} a_{kj}|x_j| = r(A)|x_k|$$

for each $k = 1, 2, \cdots, n$. Thus,

$$\left| \sum_{j=1}^{n} a_{kj}x_j \right| = \sum_{j=1}^{n} a_{kj}|x_j|, \quad k = 1, 2, \cdots, n.$$

The only possibility for the above equalities to be valid is that there is a real number θ such that $\exp(i\theta)a_{kj}x_j > 0$ for all $k, j = 1, \cdots, n$. Since all $a_{kj} > 0$, we must have $z \equiv \exp(i\theta)x > 0$. Since $Ax = \lambda x$, $Az = \lambda z$, so $\lambda = r(A)$ by Theorem 2.1(iii). $\square$

The Perron vector and the left Perron vector of a positive square matrix A basically give the asymptotic behavior of the powers A^k as k goes to infinity, which is precisely stated by the following theorem.

Theorem 2.3. *Suppose that a matrix $A \in \mathbb{R}^{n \times n}$ is positive. Let two vectors $x, y \in \mathbb{R}^n$ be such that*

$$Ax = r(A)x, \quad y^T A = r(A)y^T, \quad x > 0, \quad y > 0, \quad y^T x = 1.$$

Then

$$\lim_{k \to \infty} \left(\frac{1}{r(A)}A \right)^k = xy^T. \tag{2.1}$$

Proof Denote $\lambda = r(A)$ and $E = xy^T$. Then

$$\left(\frac{A}{\lambda} \right)^k = \frac{1}{\lambda^k} \left(A^k - \lambda^k E \right) + E.$$

To prove (2.1), it is enough to show the equality $A^k - \lambda^k E = (A - \lambda E)^k$ and the inequality $r(A - \lambda E) < \lambda$.

A direct calculation shows that $E^2 = xy^T xy^T = xy^T = E$, $EA = xy^T A = \lambda xy^T = \lambda E$, and $AE = Axy^T = \lambda xy^T = \lambda E$. So $E(A - \lambda E) = (A - \lambda E)E = 0$, and $EA^k = A^k E = \lambda^k E$ for any k. Let $\gamma \neq 0$ be an eigenvalue of $A - \lambda E$ with an associated eigenvector v. Then $0 = 0v = E(A - \lambda E)v = \gamma Ev$. Thus $Ev = 0$, so $Av = (A - \lambda E)v = \gamma v$. Namely, every nonzero eigenvalue of $A - \lambda E$ is an eigenvalue of A. We claim that λ is not an eigenvalue of $A - \lambda E$. If v were an eigenvector of $A - \lambda E$ corresponding to eigenvalue λ, then $Ev = 0$ from the above, so v would be an eigenvector of A with the same eigenvalue. By Theorem 2.1(ii), $v = ax$ for some number $a \neq 0$. But then

$$\lambda v = (A - \lambda E)v = (A - \lambda E)ax = a\lambda x - \lambda ax = 0,$$

which implies that $v = 0$. In summary, because of Theorem 2.2, we have proved that $r(A - \lambda E) < \lambda$, and consequently

$$\lim_{k \to \infty} \left(\frac{A - \lambda E}{\lambda} \right)^k = 0.$$

Now, for $k = 1, 2, \cdots$,

$$(A - \lambda E)^k = \sum_{j=0}^{k} \binom{k}{j} A^{k-j}(-\lambda E)^j = A^k + \sum_{j=1}^{k} \binom{k}{j}(-\lambda)^j A^{k-j} E$$

$$= A^k + \sum_{j=1}^{k} \binom{k}{j}(-\lambda)^j \lambda^{k-j} E = A^k + \lambda^k \sum_{j=1}^{k}(-1)^j \binom{k}{j} E$$

$$= A^k + \lambda^k \left(\sum_{j=0}^{k}(-1)^j \binom{k}{j} - 1 \right) E = A^k - \lambda^k E.$$

It follows that

$$\left(\frac{A}{\lambda} \right)^k = \frac{A^k - \lambda^k E}{\lambda^k} + E = \left(\frac{A - \lambda E}{\lambda} \right)^k + E \to E$$

as $k \to \infty$. This proves the theorem. $\square$

Remark 2.1. The matrix E is actually a projection onto the null space of the matrix $r(A)I - A$ along the range of $r(A)I - A$, and we have the following *direct sum decomposition* of $\mathbb{R}^n$:

$$\mathbb{R}^n = N(r(A)I - A) \oplus R(r(A)I - A). \tag{2.2}$$

In Section 2.3, we can see that (2.2) is still valid for a wider class of nonnegative square matrices.

The proof of the above limit theorem can be used to show that the spectral radius of a positive square matrix is an algebraically simple eigenvalue of the matrix. But we need a preliminary result, whose proof is based on the following *determinant identity* [Bernstein (2005)]

$$\det(A - UV^T) = \det A \cdot \det(I - V^T A^{-1} U), \tag{2.3}$$

where $A \in \mathbb{C}^{n \times n}$ is nonsingular and $U, V \in \mathbb{C}^{n \times k}$.

Lemma 2.2. *Let $A \in \mathbb{C}^{n \times n}$ with all the eigenvalues $\lambda_1, \cdots, \lambda_n$, counting algebraic multiplicity, and for $1 \le k \le n$ let $U = [u_1, \cdots, u_k]$ and $V = [v_1, \cdots, v_k]$ be in $\mathbb{C}^{n \times k}$ such that the vectors $u_1, \cdots, u_k$ are eigenvectors of A corresponding to the eigenvalues $\lambda_1, \cdots, \lambda_k$, respectively. Then the eigenvalues of the matrix $A + UV^T$ are*

$$\{\gamma_1, \cdots, \gamma_k, \lambda_{k+1}, \cdots, \lambda_n\},$$

counting algebraic multiplicity, where $\gamma_1, \cdots, \gamma_k$ are the eigenvalues of the $k \times k$ matrix $\operatorname{diag}(\lambda_1, \cdots, \lambda_k) + V^T U$, counting algebraic multiplicity.

Proof Let λ be any complex number such that $\lambda \notin \sigma(A)$. Since the matrix $\lambda I - A$ is nonsingular, formula (2.3) implies that

$$\det[\lambda I - (A + UV^T)] = \prod_{j=1}^{n} (\lambda - \lambda_j) \cdot \det[I - V^T(\lambda I - A)^{-1} U]. \tag{2.4}$$

Denote $\Lambda = \operatorname{diag}(\lambda_1, \cdots, \lambda_k)$. Then the assumption of the lemma gives $(\lambda I - A)U = U(\lambda I - \Lambda)$, which implies

$$(\lambda I - A)^{-1} U = U(\lambda I - \Lambda)^{-1}.$$

It follows that

$$\det[I - V^T(\lambda I - A)^{-1} U] = \det[I - V^T U(\lambda I - \Lambda)^{-1}]$$

$$= \det[\lambda I - (\Lambda + V^T U)] \det(\lambda I - \Lambda)^{-1} = \frac{\det[\lambda I - (\Lambda + V^T U)]}{\prod_{j=1}^{k}(\lambda - \lambda_j)}.$$

Thus, from (2.4),

$$\det[\lambda I - (A + UV^T)] \cdot \prod_{j=1}^{k} (\lambda - \lambda_j) = \prod_{j=1}^{n} (\lambda - \lambda_j) \cdot \prod_{j=1}^{k} (\lambda - \gamma_j),$$

and it follows that

$$\det[\lambda I - (A + UV^T)] = \prod_{j=1}^{k} (\lambda - \gamma_j) \cdot \prod_{j=k+1}^{n} (\lambda - \lambda_j). \qquad \square$$

Theorem 2.4. *Let $A \in \mathbb{R}^{n \times n}$ be positive. Then $r(A)$ is an eigenvalue of A with algebraic multiplicity 1; that is, $r(A)$ is a simple root of the characteristic polynomial of A.*

Proof Let $\lambda_1 = r(A), \lambda_2, \cdots, \lambda_n$ be the eigenvalues of A, counting algebraic multiplicity. For the projection matrix xy^T given in Theorem 2.3, it was shown there that λ_1 is not an eigenvalue of $A - \lambda_1 xy^T$. Since x in Theorem 2.3 is an eigenvector of A corresponding to eigenvalue λ_1, all the conditions of Lemma 2.2 are satisfied with $k = 1, u_1 = -\lambda_1 x$, and $v_1 = y$. Thus, the eigenvalues of $A - \lambda_1 xy^T$ are, counting algebraic multiplicity, $\gamma_1, \lambda_2, \cdots, \lambda_n$, where $\gamma_1 = \lambda_1 + y^T(-\lambda_1 x) = \lambda_1 - \lambda_1 = 0$. Therefore, $\lambda_1 \neq \lambda_i$ for all $i = 2, 3, \cdots, n$. □

2.2 Frobenius' Theorem for Irreducible Nonnegative Matrices

Most nonnegative matrices in applications are not positive, so it is necessary to develop the spectral theory for them. This theory is named after Frobenius since he [Frobenius (1912)] extended Perron's results to this more general case in 1912. We begin our discussion with the following simple result about the *resolvent* $R(\lambda, A) \equiv (\lambda I - A)^{-1}$ of A, which is useful in the study of nonnegative matrices and more general positive operators.

Proposition 2.1. *Let $A \in \mathbb{R}^{n \times n}$ and suppose that $A \geq 0$. Then*

$$R(\lambda, A) \geq 0$$

for all $\lambda > r(A)$.

Proof By Corollary A.7, the Laurent series (A.7) for $R(\lambda, A)$ converges for $|\lambda| > r(A)$. Since each term of the series is nonnegative for $\lambda > r(A)$, it is clear that $R(\lambda, A) \geq 0$. □

Our first spectral result below shows that the spectral radius is still an eigenvalue for an arbitrary nonnegative square matrix, and its proof is based on Perron's theorem and a limit argument.

Theorem 2.5. *Let $A \in \mathbb{R}^{n \times n}$ and suppose that $A \geq 0$. Then $r(A)$ is an eigenvalue of A, and there is a nonnegative eigenvector x and a nonnegative left eigenvector y associated with $r(A)$.*

Proof Let $A(\epsilon) = A + \epsilon ee^T = (a_{ij} + \epsilon) > 0$, where ϵ is an arbitrary positive number. Denote by $x(\epsilon) > 0$ the Perron vector of $A(\epsilon)$, so that $e^T x(\epsilon) = 1$. Since all such vectors $x(\epsilon)$ are contained in the closed unit ball of $\mathbb{R}^n$ under the vector 1-norm, by the Bolzano-Weierstrass theorem in real analysis [Roydon (1968)], there is a monotonically decreasing sequence $\{\epsilon_k\}$ with $\lim_{k\to\infty} \epsilon_k = 0$ such that $\lim_{k\to\infty} x(\epsilon_k) = x \geq 0$ exists. Moreover, we have $e^T x = \lim_{k\to\infty} e^T x(\epsilon_k) = 1$. Since $A(\epsilon_k) \geq A(\epsilon_{k+1}) \geq \cdots \geq A$, Corollary 1.1 ensures that the sequence of the positive numbers $r(A(\epsilon_k))$ is monotonically decreasing, hence the limit $r \equiv \lim_{k\to\infty} r(A(\epsilon_k))$ exists and clearly $r \geq r(A)$. Taking limit from the equality

$$A(\epsilon_k)x(\epsilon_k) = r(A(\epsilon_k))x(\epsilon_k),$$

we obtain that $Ax = rx$, so r is an eigenvalue of A. Since $r \leq r(A)$ and $r \geq r(A)$, it must be that $r = r(A)$.

Applying the above argument to A^T gives the existence of a nonnegative left eigenvector of A corresponding to $r(A)$. □

The spectral radius $r(A)$ of a nonnegative matrix $A \in \mathbb{R}^{n \times n}$ is sometimes called the *Perron root* of A, but the terms of the Perron vector and the left Perron vector are not applicable to general nonnegative square matrices since a normalized nonnegative eigenvector associated with the Perron root is not necessarily unique for arbitrary nonnegative matrices. A discussion of the peripheral spectrum for a nonnegative square matrix will be performed in the last part of the section.

We shall use the properties of irreducible nonnegative matrices introduced in the previous chapter to provide more information about their spectral properties. The following result is the main part of the so-called Perron-Frobenius theorem which extends Perron's theorem from positive matrices to irreducible nonnegative matrices.

Theorem 2.6. *(The Perron-Frobenius theorem) Suppose that $A \in \mathbb{R}^{n \times n}$ is nonnegative and irreducible. Then*

(i) *the spectral radius $r(A)$ of A is a positive eigenvalue with a positive eigenvector x and a positive left eigenvector y;*

(ii) *the eigenvalue $r(A)$ is geometrically simple;*

(iii) *any nonnegative eigenvector of A is a positive scalar multiple of x corresponding to $r(A)$;*

(iv) *the maximal eigenvalue $r(A)$ is algebraically simple.*

Proof By Theorem 2.5, $r(A)$ is an eigenvalue with a nonnegative eigenvector x and left eigenvector y. Since $Ax = r(A)x$,

$$(I + A)^{n-1}x = [1 + r(A)]^{n-1}x.$$

From Theorem 1.5, $(I + A)^{n-1} > 0$. Then $(I + A)^{n-1}x > 0$, so

$$x = \frac{1}{[1 + r(A)]^{n-1}}(I + A)^{n-1}x > 0.$$

It follows that $r(A) = (e^T x)^{-1} e^T Ax > 0$ is a positive eigenvalue of A with a positive eigenvector x. The same reason applied to A^T proves the positivity of y. This proves (i). (ii) and (iii) can be proved the same way as for (ii) and (iii) of Theorem 2.1, and the demonstration of (iv) is basically a re-statement of the proof to Theorem 2.4. □

Corollary 2.2. *Let $A \in \mathbb{R}^{n \times n}$ be nonnegative and irreducible. Then there exists a unique vector x such that $Ax = r(A)x$, $x > 0$, and $e^T x = 1$, and there exists a unique vector y such that $y^T A = r(A)y^T$, $y > 0$, and $y^T e = 1$.*

The unique normalized eigenvector x and left eigenvector y in the above corollary for any irreducible nonnegative square matrix A are again called the *Perron vector* and the *left Perron vector* of A, respectively.

We now study *variational inequalities* for the purpose of estimating the maximal eigenvalue $r(A)$ of a nonnegative square matrix A. First, we give a simple but useful result.

Proposition 2.2. *Let $A \in \mathbb{R}^{n \times n}$ be a nonnegative matrix and let $v \in \mathbb{R}^n$ be a positive vector. If $av \le Av$ for some number $a \ge 0$, then $a \le r(A)$, and if there is a number $b \ge 0$ such that $Av \le bv$, then $r(A) \le b$. Moreover, if $av < Av$, then $a < r(A)$, and if $Av < bv$, then $r(A) < b$.*

Proof. If $av \le Av$, then
$$ay^T v \le y^T Av = r(A)y^T v,$$
where y is the left nonnegative eigenvector of A corresponding to $r(A)$ from Theorem 2.5. $y^T v > 0$ since $y \ge 0$, $y \ne 0$, and $v > 0$. Therefore $a \le r(A)$. If $av < Av$, then there is a number $a' > a$ such that $a'v \le Av$. Thus, $a < a' \le r(A)$. The upper bounds can be proved similarly. □

Corollary 2.3. *Let $A \in \mathbb{R}^{n \times n}$ be a nonnegative matrix. If $Ax = \lambda x$ and $x > 0$, then $\lambda = r(A)$.*

Proof Since $\lambda x \le Ax \le \lambda x$, Proposition 2.2 ensures that $\lambda \le r(A) \le \lambda$. Hence, $\lambda = r(A)$. □

Corollary 2.4. *Let $A \in \mathbb{R}^{n \times n}$ be nonnegative. Then*

$$\min_{1 \le i \le n} \sum_{j=1}^{n} a_{ij} \le r(A) \le \max_{1 \le i \le n} \sum_{j=1}^{n} a_{ij} \tag{2.5}$$

and

$$\min_{1 \le j \le n} \sum_{i=1}^{n} a_{ij} \le r(A) \le \max_{1 \le j \le n} \sum_{i=1}^{n} a_{ij}. \tag{2.6}$$

Proof We prove the inequality (2.5) only since (2.6) can be proved by applying (2.5) to A^T directly. Let

$$a = \min_{1 \leq i \leq n} \sum_{j=1}^{n} a_{ij} \quad \text{and} \quad b = \max_{1 \leq i \leq n} \sum_{j=1}^{n} a_{ij}.$$

Then $ae \leq Ae \leq be$, so Proposition 2.2 implies (2.5). $\qquad\square$

Since $\sigma(P^{-1}AP) = \sigma(A)$ for any nonsingular matrix $P \in \mathbb{R}^{n \times n}$, which ensures that $r(P^{-1}AP) = r(A)$, applying the above corollary to the matrix $X^{-1}AX = (x_i^{-1}a_{ij}x_j)$ for any positive vector $x = (x_1, \cdots, x_n)^T \in \mathbb{R}^n$, where $X = \text{diag}(x_1, \cdots, x_n)$, we obtain the following more general result.

Theorem 2.7. *Let $A \in \mathbb{R}^{n \times n}$ be a nonnegative matrix. Then for any positive vector $x \in \mathbb{R}^n$,*

$$\min_{1 \leq i \leq n} \frac{1}{x_i} \sum_{j=1}^{n} a_{ij}x_j \leq r(A) \leq \max_{1 \leq i \leq n} \frac{1}{x_i} \sum_{j=1}^{n} a_{ij}x_j$$

and

$$\min_{1 \leq j \leq n} x_j \sum_{i=1}^{n} \frac{a_{ij}}{x_i} \leq r(A) \leq \max_{1 \leq j \leq n} x_j \sum_{i=1}^{n} \frac{a_{ij}}{x_i}.$$

Corollary 2.5. *Let $A \in \mathbb{R}^{n \times n}$ be a nonnegative matrix. If A has a positive eigenvector, then*

$$r(A) = \max_{x > 0} \min_{1 \leq i \leq n} \frac{1}{x_i} \sum_{j=1}^{n} a_{ij}x_j = \min_{x > 0} \max_{1 \leq i \leq n} \frac{1}{x_i} \sum_{j=1}^{n} a_{ij}x_j. \qquad (2.7)$$

Similarly, if A has a positive left eigenvector, then

$$r(A) = \max_{x > 0} \min_{1 \leq j \leq n} x_j \sum_{i=1}^{n} \frac{a_{ij}}{x_i} = \min_{x > 0} \max_{1 \leq j \leq n} x_j \sum_{i=1}^{n} \frac{a_{ij}}{x_i}. \qquad (2.8)$$

Proof By Corollary 2.3, $r(A)$ is the eigenvalue associated with a positive eigenvector x^*. Clearly

$$r(A) = \frac{1}{x_i^*} \sum_{j=1}^{n} a_{ij}x_j^*, \quad \forall\, i = 1, \cdots, n.$$

Thus (2.7) follows from the first formula of Theorem 2.7. Applying the same argument to A^T gives (2.8). $\qquad\square$

The following proposition indicates that the powers $\left(r(A)^{-1}A\right)^k$ of the normalized matrix $r(A)^{-1}A$ are uniformly bounded if a nonnegative square matrix A has a positive eigenvector, a useful property for proving the mean ergodicity of such matrices in the next section.

Proposition 2.3. *Let $A \in \mathbb{R}^{n \times n}$ be a nonnegative matrix. If A has a positive eigenvector x, then for all $k = 1, 2, \cdots$,*

$$\frac{\min_{1 \leq i \leq n} x_i}{\max_{1 \leq i \leq n} x_i} e \leq \left(\frac{1}{r(A)} A \right)^k e \leq \frac{\max_{1 \leq i \leq n} x_i}{\min_{1 \leq i \leq n} x_i} e. \tag{2.9}$$

Proof Since $Ax = r(A)x$ from Corollary 2.3, $A^k x = r(A)^k x$ for all positive integers k. Denote $m = \min_{1 \leq i \leq n} x_i$ and $M = \max_{1 \leq i \leq n} x_i$. Then we have $me \leq x \leq Me$. Since $A^k \geq 0$,

$$mA^k e \leq A^k x = r(A)^k x \leq r(A)^k Me,$$

which implies the right-hand side inequality of (2.9). The left-hand side one of the same formula follows from

$$r(A)^k me \leq r(A)^k x = A^k x \leq MA^k e. \qquad \square$$

Using a limit argument, we are able to extend part of Proposition 2.2 from positive vectors to nonnegative ones.

Theorem 2.8. *Let $A \in \mathbb{R}^{n \times n}$ and $x \in \mathbb{R}^n$. Suppose that $A \geq 0$, $x \geq 0$, and $x \neq 0$. If $ax \leq Ax$ for some $a \geq 0$, then $a \leq r(A)$.*

Proof Let $A(\epsilon) = (a_{ij} + \epsilon)$ with $\epsilon > 0$, and let $y(\epsilon) > 0$ be the left Perron vector of $A(\epsilon)$. Since $ax \leq Ax < A(\epsilon)x$, we have

$$ay(\epsilon)^T x < y(\epsilon)^T A(\epsilon)x = r(A(\epsilon))y(\epsilon)^T x.$$

It follows from $y(\epsilon)^T x > 0$ that $a < r(A(\epsilon))$ for all $\epsilon > 0$, which gives rise to $a \leq \lim_{\epsilon \to 0} r(A(\epsilon)) = r(A)$. $\qquad \square$

As a corollary of the above theorem, we have the following variational result which extends part of Corollary 2.5.

Corollary 2.6. *Let $A \in \mathbb{R}^{n \times n}$ be nonnegative. Then*

$$r(A) = \max_{x \geq 0, x \neq 0} \min_{1 \leq i \leq n, x_i \neq 0} \frac{1}{x_i} \sum_{j=1}^{n} a_{ij} x_j.$$

Proof For any nonzero vector $x \geq 0$, define

$$a(x) = \min_{1 \leq i \leq n, x_i \neq 0} \frac{1}{x_i} \sum_{j=1}^{n} a_{ij} x_j.$$

Then $a(x)x \leq Ax$, so $a(x) \leq r(A)$ by Theorem 2.8. Let $x^* \geq 0$ be an eigenvector of A associated with $r(A)$, the existence of which is guaranteed by Theorem 2.5. Then $a(x^*) = r(A)$. $\qquad\square$

The following proposition provides a sufficient condition for a vector inequality to become an equality.

Proposition 2.4. *Let $A \in \mathbb{R}^{n \times n}$. Suppose that $A \geq 0$ with a positive left eigenvector. If $x \geq 0$ is such that $x \neq 0$ and either $Ax \geq r(A)x$ or $Ax \leq r(A)x$, then $Ax = r(A)x$.*

Proof Let y be a positive left eigenvector of A. Then the corresponding eigenvalue must be $r(A)$ by Corollary 2.3 applied to A^T. So $y^T A = r(A)y^T$, which implies that

$$y^T [Ax - r(A)x] = r(A)y^T x - r(A)y^T x = 0.$$

The assumption that $y > 0$ and either $Ax - r(A)x \geq 0$ or $Ax - r(A)x \leq 0$ then imply $Ax - r(A)x = 0$. $\qquad\square$

In the last part of the section we turn our attention to the structure analysis of the peripheral spectrum for a given nonnegative square matrix. When A is a positive matrix, Perron's theorem guarantees that $r(A)$ is the unique eigenvalue of A of maximal modulus. If A is only nonnegative, there may be more than one eigenvalue of maximal modulus, but the peripheral spectrum of A exhibits a nice group property. The peripheral spectrum of an irreducible nonnegative matrix A turns out to be the roots of 1 multiplied by the spectral radius of A. In order to determine this fine structure of the peripheral spectrum, we need the following remarkable result for general complex square matrices.

Lemma 2.3. (Wielandt) *Let a real matrix $A \in \mathbb{R}^{n \times n}$ and a complex matrix $B \in \mathbb{C}^{n \times n}$ be such that A is irreducible and $|B| \leq A$. Then $\lambda = r(A)\exp(i\phi)$ is an eigenvalue of B if and only if*

$$B = \exp(i\phi)DAD^{-1}, \tag{2.10}$$

where the diagonal matrix D satisfies $|D| = I$.

Proof Theorem 1.1 ensures $r(B) \leq r(A)$. Suppose that (2.10) is satisfied. Then $B \sim \exp(i\phi)A$, that is, B is similar to $\exp(i\phi)A$. Since $\exp(i\phi)r(A)$ is an eigenvalue of $\exp(i\phi)A$, it is an eigenvalue of B.

Conversely, let $r(B) = r(A)$ and $\lambda = r(A)\exp(i\phi)$ be an eigenvalue of B. Then there exists $x \neq 0$ with $Bx = \lambda x$, hence

$$r(A)|x| = |\lambda||x| = |\lambda x| = |Bx| \leq |B||x| \leq A|x|.$$

Since A is irreducible, A has a positive left eigenvector, so we see from Proposition 2.4 that $A|x| = r(A)|x|$. Thus, $A|x| = |B||x| = |Bx|$, which implies $(A - |B|)|x| = 0$. On the other hand, $|x| > 0$ by Theorem 2.6, which implies $A = |B|$. Define a diagonal matrix

$$D = \operatorname{diag}\left(\frac{x_1}{|x_1|}, \frac{x_2}{|x_2|}, \cdots, \frac{x_n}{|x_n|}\right).$$

Then $x = D|x|$, therefore

$$\exp(-i\phi)D^{-1}BD|x| = r(A)D^{-1}x = r(A)|x| = A|x|.$$

Since $|\exp(-i\phi)D^{-1}BD| = |B| = A$, we have

$$|\exp(-i\phi)D^{-1}BD||x| = \exp(-i\phi)D^{-1}BD|x|.$$

It follows that $\exp(-i\phi)D^{-1}BD = |\exp(-i\phi)D^{-1}BD| = A$ since $|x| > 0$. In other words, (2.10) is satisfied. □

It is easy to see that Theorem 1.1 is a corollary of Lemma 2.3 by a continuity argument. However, if A is reducible, the condition (2.10) for the equality $r(B) = r(A)$ is not necessary.

Corollary 2.7. *Let $A \in \mathbb{R}^{n \times n}$ be irreducible. If $A \geq B \geq 0$ and $A \neq B$, then $r(A) > r(B)$.*

Lemma 2.3 implies the following result on the structure of the peripheral spectrum of an irreducible nonnegative matrix.

Theorem 2.9. *Let $A \in \mathbb{R}^{n \times n}$ be irreducible and nonnegative with index of imprimitivity s. Then all the eigenvalues of A of modulus $r(A)$ are the distinct sth roots of $r(A)^s$.*

Proof Suppose that $\lambda_j = r(A)\exp(i\phi_j), j = 1, 2, \cdots, s$, are the elements of the peripheral spectrum of A. Since $|\lambda_j| = r(A)$, equality (2.10) in Lemma 2.3 with $B = A$ and $\lambda = \lambda_j$ implies that

$$A = \exp(i\phi_j)D_j A D_j^{-1}, \quad j = 1, 2, \cdots, s. \tag{2.11}$$

Hence, $A \sim \exp(i\phi_j)A$. Since $r(A)$ is a simple eigenvalue of A, it follows that $\exp(i\phi_j)r(A)$ is a simple eigenvalue of $\exp(i\phi_j)A$, and thus of A, for each j. By (2.11), for all j and k,

$$\begin{aligned} A &= \exp(i\phi_j)D_j[\exp(i\phi_k)D_k A D_k^{-1}]D_j^{-1} \\ &= \exp(i(\phi_j + \phi_k))(D_j D_k)A(D_j D_k)^{-1}, \end{aligned}$$

namely $A \sim \exp(i(\phi_j + \phi_k))A$, so $\exp(i(\phi_j + \phi_k))r(A)$ is an eigenvalue of A. Hence, for all j, $k = 1, 2, \cdots, s$, the numbers $\exp(i(\phi_j + \phi_k))r(A)$ are among the eigenvalues

$$\exp(i\phi_1)r(A), \ \exp(i\phi_2)r(A), \ \cdots, \ \exp(i\phi_s)r(A).$$

In other words, the set of the distinct numbers

$$\{\exp(i\phi_1), \ \exp(i\phi_2), \ \cdots, \ \exp(i\phi_s)\}$$

is closed under multiplication, so they are the sth roots of 1. □

The next theorem shows that not only the peripheral spectrum, but also the whole spectrum of any irreducible and nonnegative square matrix has the specified cyclic property.

Theorem 2.10. *The spectrum of an irreducible nonnegative matrix $A \in \mathbb{R}^{n \times n}$ with index s is invariant under a rotation of angle $2\pi/s$, but not so under a rotation of any smaller positive angle.*

Proof Let $\sigma(A) = \{\lambda_1, \lambda_2, \cdots, \lambda_n\}$ be the spectrum of A, counting algebraic multiplicity. Then, since A and $\exp(i2\pi/s)A$ are similar from the proof of the previous theorem,

$$\sigma(A) = \left\{ \lambda_1 \exp\left(i\frac{2\pi}{s}\right), \lambda_2 \exp\left(i\frac{2\pi}{s}\right), \cdots, \lambda_n \exp\left(i\frac{2\pi}{s}\right) \right\}.$$

This proves the first part, and the second part also follows from the same theorem since the peripheral spectrum of A cannot be preserved under a rotation of positive angle less than $2\pi/s$. □

The proof of the next structure theorem for the peripheral spectrum of a general nonnegative square matrix is based on the following preliminary result, which can also be extended to infinite dimensional positive operators of Banach lattices in Chapter 6.

Lemma 2.4. *Let $A \in \mathbb{R}^{n \times n}$ be nonnegative. If $Ax = \lambda x$ with $|\lambda| = 1$ and $A|x| = |x|$, then for all integers k,*

$$\left(|x_1|(\operatorname{sgn} x_1)^k, \ |x_2|(\operatorname{sgn} x_2)^k, \ \cdots, \ |x_n|(\operatorname{sgn} x_n)^k \right)^T$$

is an eigenvector of A corresponding to the eigenvalue λ^k.

Proof Let $v = (\operatorname{sgn} x_1, \cdots, \operatorname{sgn} x_n)^T$. Then $x_i = |x_i| v_i$ for all i. The equalities $Ax = \lambda x$ and $A|x| = |x|$ can be written as

$$\sum_{j=1}^{n} a_{ij} |x_j| v_j = \lambda |x_i| v_i \text{ and } \sum_{j=1}^{n} a_{ij} |x_j| = |x_i|, \quad i = 1, 2, \cdots, n.$$

Hence $\lambda v_i = v_j$ for all pairs (i, j) with $a_{ij} |x_j| > 0$. Then $\lambda^k v_i^k = v_j^k$ for these same (i, j), which proves the lemma. □

Theorem 2.11. *The peripheral spectrum of any nonnegative square matrix $A \in \mathbb{R}^{n \times n}$ must be cyclic.*

Proof Without loss of generality, assume $r(A) = 1$. Let $Ax = \lambda x$, $x \neq 0$, and $|\lambda| = 1$. Then $|x| = |\lambda x| = |Ax| \leq A|x|$. Denote by N the smallest ideal of $\mathbb{C}^n$ containing $A^k |x|$ for all $k \geq 0$.

Suppose first that $N = \mathbb{C}^n$. By Theorem 2.5, there is a nonzero vector $y \geq 0$ such that $y^T A = y^T$. We must have $y^T |x| > 0$, since otherwise $y^T A^k |x| = y^T |x| = 0$ for all k and so $y = 0$. Let $M = \{z \in \mathbb{C}^n : y^T |z| = 0\}$. Then M is an ideal and $AM \subset M$. Since $x \notin M$, we see that $\hat{x} \equiv x + M \neq 0$ in the quotient space $\mathbb{C}^n / M$. If we denote $A_M : \mathbb{C}^n / M \to \mathbb{C}^n / M$ the induced linear operator, then $Ax = \lambda x$ implies $A_M \hat{x} = \lambda \hat{x}$ and the fact that $A|x| - |x| \in M$ implies $A_M |\hat{x}| = |\hat{x}|$. Now Lemma 2.4 ensures that $\lambda^k \in \sigma(A_M)$ for all positive integers k, hence $\lambda^k \in \sigma(A)$ since $\sigma(A_M) \subset \sigma(A)$ from the formula (A.1) in Appendix A.2.

If $N \neq \mathbb{C}^n$, then the above argument can be applied to the restriction $A|_N$ of A to N since $r(A|_N) = 1$. Thus, $\lambda^k \in \sigma(A|_N) \subset \sigma(A)$. This proves that the spectrum $\sigma(A)$ is cyclic. □

Proposition 2.5. *Let $A \in \mathbb{R}^{n \times n}$ be nonnegative. Then its maximal eigenvalue $r(A)$ is a pole of $R(\lambda, A)$ of the maximal order among the elements of the peripheral spectrum of A.*

Proof Let γ be any point in the peripheral spectrum of A and set $\lambda = t\gamma$, where $t > 1$. Then from the Laurent series representation given by Corollary A.7 of Appendix A.4, we see that

$$|(\lambda - \gamma)^k R(\lambda, A)| \leq (t - 1)^k R(t, A)$$

for each k. This proves the assertion of the theorem. □

The last theorem of this section gives a relation between the index of imprimitivity of a matrix and its characteristic polynomial.

Theorem 2.12. *Let $A \in \mathbb{R}^{n \times n}$ be nonnegative and irreducible with index s. If the characteristic polynomial of A is*

$$\lambda^n + a_1 \lambda^{n_1} + \cdots + a_k \lambda^{n_k},$$

where $n > n_1 > \cdots > n_k \geq 0$ and $a_j \neq 0$ for all j, then

$$s = \text{g.c.d.} \ (n - n_1, n_1 - n_2, \cdots, n_{k-1} - n_k). \tag{2.12}$$

Proof Since the two matrices A and $\exp(i2\pi/s)A$ have exactly the same characteristic polynomials, we obtain the equality

$$\lambda^n + a_1 \lambda^{n_1} + a_2 \lambda^{n_2} + \cdots + a_k \lambda^{n_k}$$
$$= \lambda^n + a_1 \theta^{n-n_1} \lambda^{n_1} + a_2 \theta^{n-n_2} \lambda^{n_2} + \cdots + a_k \theta^{n-n_k} \lambda^{n_k},$$

where $\theta = \exp(i2\pi/s)$. It follows that

$$a_j = a_j \theta^{n-n_j}, \quad j = 1, 2, \cdots, k,$$

and therefore s divides each of $n - n_1, n_1 - n_2, \cdots, n_{k-1} - n_k$. Now, by Theorem 2.10, A and $\exp(i2\pi/t)A$ have different characteristic polynomials for $t > s$. Thus (2.12) follows. $\square$

Corollary 2.8. *Any irreducible and nonnegative matrix $A \in \mathbb{R}^{n \times n}$ with a positive trace is primitive.*

Proof This is true because

$$\text{g.c.d.}\{n - (n - 1), \cdots\} = 1. \qquad \square$$

Corollary 2.9. *If a square matrix $A \geq 0$ is irreducible and $B \geq 0$ of the same size has a positive trace, then $A + B$ is primitive.*

Proof Since $A + B$ is irreducible and $\text{tr}(A + B) \geq \text{tr}B > 0$, Corollary 2.8 gives the conclusion. $\square$

Remark 2.2. Corollary 2.9 is identical to Corollary 1.5.

2.3 Ergodic Theorems for Nonnegative Matrices

In many applications of nonnegative matrices, for example, in the theory of stochastic matrices of the next chapter and its application to stationary Markov chains with finitely many states in Chapter 4, it is important to know the asymptotic behavior of the powers A^k of a nonnegative square matrix A as k approaches infinity. It is well-known that $\lim_{k\to\infty} A^k = 0$

if and only if $r(A) < 1$ and that the sequence $\{A^k\}$ is unbounded and so $\lim_{k \to \infty} A^k$ cannot exist if $r(A) > 1$. Consequently, the most interesting situation in the convergence theory of nonnegative matrices is the eventual property of A^k as k approaches infinity in the case of $r(A) = 1$.

Unlike the case of positive matrices, the sequence of normalized powers $\left(r(A)^{-1}A\right)^k$ of an irreducible nonnegative matrix may not converge as k goes to infinity. For example, if

$$A = \begin{bmatrix} 0 & 1 \\ 1 & 0 \end{bmatrix},$$

then the limit $\lim_{k \to \infty} A^k$ does not exist since $A^k = I$ when k is even and $A^k = A$ when k is odd.

A less restrictive requirement is the *convergence in the mean* or of the *Cesáro averages*, in other words, the convergence for the sequence of the average matrices of $I, A, A^2, \cdots, A^{k-1}$ for all natural numbers k. Let $A \in \mathbb{C}^{n \times n}$. The *Cesáro averages* of the powers A^k of A is defined as

$$A_k \equiv \frac{1}{k} \sum_{j=0}^{k-1} A^j, \quad k = 1, 2, \cdots.$$

Clearly, if the sequence of the powers of A converges, so does the averages sequence. As will be shown in this section, the averages sequence of the powers of the above 2×2 permutation matrix does converge. The following Theorem 2.13 is a general limit theorem for irreducible nonnegative matrices; its proof is based on the *mean ergodic theorem*, Lemma 2.5 below, for general complex square matrices of spectral radius 1.

Definition 2.1. A matrix $A \in \mathbb{C}^{n \times n}$ with $r(A) = 1$ is said to be *mean ergodic* if $\lim_{k \to \infty} A_k$ exists.

In the proof of Lemma 2.5 below, we need the simple identity

$$AA_k - A_k = \frac{1}{k}(A^k - I) \tag{2.13}$$

for A_k, which can be verified by a direct computation.

Lemma 2.5. (The mean ergodic theorem for matrices) *Let $A \in \mathbb{C}^{n \times n}$ be such that $r(A) = 1$. If the sequence $\{A^k\}$ is uniformly bounded, then the matrix A is mean ergodic.*

Proof Let $\|A\|$ be any matrix norm of A. Since the sequence $\{\|A^k\|\}$ is uniformly bounded, so is the sequence $\{\|A_k\|\}$. The Balzano-Weierstrass theorem implies that there exists a subsequence $\{A_{k_j}\}$ of $\{A_k\}$ and a matrix $E \in \mathbb{C}^{n \times n}$ such that

$$\lim_{j \to \infty} A_{k_j} = E.$$

We show that $\lim_{k \to \infty} A_k = E$. Suppose that there is a subsequence $\{A_{k'_j}\}$ of $\{A_k\}$ with $\lim_{j \to \infty} A_{k'_j} = E'$. Then (2.13) implies $AE = E$ and $AE' = E'$, so $A_k E = E$ and $A_k E' = E'$ for all positive integers k. Since

$$\|E - E'\| = \|A_{k'_j}(E - A_{k_j}) + A_{k_j}(A_{k'_j} - E')\|$$

$$\leq \|A_{k'_j}\| \|E - A_{k_j}\| + \|A_{k_j}\| \|A_{k'_j} - E'\| \to 0$$

as $j \to \infty$, we see that $E = E'$. Since all convergent subsequences of A_k converge to E, so $\lim_{k \to \infty} A_k = E$. $\qquad\square$

Remark 2.3. The limit matrix E satisfies $AE = EA = E$ and $E^2 = E$. It follows that $R(E) = N(I - A)$ and $N(E) = R(I - A)$. In other words, E is a projection from $\mathbb{C}^n$ onto the *fixed point space* of A along the range of $I - A$; see Remark 2.1.

Theorem 2.13. *Let $A \geq 0$ be $n \times n$ and irreducible. Then*

$$\lim_{k \to \infty} \frac{1}{k} \sum_{j=0}^{k-1} \left(\frac{1}{r(A)} A \right)^j = E, \qquad (2.14)$$

where $E = xy^T$, $x > 0$, $y > 0$, and x, y satisfy the equalities

$$Ax = r(A)x, \quad y^T A = r(A)y^T, \quad y^T x = 1.$$

Proof The existence of x and y is guaranteed by Theorem 2.6. Since $r(A) > 0$, the spectral radius of the matrix $r(A)^{-1}A$ is 1. By Proposition 2.3, the entries of the sequence of the matrices $\left(r(A)^{-1}A \right)^k$ are uniformly bounded for all $k = 1, 2, \cdots$, so Lemma 2.5 implies that $r(A)^{-1}A$ is mean ergodic. Thus, $\lim_{k \to \infty} (r(A)^{-1}A)_k = E$ for some matrix E which is a projection from $\mathbb{R}^n$ onto $N(I - A)$ that is spanned by x. So $E = xz^T$ for some vector z. Applying the same conclusion to A^T gives $E^T = yw^T$ for some vector w. Since $E^T = zx^T$ and $E^2 = E$, we immediately see that $z = y$ and $w = x$. Therefore, (2.14) is valid. $\qquad\square$

Theorem 2.13 gives a sufficient condition for the mean ergodicity of nonnegative matrices. Now we further study equivalent conditions for the mean ergodicity of $A \in \mathbb{R}_+^{n \times n}$. The existence of $\lim_{\lambda \to 1}(\lambda - 1)R(\lambda, A)$ does not, in general, imply the existence of the limit of A_k. However, if 1 is a *simple* pole of $R(\lambda, A)$, we can prove the following decomposition result for general matrices, which will be applied to prove Theorem 2.14 below.

Lemma 2.6. *Let $A \in \mathbb{C}^{n \times n}$ with $r(A) = 1$. If 1 is a simple pole of the resolvent of A and the only eigenvalue of modulus 1, then $A = E + B$, where $E = \lim_{\lambda \to 1} (\lambda - 1) R(\lambda, A)$ is a projection onto the null space of $I - A$, $EB = BE = 0$, and $r(B) < 1$.*

Proof Since 1 is a simple pole of $R(\lambda, A)$ by assumption, the limit $E = \lim_{\lambda \to 1} (\lambda - 1) R(\lambda, A)$ exists. Then, because of the Laurent series expansion $R(\lambda, A) = \sum_{k=0}^{\infty} \lambda^{-(k+1)} A^k$,

$$E = \lim_{\lambda \to 1} \sum_{k=1}^{\infty} \frac{A^{k-1}(A - I)}{\lambda^k} + I.$$

Therefore,

$$\begin{aligned}
AE &= A \left[\lim_{\lambda \to 1} \sum_{k=1}^{\infty} \frac{A^{k-1}(A - I)}{\lambda^k} + I \right] \\
&= \lim_{\lambda \to 1} \left(\lambda \sum_{k=1}^{\infty} \frac{A^k(A - I)}{\lambda^{k+1}} \right) + A \\
&= \lim_{\lambda \to 1} \left[\sum_{k=1}^{\infty} \frac{A^{k-1}(A - I)}{\lambda^k} - (A - I) \right] + A \\
&= \lim_{\lambda \to 1} \sum_{k=1}^{\infty} \frac{A^{k-1}(A - I)}{\lambda^k} + I = E.
\end{aligned}$$

Similarly, $EA = E$. So, $(\lambda I - A)E = (\lambda - 1)E$, which implies

$$E = (\lambda - 1)R(\lambda, A)E, \quad \forall \, \lambda \in \rho(A).$$

Letting $\lambda \to 1$ in the above equality gives $E^2 = E$. It is obvious now that $Ax = x$ if and only if $Ex = x$. Define $B = A - E$, and we see that $EB = BE = 0$. To show that $r(B) < 1$, assume $Bx = \lambda x$ for some complex number λ with $|\lambda| \geq 1$. Then $Ex = 0$ since $\lambda Ex = EBx = 0$, hence $Ax = \lambda x$. If $\lambda = 1$, then $x = Ex = 0$, and if $\lambda \neq 1$, then $x = 0$ since 1 is the only peripheral spectral point of A. $\qquad \square$

Theorem 2.14. *Let $A \in \mathbb{R}^{n \times n}$ be nonnegative with $r(A) = 1$. Then the following are equivalent:*

(i) *A is mean ergodic.*

(ii) *$\lambda = 1$ is a simple pole of $R(\lambda, A)$.*

(iii) *The sequence $\{A^k\}$ is uniformly bounded.*
Moreover, the existence of either one of the two limits, $\lim_{\lambda \to 1}(\lambda - 1)R(\lambda, A)$ and $\lim_{k \to \infty} k^{-1}\sum_{j=0}^{k-1} A^j$, will imply the other with the same value.

Proof (i) $\Rightarrow$ (ii): Let $E = \lim_{k \to \infty} k^{-1}\sum_{j=0}^{k-1} A^j$. In the Laurent series expansion of $R(\lambda, A)$ at $\lambda = 1$, the coefficient of $(\lambda - 1)^{-m}$ with $m \geq 1$ is given by $Q_m = (I - A)^{m-1}Q_1$, where $Q_1 = \lim_{\lambda \to 1}(\lambda - 1)R(\lambda, A)$ is the *residuum* at $\lambda = 1$. Since $EA = E$, $EQ_m = 0$ for all integers $m > 1$. On the other hand, if m_0 is the order of the pole, $Q_{m_0}x$ is a fixed vector of A for any $x \in \mathbb{C}^n$, hence $EQ_{m_0} = Q_{m_0}$. Therefore $m_0 = 1$.

(ii) $\Rightarrow$ (iii): The peripheral spectrum of A consists only of the roots of 1, so there exists an integer $l > 0$ such that the peripheral spectrum of A^l is $\{1\}$. Let $\lambda \in (1, 2]$ and $\gamma = \lambda^l$. Then, since the matrix A is nonnegative,

$$0 \leq (\gamma - 1)R(\gamma, A^l) \leq (\lambda^l - 1)R(\gamma, A)$$
$$\leq l2^{l-1}(\lambda - 1)R(\lambda, A).$$

When $\gamma \to 1^+$, $(\gamma - 1)R(\gamma, A^q)$ is bounded for $(\lambda - 1)R(\lambda, A)$ is bounded. Thus, $\gamma = 1$ is a simple pole of $R(\gamma, A^l)$. Applying Lemma 2.6 to A^l gives $A^l = E + B$, where $E^2 = E, EB = BE = 0$, and $r(B) < 1$. It follows that $B^k \to 0$ as $k \to \infty$. Since

$$A^{lk} = (A^l)^k = E + B^k, \ \forall \, k \geq 1,$$

$A^{lk} \to E$ as $k \to \infty$, which implies that the sequence $\{A^{lk}\}$ is bounded. Now the set $\{A^k\}$ is the union of the l bounded sets $A^j\{A^{lk}\}$, $j = 0, 1, \cdots, l - 1$, hence it itself is bounded.

(iii) $\Rightarrow$ (i): It is immediate due to Lemma 2.5.

Finally, if either limit mentioned in the theorem exists, then so does the other by the preceding arguments. Suppose so, then we have $AE = EA = E$ and $AQ_1 = Q_1A = Q_1$. Thus,

$$Q_1E = \lim_{k \to \infty} Q_1A_k = Q_1$$

and

$$EQ_1 = \lim_{\lambda \to 1}(\lambda - 1)ER(\lambda, A) = E.$$

Since E and Q_1 commute, $Q_1 = E$. $\qquad \square$

In general, the limit theorem, Theorem 2.3, for positive square matrices is no longer valid for general irreducible nonnegative square matrices, but it can directly be extended to the subclass of primitive matrices with exactly the same proof because of the key fact that $r(A - r(A)xy^T) < r(A)$ for primitive matrices A, where x and y are as in the following theorem.

Theorem 2.15. *If $A \in \mathbb{R}^{n \times n}$ is primitive, then*

$$\lim_{k \to \infty} \left(\frac{1}{r(A)} A \right)^k = xy^T,$$

where $Ax = r(A)x$, $y^T A = r(A)y^T$, $x > 0$, $y > 0$, and $y^T x = 1$.

More generally, we have the following equivalent condition for the existence of $\lim_{k \to \infty} A^k$ for arbitrary square matrices with spectral radius 1, which also implies Theorem 2.15 directly.

Theorem 2.16. *Let $A \in \mathbb{C}^{n \times n}$ with $r(A) = 1$. Then $\lim_{k \to \infty} A^k$ exists if and only if $\lambda = 1$ is a simple pole of the resolvent $R(\lambda, A)$ of A and the only eigenvalue of modulus 1.*

Proof If the sufficient condition is satisfied, then Lemma 2.5 implies that $A^k = E^k + B^k$ for all $k \geq 1$, where $\lim_{k \to \infty} B^k = 0$. So $\lim_{k \to \infty} A^k = E$. Conversely, let the sequence $\{A^k\}$ converge. If λ is an eigenvalue of A with $|\lambda| = 1$ and $x \neq 0$ is a corresponding eigenvector, then $A^k x = \lambda^k x$ for all $k \geq 1$. Since the vector sequence $\{A^k x\}$ converges, so does the number sequence $\{\lambda^k\}$, which implies that $\lambda = 1$. Since $\{A^k\}$ is a bounded sequence, a simple estimate shows that the family of matrices

$$(\lambda - 1)R(\lambda, A) = (\lambda - 1) \sum_{k=0}^{\infty} \frac{A^k}{\lambda^{k+1}}$$

is uniformly bounded for $\lambda > 1$, thus the order of the pole $\lambda = 1$ of $R(\lambda, A)$ is 1. In other words, the eigenvalue 1 is simple.   □

2.4   Characterizations of Irreducibility and Primitivity

So far we have been interested in the spectral analysis of irreducible nonnegative matrices. In this section we present a few characterizations of irreducibility and primitivity.

The first theorem in the following gives several spectral properties which are equivalent to irreducibility.

Theorem 2.17. *Let $A \in \mathbb{R}^{n \times n}$ be a nonnegative matrix. The following statements are equivalent:*

(i) *A is irreducible.*

(ii) A^T is irreducible.

(iii) $r(A)$ is an algebraically simple eigenvalue of A with a positive eigenvector and a positive left eigenvector.

(iv) $R(\lambda, A)v > 0$ for any nonzero and nonnegative vector $v \in \mathbb{R}^n$ and any real number $\lambda > r(A)$.

Proof The equivalence of (i) and (ii) is obvious from the definition of irreducibility, and that (i) $\Rightarrow$ (iii) is from Theorem 2.6. So we need to show that (iii) $\Rightarrow$ (iv) and (iv) $\Rightarrow$ (i) only.

(iii) $\Rightarrow$ (iv). The assumption guarantees that there exist two vectors $x > 0$ and $y > 0$ such that

$$Ax = r(A)x, \ y^T A = r(A)y^T, \ y^T x = 1.$$

Denote by $E = xy^T$ the projection onto the null space $N(r(A)I - A)$. Then, as the proof of Theorem 2.3 shows, $r(A - r(A)E) < r(A)$. Hence

$$\lim_{k \to \infty} \left(\frac{A - r(A)E}{r(A)} \right)^k = 0,$$

from which it follows that for any nonzero vector $v \geq 0$,

$$\lim_{k \to \infty} \left[Ev + \left(\frac{A - r(A)E}{r(A)} \right)^k v \right] = Ev.$$

On the other hand, $Ev = (y^T v)x > 0$. Therefore we can find a positive integer k_0 such that

$$Ev + \left(\frac{A - r(A)E}{r(A)} \right)^k v > 0, \ \forall k > k_0.$$

We now have, for any $\lambda > r(A)$,

$$\lambda R(\lambda, A)v = \sum_{k=0}^{\infty} \left(\frac{A}{\lambda} \right)^k v = \sum_{k=0}^{k_0} \left(\frac{A}{\lambda} \right)^k v + \sum_{k=k_0+1}^{\infty} \left(\frac{A}{\lambda} \right)^k v$$

$$\geq \sum_{k=k_0+1}^{\infty} \frac{1}{\lambda^k} \left[A^k Ev + A^k (I - E)v \right]$$

$$= \sum_{k=k_0+1}^{\infty} \frac{1}{\lambda^k} \left\{ r(A)^k Ev + \left[A^k - r(A)^k E \right] v \right\}$$

$$= \sum_{k=k_0+1}^{\infty} \left(\frac{r(A)}{\lambda} \right)^k \left[Ev + \left(\frac{A - r(A)E}{r(A)} \right)^k v \right] > 0.$$

Therefore, $R(\lambda, A)v > 0$.

(iv) $\Rightarrow$ (i). Denote $A^k = (a_{ij}^{(k)})$. Thanks to Corollary 1.4 and Theorem 1.4, we just need to show that for any pair (i,j), there is a positive integer k such that $a_{ij}^{(k)} > 0$. Let $\lambda > r(A)$. Since

$$R(\lambda, A)e_j = \sum_{k=0}^{\infty} \frac{A^k}{\lambda^{k+1}} e_j > 0$$

by the assumption, its ith component

$$\sum_{k=0}^{\infty} \frac{a_{ij}^{(k)}}{\lambda^{k+1}} > 0.$$

Therefore, $a_{ij}^{(k)} > 0$ for at least one k. $\square$

The following result characterizes primitivity of a matrix.

Theorem 2.18. *Let $A \in \mathbb{R}^{n \times n}$ be nonnegative and irreducible. Then the following statements are equivalent:*
  (i) *A is primitive.*
  (ii) *A^T is primitive.*
  (iii) *$\lim_{k \to \infty} [r(A)^{-1} A]^k$ exists.*
  (iv) *$A^k > 0$ for some $k \geq 1$.*
  (v) *A^k is irreducible for all $k \geq 1$.*

Proof It is obvious that the statements (i) and (ii) are equivalent. Theorem 2.15 means that (i) $\Rightarrow$ (iii). What remains for us to show is the implications (iii) $\Rightarrow$ (iv) $\Rightarrow$ (v) $\Rightarrow$ (i).

(iii) $\Rightarrow$ (iv). Since $A \geq 0$ is irreducible, (iii) implies that A is mean ergodic. Moreover, $\lim_{k \to \infty} [r(A)^{-1} A]^k = xy^T > 0$, where $x > 0$ and $y > 0$ are the positive eigenvector and left eigenvector of A corresponding to $r(A)$ with $y^T x = 1$ (see Theorem 2.13 and its proof). Therefore, there is an integer $k \geq 1$ such that $[r(A)^{-1} A]^k > 0$.

(iv) $\Rightarrow$ (v). Suppose that $A^k > 0$ for some $k \geq 1$. Then $A^{k+l} > 0$ for all $l \geq 0$. If $A^{k'}$ were reducible for some k', then so would be $A^{jk'}$ for all $j \geq 1$. This conflicts with the fact that $A^j > 0$ for all $j \geq k$.

(v) $\Rightarrow$ (i). Suppose that the matrix A is not primitive. Then there exists an eigenvalue $ar(A) \neq r(A)$, where the complex number a satisfies $|a| = 1$. Since $a^s = 1$ for some $s \geq 2$ by Theorem 2.9, A^s has at least two linearly independent eigenvectors associated with eigenvalue $r(A)$, which contradicts the irreducibility of A^s. This completes the proof. $\square$

The next result shows that an irreducible nonnegative matrix with all positive diagonal entries must be primitive.

Theorem 2.19. *Let $A \in \mathbb{R}^{n \times n}$ be nonnegative and irreducible. If $a_{ii} > 0$ for all i, then $A^{n-1} > 0$, and so A is primitive.*

Proof Let $B = A - \text{diag}(a_{11}, a_{22}, \cdots, a_{nn}) \geq 0$. Then B is irreducible. Define $a = \min_{i=1}^{n} a_{ii} > 0$. Then $a^{-1}B \geq 0$ and is also irreducible. Thus, $(I + a^{-1}B)^{n-1} > 0$ by Theorem 1.5. Since $A \geq aI + B = a[I + a^{-1}B]$,

$$A^{n-1} \geq a^{n-1} \left(I + \frac{1}{a} B \right)^{n-1} > 0.$$

Therefore A is primitive from Theorem 2.18. □

Finally, we give two structure theorems of nonnegative matrices.

Theorem 2.20. (The Frobenius standard form for irreducible matrices) *Let $A \in \mathbb{R}^{n \times n}$ be nonnegative and irreducible with index of imprimitivity $s \geq 2$. Then there exists an $n \times n$ permutation matrix P such that*

$$P^T A P = \begin{bmatrix} 0 & A_{12} & 0 & \cdots & 0 & 0 \\ 0 & 0 & A_{23} & \cdots & 0 & 0 \\ \vdots & \vdots & \vdots & & \vdots & \vdots \\ 0 & 0 & 0 & \cdots & 0 & A_{s-1,s} \\ A_{s1} & 0 & 0 & \cdots & 0 & 0 \end{bmatrix}, \tag{2.15}$$

where the zero blocks along the main diagonal are square.

Proof From Theorem 2.9 and its proof,

$$\lambda_j = r(A) \exp \left(i \frac{2\pi j}{s} \right), \quad j = 0, 1, \cdots, s - 1$$

are the eigenvalues of A of modulus $r(A)$, and

$$A = \exp \left(i \frac{2\pi j}{s} \right) D_j A D_j^{-1}, \quad j = 0, 1, \cdots, s - 1, \tag{2.16}$$

where $|D_j| = I$. Without loss of generality, we may assume that the $(1,1)$-entry of each D_j is 1. Let x be the Perron vector of A and let $x^j = D_j x$ for $j = 0, 1, \cdots, s - 1$. Then (2.16) implies that

$$Ax^j = \exp \left(i \frac{2\pi j}{s} \right) D_j A D_j^{-1} D_j x = \exp \left(i \frac{2\pi j}{s} \right) D_j r(A) x = \lambda_j x^j.$$

Thus, x^j is an eigenvector of A corresponding to eigenvalue λ_j. If we apply (2.16) twice, then we have

$$A = \exp\left(i\frac{2\pi j}{s}\right) D_j \left(\exp\left(i\frac{2\pi k}{s}\right) D_k A D_k^{-1}\right) D_j^{-1}$$

$$= \exp\left(i\frac{2\pi(k+j)}{s}\right)(D_j D_k)A(D_j D_k)^{-1}.$$

The same argument as above for $D_j x$ shows that $D_j D_k x$ is an eigenvector corresponding to $r(A)\exp(i2\pi(k+j)/s)$. In particular, $D_1^s x$ is an eigenvector of A corresponding to $r(A)\exp(i2\pi s/s) = r(A)$. Therefore, $D_1^s x = ax$ for some number a, since $r(A)$ is a simple eigenvalue of A. But $a = 1$ from the fact that the $(1,1)$-entry of the diagonal matrix D_1^s is 1. It follows that $D_1^s = I$, and so the main diagonal entries of D_1 are the sth roots of 1. We can rearrange such main diagonal entries so that the same numbers are together. That is, there is a permutation matrix P such that $P^T D_1 P$ is a $k \times k$ block diagonal matrix with the main diagonal blocks

$$\exp\left(i\frac{m_1 2\pi}{s}\right) I_{n_1}, \ \exp\left(i\frac{m_2 2\pi}{s}\right) I_{n_2}, \ \cdots, \ \exp\left(i\frac{m_k 2\pi}{s}\right) I_{n_k},$$

where I_{n_j} is the $n_j \times n_j$ identity matrix, and $0 = m_1 < m_2 < \cdots < m_k \le s - 1$. Partition the matrix

$$P^T AP = \begin{bmatrix} A_{11} & A_{12} & \cdots & A_{1k} \\ A_{21} & A_{22} & \cdots & A_{2k} \\ \vdots & \vdots & & \vdots \\ A_{k1} & A_{k2} & \cdots & A_{kk} \end{bmatrix}$$

the same way as $P^T D_1 P$ above, so each A_{pq} is $n_p \times n_q$, $p, q = 1, 2, \cdots, k$. The block matrix multiplication of the right-hand side of the equality

$$P^T AP = \exp\left(i\frac{2\pi}{s}\right)(P^T D_1 P)(P^T AP)(P^T D_1^{-1} P)$$

gives

$$A_{pq} = \exp\left(i\frac{(1 + m_p - m_q)\pi}{s}\right) A_{pq}$$

for all (p, q). Therefore, for each pair (p, q), either

$$A_{pq} = 0 \tag{2.17}$$

or

$$m_q - m_p \equiv 1 \ (\text{mod } s). \tag{2.18}$$

If $p = q$, then (2.18) cannot be satisfied, so $A_{pp} = 0$ for all $p = 1, 2, \cdots, k$. On the other hand, the assumption that A is irreducible rules out the possibility that $A_{pq} = 0$ for some p and all q or for some q and all p.

If $p = 1$, then the condition (2.18) is

$$m_q \equiv 1 \pmod{s}, \tag{2.19}$$

and since $0 < m_2 < \cdots < m_s < s$, the only solution of (2.19) is $m_2 = 1$. So $A_{1q} = 0$ for all $q \neq 2$. Next, for $p = 2$ the condition (2.18) becomes

$$m_q \equiv 2 \pmod{s}.$$

Then the only solution is $m_3 = 2$, and therefore $A_{2q} = 0$ for all $q \neq 3$. Continuing in the same manner, we conclude that $m_{p+1} = p$ and $A_{pq} = 0$ for all $q \neq p + 1$ for $p = 1, 2, \cdots, k - 1$.

For the last case $p = k$, the conditions (2.17) and (2.18) state that for each q either $A_{kq} = 0$ or $m_q - m_k \equiv 1 \pmod{s}$, that is, $m_q \equiv k \pmod{s}$. Now, A_{k1} cannot be zero since all other $A_{p1} = 0$. Hence, we must have

$$m_1 \equiv k \pmod{s},$$

which implies that $k = s$, and thus $m_q \not\equiv k \pmod{s}$ for all $q \neq 1$. Hence, $A_{kq} = 0$ for all $q \neq 1$. This completes the proof. $\qquad\square$

The Frobenius standard form for irreducible nonnegative matrices implies the following *normal form* for a reducible matrix.

Theorem 2.21. (The normal form for reducible matrices) *Let $A \in \mathbb{R}^{n \times n}$ be nonnegative and reducible. Then there is a permutation matrix P such that $P^T A P$ is of the form*

$$
\begin{bmatrix}
A_{11} & 0 & \cdots & 0 & 0 & \cdots & 0 \\
0 & A_{22} & \cdots & 0 & 0 & \cdots & 0 \\
\vdots & \vdots & & \vdots & \vdots & & \vdots \\
0 & 0 & \cdots & A_{kk} & 0 & \cdots & 0 \\
A_{k+1,1} & A_{k+1,2} & \cdots & A_{k+1,k} & A_{k+1,k+1} & 0 & \cdots & 0 \\
\vdots & \vdots & & \vdots & \vdots & & \vdots \\
A_{r1} & A_{r2} & \cdots & A_{rk} & A_{r,k+1} & \cdots & A_{rr}
\end{bmatrix}, \tag{2.20}
$$

where the blocks A_{ij} are $n_i \times n_j$ for $i, j = 1, \cdots, r$, the blocks $A_{11}, \cdots, A_{kk}$ are irreducible, and

$$A_{i1} + A_{i2} + \cdots + A_{i,i-1} \neq 0, \quad i = k+1, \cdots, r.$$

Proof There is a permutation matrix P such that

$$P^T A P = \begin{bmatrix} B & 0 \\ C & D \end{bmatrix},$$

where B and D are square sub-matrices. If both of them are irreducible, the proof is done. Otherwise, we can continue the process until we obtain a matrix of the form

$$\begin{bmatrix} A_{11} & 0 & 0 & \cdots & 0 \\ A_{21} & A_{22} & 0 & \cdots & 0 \\ A_{31} & A_{32} & A_{33} & \cdots & 0 \\ \vdots & \vdots & \vdots & \cdots & \vdots \\ A_{r1} & A_{r2} & A_{r3} & \cdots & A_{rr} \end{bmatrix}, \qquad (2.21)$$

where the A_{ii} are square irreducible sub-matrices. If all the off-diagonal blocks are zero in rows $1 < i_1, \cdots, i_{k-1}, i_k$ of (2.21), then permute the rows and the columns the same way so that the blocks $A_{i_t i_t}$, $t = 1, 2, \cdots, k-1$, can be brought to the places 2 through k along the main diagonal. Thus we obtain a matrix of the form (2.20) via exactly the same type of row and column permutations to the original matrix A. □

2.5   Notes and Remarks

1. The classic theory of nonnegative matrices is named after German mathematician Oskar Perron (1880-1975), who proved the very first major spectral theorem in [Perron (1907)] for positive matrices, and his countryman and Weierstrass' student Ferdinand Georg Frobenius (1849-1917), who extended Perron's main result to irreducible nonnegative matrices five years later [Frobenius (1912)]. Frobenius also gave the first full proof for the Cayley-Hamilton theorem (Theorem A.3 in Appendix A).

2. The proof of Theorem 2.1(i), based on Brouwer's fixed point theorem, follows the idea of [Bapat and Raghavan (1997)], which is an analytic one instead of algebraic ones adopted by most textbooks on nonnegative matrices. This approach is from the consideration that the study of positive operators in the sequel is mainly based on functional analysis. We refer to [MacCluer (2000)] and references cited therein for other proofs. In particular, the approach based on the Collatz-Wielandt function is contained in [Minc (1988)], and will be discussed in the exercise set below.

3. Theorem 2.4 was proved by means of Lemma 2.2 which was proved by the authors in [Ding and Zhou (2008)]. This proof seems much shorter than some other approaches in the literature, such as those in [Minc (1988); Bapat and Raghavan (1997)] based on derivative arguments.

4. The proof of Theorem 2.20 on the Frobenius standard form is due to [Wielandt (1950)] and follows the presentation of [Minc (1988)].

5. Applied fields involving nonnegative matrices include dynamical systems [Berman *et al.* (1989)], and game theory and economics [Bapat and Raghavan (1997)]. Chapter 4 is concerned with some other applications.

Exercises

2.1 Show that the matrix
$$A = \begin{bmatrix} 1 & 1 \\ 0 & 1 \end{bmatrix}$$
has spectral radius 1, but that the sequence $\{A^k\}$ is unbounded.

2.2 Let $A \in \mathbb{R}_+^{n \times n}$ have a positive eigenvector. Show that
$$A^k \leq r(A)^k C, \ \forall \, k \geq 1,$$
where C is a constant matrix.

2.3 Let a nonnegative matrix $A \in \mathbb{R}^{n \times n}$ have a positive eigenvector. Show that A is similar to some nonnegative matrix $B \in \mathbb{R}^{n \times n}$ with constant row sums. What is the constant?

2.4 Find a reducible matrix which has a positive eigenvector.

2.5 Let x be the Perron vector of a positive matrix $A \in \mathbb{R}^{n \times n}$. Prove
$$r(A) = \sum_{i,j=1}^{n} a_{ij} x_j.$$

2.6 Show that the inverse of a nonsingular positive matrix cannot be nonnegative. How about a nonnegative matrix?

2.7 Show that there is a positive matrix $B \in \mathbb{R}^{n \times n}$ that commutes with a given nonnegative matrix $A \in \mathbb{R}^{n \times n}$ if and only if A has positive eigenvectors and left eigenvectors.

2.8 Let A be an $n \times n$ nonnegative and irreducible matrix, and let $K = \{x \in \mathbb{R}^n : x \geq 0, e^T x = 1\}$ be the standard simplex in $\mathbb{R}^n$. The function $f_A : K \to \mathbb{R}_+$ defined by
$$f_A(x) = \min_{x_i \neq 0} \frac{(Ax)_i}{x_i}, \ \forall \, x \in K$$

is called the *Collatz-Wielandt function* [Wielandt (1950)] associated with A. Assume $x \in K$. Show that

(i) if r is the largest real number for which the inequality $Ax \geq rx$ is satisfied, then $r = f_A(x)$;

(ii) if $y = (I + A)^{n-1}x$, then $f_A(y) \geq f_A(x)$.

2.9 Show that the above function f_A is bounded on K.

2.10 Is the Collatz-Wielandt function f_A continuous on K?

2.11 Use Theorem 1.5 to show that the Collatz-Wielandt function f_A attains its maximal value on K.

2.12 Use the Collatz-Wielandt function f_A and its above results to provide another proof of Theorem 2.6(i).

2.13 Find a necessary and sufficient condition on A for the Collatz-Wielandt function f_A to be continuous on K.

2.14 Let A be an $n \times n$ nonnegative and irreducible matrix. Define the *dual Collatz-Wielandt function* $g_A : K \to \mathbb{R}_+$ by

$$g_A(x) = \max_{x_i \neq 0} \frac{(Ax)_i}{x_i}, \quad \forall \, x \in K.$$

Assume $x \in K$ such that $x_i > 0$ whenever $(Ax)_i > 0$. Show that

(i) if r is the smallest real number for which the inequality $Ax \leq rx$ is satisfied, then $r = g_A(x)$;

(ii) if $y = (I + A)^{n-1}x$, then $g_A(y) \leq g_A(x)$.

2.15 Show that the dual Collatz-Wielandt function g_A attains its minimal value on K at a point with all positive components.

2.16 Show that $r(A) = \min\{g_A(x) : x \in K\}$. Thus, for any irreducible matrix $A \in \mathbb{R}_+^{n \times n}$, its spectral radius

$$r(A) = \max_{x \in K} f_A(x) = \min_{x \in K} g_A(x).$$

2.17 Let the spectral radius $r(A)$ of an $n \times n$ matrix $A \geq 0$ be a simple eigenvalue of A. Show that A is irreducible if and only if A and A^T have positive eigenvectors associated with $r(A)$.

2.18 Use the maximal eigenvalue estimates in Section 2.2 to find several bounds for $r(A)$ when

$$A = \begin{bmatrix} 3 & 1 & 7 & 4 \\ 1 & 5 & 3 & 2 \\ 1 & 4 & 7 & 1 \\ 1 & 1 & 4 & 8 \end{bmatrix}.$$

2.19 Evaluate $f_A(x)$ and $g_A(x)$ with

$$A = \begin{bmatrix} 1 & 2 & 3 \\ 4 & 5 & 6 \\ 7 & 8 & 9 \end{bmatrix}$$

for the three vectors $x = (1,1,1)^T, (1,0,2)^T$, and $(1,2,3)^T$ in $\mathbb{R}^3$. Then deduce a lower and upper bound for $r(A)$.

2.20 Let an $n \times n$ matrix A be nonnegative and let $x \in \mathbb{R}^n$ be a nonnegative vector which satisfies $e^T x = 1$. Show that

$$\min_{j=1}^{n} (e^T A)_j \le e^T A x \le \max_{j=1}^{n} (e^T A)_j.$$

2.21 Verify the equality (2.13).

2.22 Use an example of 2×2 matrix A to show that the existence of the limit $\lim_{\lambda \to 1} (\lambda - 1) R(\lambda, A)$ does not necessarily imply the existence of the limit of the sequence $\{A_k\}$ of the Cesáro averages of the powers A^k.

2.23 Provide a detailed proof of Theorem 2.15.

2.24 Let $A \in \mathbb{R}^{n \times n}$ be nonnegative. Show that if $R(\lambda, A)$ exists and is nonnegative, then λ is real and $\lambda > r(A)$.

2.25 Let $A \in \mathbb{R}^{n \times n}$ be nonnegative. Show that if A is mean ergodic and $\sigma(A) \subset \partial \mathbb{D}^2$, then the minimal polynomial of A is $p_m(\lambda) = \lambda^m - 1$ and

$$\lim_{k \to \infty} A_k = \lim_{k \to \infty} \frac{1}{k} \sum_{j=0}^{k-1} A^j = \frac{1}{m} \sum_{j=0}^{m-1} A^j.$$

2.26 Apply the result of Exercise 2.25 to the 2×2 permutation matrix $A = [e_2, e_1]$ and compare with Theorem 2.13.

2.27 Let $A \in \mathbb{R}^{n \times n}$ be a nonnegative matrix with $r(A) = 1$ being the only peripheral spectral point of A which has n linearly independent eigenvectors $v_1, \cdots, v_r, v_{r+1}, \cdots, v_n$ such that $v_1, \cdots, v_r$ are the eigenvectors associated with the same eigenvalue 1. Show that

(i) $A = VDV^{-1}$, where the matrices $V = [v_1, v_2, \cdots, v_n]$ and $D = \mathrm{diag}(1, \cdots, 1, \lambda_{r+1}, \cdots, \lambda_n)$ in which $|\lambda_j| < 1$, $j = r+1, \cdots, n$;

(ii) $\lim_{n \to \infty} A^n = V \mathrm{diag}(1, \cdots, 1, 0, \cdots, 0) V^{-1}$.

2.28 Let $A \in \mathbb{R}_+^{n \times n}$ with $r(A) = 1$ be such that $\lim_{n \to \infty} A^n = E$ exists. Show that the matrix E is a projection from $\mathbb{R}^{n \times n}$ onto the null space $N(I - A)$ along the range $R(I - A)$.

2.29 Find 3×3 reducible matrices A and B such that

(i) $AB > 0$;

(ii) AB is primitive but not positive;

(iii) AB is imprimitive;

2.30 Construct a 3×3 imprimitive matrix A and a 3×3 reducible real matrix B such that

(i) AB is primitive;

(ii) AB is reducible.

2.31 Show that a symmetric imprimitive matrix must have index 2.

2.32 Let $A \in \mathbb{R}^{n \times n}$ be nonnegative and irreducible. Show that the matrix $A + \epsilon I$ is primitive for any real number $\epsilon > 0$.

2.33 Give an example of a 3×3 irreducible matrix A with $r(A) = 1$ such that $\lim_{k \to \infty} A^k$ does not exist.

2.34 Let $A \in \mathbb{R}_+^{n \times n}$ be reducible. Give an example to show that the condition (2.10) is not necessary for $r(B) = r(A)$.

2.35 Let a nonnegative matrix $A \in \mathbb{R}^{n \times n}$ be irreducible and nonsingular. Assume that the index of imprimitivity of A is s. How many real eigenvalues at most can A have?

2.36 Let A and B be two 3×3 imprimitive matrices. Find examples for each of the following statements, or show that no matrices exist to satisfy the given statement.

(i) AB is primitive;

(ii) AB is primitive and BA is also primitive;

(iii) AB is imprimitive;

(iv) AB is reducible.

2.37 Let a nonnegative matrix $A \in \mathbb{R}^{n \times n}$ be irreducible with the index of imprimitivity s. Show that the matrix A^k is irreducible if and only if the two positive integers s and k are relatively prime.

2.38 Let $A = (a_{ij}) = [e_n, e_1, \cdots, e_{n-1}] \in \mathbb{R}^{n \times n}$. Show that the matrix B obtained from A by adding 1 to a_{13} is primitive.

2.39 Show that the 4×4 matrix

$$A = \begin{bmatrix} 0 & 0 & 0 & 1 \\ 0 & 0 & 1 & 1 \\ 1 & 1 & 0 & 0 \\ 1 & 0 & 0 & 0 \end{bmatrix}$$

is of the form (2.20), but it is reducible.

2.40 Let $A, B \in \mathbb{C}^{n \times n}$. Show that the matrices AB, BA have the same nonzero eigenvalues.

2.41 Let $A \in \mathbb{C}^{m \times n}$, $B \in \mathbb{C}^{n \times k}$, $C \in \mathbb{C}^{k \times m}$. Show that the matrices ABC, BCA, CAB have the same nonzero eigenvalues.

2.42 Can you extend the conclusion of Exercise 2.41 to k matrices $A_1 \in \mathbb{C}^{m \times n_1}$, $A_2 \in \mathbb{C}^{n_1 \times n_2}, \cdots, A_k \in \mathbb{C}^{n_{k-1} \times n}$?

2.43 Let $A \in \mathbb{R}^{n \times n}$ be partitioned as

$$A = \begin{bmatrix} 0 & A_{12} & 0 \\ 0 & 0 & A_{23} \\ A_{31} & 0 & 0 \end{bmatrix},$$

where A_{12}, A_{23}, A_{31} are square matrices. Show that the algebraic multiplicity of eigenvalue zero is a multiple of 3.

Chapter 3

Stochastic Matrices

In this chapter we continue the study of nonnegative matrices, but now our focus is on more special ones called stochastic matrices. The theory of stochastic matrices has found many new applications in modern science and technology, such as the PageRank computation in the Google search engine [Bryan and Leise (2006); Langville and Meyer (2006)]. In the structure-preserving finite dimensional approximation of infinite dimensional Markov operators that is a special class of positive operators to be studied in Chapter 7, which will be the main topic of Chapter 10, the resulting square matrices are usually stochastic ones.

Section 3.1 will be devoted to the general properties of stochastic matrices. In Section 3.2 we explore the spectral properties of doubly stochastic matrices. Some interesting inequalities related to doubly stochastic matrices will be presented in Section 3.3. In Section 3.4 we shall explore the ergodic properties of stochastic matrices.

3.1 Stochastic Matrices

A matrix is called *quasi-stochastic* if each of its row sums is 1 and *column quasi-stochastic* if all its column sums are 1. A matrix which is both quasi-stochastic and column quasi-stochastic is called *doubly quasi-stochastic*. A quasi-stochastic matrix is said to be *stochastic* if it is square and nonnegative, and a column quasi-stochastic matrix is called *column stochastic* if it is also a nonnegative square matrix. A *doubly stochastic matrix* is a nonnegative and doubly quasi-stochastic matrix.

Every row of a stochastic matrix can be thought of as a discrete probability distribution, and each column of a column stochastic matrix is a probability vector. These matrices arise in various applied fields, such as

Markov chains and probability modeling. They are also used in the numerical analysis of Markov operators that will be studied in later chapters.

The definition of a quasi-stochastic (or stochastic) matrix can be restated as follows. A matrix (or nonnegative matrix) $A \in \mathbb{R}^{n \times n}$ is quasi-stochastic (or stochastic) if and only if $Ae = e$. The latter condition simply says that e is an eigenvector of A corresponding to the eigenvalue 1. If A and B are $n \times n$ quasi-stochastic (or stochastic) matrices, then, due to $ABe = Ae = e$, so is their product AB. Similarly, a column quasi-stochastic (or column stochastic) matrix is a matrix (or nonnegative matrix) $A \in \mathbb{R}^{n \times n}$ such that $e^T A = e^T$. A doubly quasi-stochastic matrix (or doubly stochastic matrix) is a matrix (or nonnegative matrix) $A \in \mathbb{R}^{n \times n}$ with $Ae = e$ and $e^T A = e^T$.

We first give elementary properties of stochastic matrices.

Lemma 3.1. *Let $A \in \mathbb{R}^{n \times n}$ be nonsingular.*
(i) *If A is quasi-stochastic, then so is A^{-1}.*
(ii) *If A is column quasi-stochastic, then so is A^{-1}.*

Proof Suppose that A is nonsingular and quasi-stochastic. Then $Ae = e$, so $e = A^{-1}Ae = A^{-1}e$. That is, A^{-1} is quasi-stochastic. This proves (i), and (ii) can be proved similarly. □

Corollary 3.1. *The inverse of a nonsingular stochastic matrix is quasi-stochastic, and the inverse of a nonsingular column stochastic matrix is column quasi-stochastic.*

Corollary 3.2. *If A and X are $n \times n$ stochastic (or column stochastic) matrices such that X is nonsingular, then the matrix XAX^{-1} is quasi-stochastic (or column quasi-stochastic).*

Let $\mathbb{S}_n$ be the set of all $n \times n$ stochastic matrices. Then $\mathbb{S}_n$ is a convex subset of $\mathbb{R}^{n \times n}$. In fact $\mathbb{S}_n$ is a *polyhedron*, since it is the intersection of the *positive orthant* $\mathbb{R}_+^{n \times n}$ with the n *hyper-planes* defined by $\sum_{j=1}^{n} a_{ij} = 1$, $i = 1, 2, \cdots, n$. Furthermore, $\mathbb{S}_n$ is closed under the matrix multiplication. Since $a_{ij} \in [0, 1]$, the set $\mathbb{S}_n$ is compact. The *dimension* of $\mathbb{S}_n$, which is defined to be the dimension of the smallest subspace of $\mathbb{R}^{n \times n}$ that contains the convex set $\mathbb{S}_n - I \equiv \{A - I : A \in \mathbb{S}_n\}$, equals $n^2 - n$.

Proposition 3.1. *Every $n \times n$ stochastic matrix is a convex combination of the n^n $n \times n$ stochastic $(0, 1)$-matrices. Moreover, a stochastic matrix A is an extreme point of $\mathbb{S}_n$ if and only if A is a stochastic $(0, 1)$-matrix.*

Proof Let $A = (a_{ij}) \in \mathbb{S}_n$. Since

$$A = \sum_{i=1}^{n} \sum_{j=1}^{n} a_{ij} S_{ij},$$

where S_{ij} is the special $(0,1)$-matrix such that only the (i,j)-entry is 1, we see that A is a convex combination of such S_{ij}.

Suppose that $A = (a_{ij})$ is an $n \times n$ stochastic $(0,1)$-matrix. If $A = aB + (1-a)C$, where $0 < a < 1$ and $B = (b_{ij})$, $C = (c_{ij})$ are stochastic, then for each index $i = 1, 2, \cdots, n$,

$$1 \text{ or } 0 = a_{ij} = a b_{ij} + (1-a) c_{ij}, \ \forall \, j = 1, 2, \cdots, n.$$

This implies that $b_{ij} = c_{ij} = 1$ or 0 depending on whether $a_{ij} = 1$ or 0, since $0 \le b_{ij} \le 1$ and $0 \le c_{ij} \le 1$. Thus $B = C = A$, which shows that A is an extreme point of $\mathbb{S}_n$.

Conversely, let $A = (a_{ij})$ be stochastic but not a $(0,1)$-matrix. Without loss of generality, assume $a_{11} \in (0,1)$ and $a_{1n} \in (0,1)$. Denote $\epsilon = \min\{a_{11}, a_{1n}\}$, and define

$$b_{11} = a_{11} + \epsilon, \ b_{1n} = a_{1n} + \epsilon, \ c_{11} = a_{11} - \epsilon, \ c_{1n} = a_{1n} - \epsilon,$$

and $b_{ij} = c_{ij} = a_{ij}$ for all other (i,j) pairs. Then $B = (b_{ij})$ and $C = (c_{ij})$ are distinct stochastic matrices satisfying $A = 2^{-1}(B+C)$. Therefore, A is not an extreme point of $\mathbb{S}_n$. □

Proposition 3.1 shows that the polyhedron $\mathbb{S}_n$ has exactly n^n extreme points since there are the same number of $n \times n$ stochastic $(0,1)$-matrices, which implies that it cannot be a simplex for $n > 1$ since $n^n > n^2 - n + 1$.

A linear operator $A : \mathbb{C}^n \to \mathbb{C}^m$ is called a *lattice homomorphism* from $\mathbb{C}^n$ into $\mathbb{C}^m$ if $|Ax| = A|x|$ for all $x \in \mathbb{C}^n$. In the case of $m = n$, it is called a lattice homomorphism of $\mathbb{C}^n$. A lattice homomorphism A of $\mathbb{C}^n$ is a *lattice isomorphism* if A^{-1} exists and is also a lattice homomorphism. A lattice homomorphism $A : \mathbb{C}^n \to \mathbb{C}$ must be a multiple of a *coordinate function*, that is, $Ax = ax_j$ for some constant a and a fixed index j, the proof of which is left as an exercise (Exercise 3.7).

Lattice homomorphisms are characterized by extreme points of $\mathbb{S}_n$, as the following result indicates.

Proposition 3.2. *Let $A \in \mathbb{R}^{n \times n}$ be a stochastic matrix. Then A defines a lattice homomorphism of $\mathbb{C}^n$ if and only if A is an extreme point of $\mathbb{S}_n$.*

Proof Suppose first that A is an extreme point of $\mathbb{S}_n$. Then A is a stochastic $(0, 1)$-matrix by Proposition 3.1. A direct verification shows that $|Ax| = A|x|$ for all vectors $x \in \mathbb{C}^n$.

Conversely, let $A \in \mathbb{S}_n$ define a lattice homomorphism of $\mathbb{C}^n$. Fix an index $i = 1, 2, \cdots, n$, and define a scalar function $s_i(x) = (Ax)_i$ with $x \in \mathbb{C}^n$. Then s_i is a lattice homomorphism from $\mathbb{C}^n$ to $\mathbb{C}$ and $s_i(e) = 1$ in addition. It follows that $s_i(x) = x_j$ for some $j = j(i)$. Thus, the (i, k)-entry of A is $(Ae_k)_i = s_i(e_k) = (e_k)_j = \delta_{jk}$, where δ_{jk} is Kronecker's notation. Hence, A is a $(0, 1)$-matrix. $\qquad\square$

Corollary 3.3. *Let $A \in \mathbb{R}^{n \times n}$ be stochastic. Then A defines a lattice isomorphism if and only if A is a permutation matrix.*

Proof The sufficiency part is obvious from the definition of lattice isomorphism. To show the necessity part, notice that a $(0, 1)$-matrix is invertible if and only if it is a permutation matrix. $\qquad\square$

The idea behind the proof of Proposition 3.2 actually proves a more general result in the next corollary.

Corollary 3.4. *Let $A \in \mathbb{C}^{n \times n}$. Then A defines a lattice homomorphism of $\mathbb{C}^n$ if and only if $A \geq 0$ and each row of A contains at most one positive entry. Similarly, the matrix A defines a lattice isomorphism if and only if $A \geq 0$ and replacement of each nonzero entry of A by 1 transforms A into a permutation matrix.*

We turn to the spectral analysis of stochastic matrices. By Corollary 2.3, if A is a stochastic matrix, then $r(A) = 1$, and so the modulus of all its eigenvalues cannot exceed 1. The special positive vector e is a common eigenvector of all stochastic matrices associated with the maximal eigenvalue 1, and $n^{-1}e$ is the common Perron vector for all stochastic matrices.

Theorem 2.5 ensures that there is a nonnegative left eigenvector for any stochastic matrix. However, because of the special structure of stochastic matrices, in other words, $Ae = e$ for any stochastic matrix A, the proof of the existence of a nonnegative left eigenvector based on Brouwer's fixed point theorem for the proof of Perron's theorem (Theorem 2.1) can be adopted here with much simplification for the particular case of stochastic matrices, even without the positivity assumption of the matrix which was essential for the proof of Perron's theorem.

Theorem 3.1. *Let $A \in \mathbb{R}^{n \times n}$ be a stochastic matrix. Then there is $y \in \mathbb{R}^n$ such that $y \geq 0$, $y^T e = 1$, and $y^T A = y^T$.*

Proof Let $K = \{x \in \mathbb{R}^n : x \geq 0, \ e^T x = 1\}$ be the standard simplex of $\mathbb{R}^n$. Then for any $x \in K$, we have $A^T x \geq 0$ and

$$(A^T x)^T e = x^T A e = x^T e = 1.$$

That is, $A^T x \in K$. Since K is compact and convex, by Brouwer's fixed point theorem, the mapping $A^T : K \to K$ has a fixed point y; in other words, $A^T y = y$. Therefore, $y^T A = y^T$. Since $y \neq 0$, it is a left eigenvector of A corresponding to eigenvalue 1. □

The following result indicates that the spectral properties of stochastic matrices are basically the same as those of other nonnegative matrices with the maximal eigenvalue equal to 1 and a corresponding positive eigenvector.

Theorem 3.2. *Let $A \in \mathbb{R}^{n \times n}$ be nonnegative. If $r(A) > 0$ with a positive eigenvector, then $r(A)^{-1} A$ is diagonally similar to a stochastic matrix.*

Proof Let $x = (x_1, \cdots, x_n)^T$ be a positive eigenvector associated with the maximal eigenvalue $r(A) > 0$. Define a diagonal matrix $D = \text{diag}(x_1, \cdots, x_n)$. Then $x = De$ and 1 is the maximal eigenvalue of $D^{-1}[r(A)^{-1} A] D = r(A)^{-1} D^{-1} A D$. Now,

$$r(A)^{-1} D^{-1} A D e = r(A)^{-1} D^{-1} A x$$
$$= r(A)^{-1} D^{-1} r(A) x = D^{-1} x = e,$$

so e is a positive eigenvector of the stochastic matrix

$$r(A)^{-1} D^{-1} A D = D^{-1}[r(A)^{-1} A] D$$

associated with the maximal eigenvalue 1. □

Remark 3.1. The existence of a positive maximal eigenvector in Theorem 3.2 is essential. For example, the matrix

$$A = \begin{bmatrix} 1 & 1 \\ 0 & 1 \end{bmatrix}$$

satisfies $r(A) = 1$ but is not similar to any stochastic matrix.

The next result presents a property of stochastic matrices which is shared by general Markov operators to be studied in Chapter 7. Although it is a special case of a general result in matrix theory that if u is a left eigenvector and v is an eigenvector corresponding to different eigenvalues of a matrix, then $u^T v = 0$, we state and prove it because of its importance for the numerical analysis of Markov operators in Chapter 10.

Proposition 3.3. *If $y \in \mathbb{C}^n$ is a left eigenvector of a stochastic matrix $A \in \mathbb{R}^{n \times n}$ corresponding to eigenvalue $\lambda \neq 1$, then $y^T e = 0$. If $x \in \mathbb{C}^n$ is an eigenvector of a column stochastic matrix $A \in \mathbb{R}^{n \times n}$ corresponding to eigenvalue $\lambda \neq 1$, then $e^T x = 0$.*

Proof For the first part of the proposition, multiplying e to the both sides of $y^T A = \lambda y^T$ from right, we have

$$y^T e = y^T A e = \lambda y^T e,$$

which implies that $y^T e = 0$ since $\lambda \neq 1$. The second part can be proved by applying the first one to A^T. □

The following eigenvalue localization result is basically a consequence of the Gershgörin disk theorem (Theorem A.8).

Proposition 3.4. *Let $A = (a_{ij}) \in \mathbb{R}^{n \times n}$ be a stochastic matrix. Then for any eigenvalue λ of A,*

$$|\lambda - a| \leq 1 - a,$$

where $a = \min_{i=1}^n a_{ii}$.

Proof Let $x = (x_1, x_2, \cdots, x_n)^T$ be an eigenvector of A associated with an eigenvalue λ, and let k be an index such that $0 < |x_k| = \max_{i=1}^n |x_i|$. Then, the equality $\lambda x = Ax$ gives

$$\lambda x_k = \sum_{j=1}^n a_{kj} x_j,$$

and therefore,

$$|\lambda - a_{kk}| = \left| \sum_{j \neq k} a_{kj} \frac{x_j}{x_k} \right| \leq \sum_{j \neq k} a_{kj} \left| \frac{x_j}{x_k} \right|$$

$$\leq \sum_{j \neq k} a_{kj} = 1 - a_{kk}.$$

Thus,

$$|\lambda - a| = |\lambda - a_{kk} + a_{kk} - a|$$
$$\leq |\lambda - a_{kk}| + |a_{kk} - a|$$
$$\leq 1 - a_{kk} + a_{kk} - a = 1 - a.$$ □

3.2 Doubly Stochastic Matrices

In the current section we focus on doubly stochastic matrices, a class of stochastic matrices that have many applications in the theory of inequalities, combinatorial matrix theory, probability, physical science, and engineering. We recall some definitions first.

Definition 3.1. A matrix $A \in \mathbb{R}^{m \times n}$ is said to be *doubly quasi-stochastic* if each of its row and column sums is 1. A nonnegative doubly quasi-stochastic matrix is called *doubly stochastic*.

A doubly quasi-stochastic matrix must be square. So, a nonnegative matrix is doubly stochastic if and only if it is stochastic and column stochastic. If A is doubly stochastic, then $r(A) = \|A\|_\infty = \|A\|_1 = \|Ae\|_\infty = \|A^T e\|_\infty = 1$. Some elementary properties of stochastic matrices can be restated for doubly stochastic ones. For example, Lemma 3.1 implies

Proposition 3.5. *Let $A \in \mathbb{R}^{m \times n}$ be doubly stochastic and nonsingular. Then its inverse is doubly quasi-stochastic.*

Applying Corollary 3.2 to doubly stochastic matrices gives

Proposition 3.6. *Let $A, X \in \mathbb{R}^{n \times n}$ be doubly stochastic such that X is nonsingular. Then XAX^{-1} is doubly quasi-stochastic.*

Of course, the matrix XAX^{-1} in the above proposition is not nonnegative in general. On the other hand, if A is doubly stochastic, even though X is nonsingular but not doubly stochastic or doubly quasi-stochastic, XAX^{-1} may be doubly stochastic. For example, if $A = I$, X may be any nonsingular matrix. However, we have the following result of Marcus, Minc, and Moyls [Marcus *et al.* (1961)] when A is irreducible.

Theorem 3.3. *Let $A, X \in \mathbb{R}^{n \times n}$ be such that A is irreducible and doubly stochastic and X is nonsingular. If $B = XAX^{-1}$ is doubly stochastic, then X is a multiple of a doubly quasi-stochastic matrix. Moreover, $YAY^{-1} = B$ for some doubly stochastic matrix Y.*

Proof If $Bv = v$, then $AX^{-1}v = X^{-1}v$. Since A is irreducible, $X^{-1}v$ must be a multiple of e, which means that v is a multiple of Xe and 1 is also a simple eigenvalue of B. Now, since $BXe = XAe = Xe$ and since $Be = e$, we deduce that $Xe = ae$ for some number a. By the same token, $e^T X = be^T$ for some number b. It follows that $a = b$ and X is a times a doubly quasi-stochastic matrix. This proves the first part of the theorem.

Let c be a number such that $a + cn > 0$ and the matrix $X + cee^T$ is positive and nonsingular. Then

$$(X + cee^T)e = Xe + cee^Te = (a + cn)e$$

and

$$e^T(X + cee^T) = e^X + ce^Tee^T = e^T(a + cn).$$

Therefore, if $Y = (a + cn)^{-1}(X + cee^T)$, then Y is a nonsingular doubly stochastic matrix, and

$$\begin{aligned} YAY^{-1} &= (X + cee^T)A(X + cee^T)^{-1} \\ &= (BX + cee^T)(X + cee^T)^{-1} \\ &= B(X + cee^T)(X + cee^T)^{-1} = B. \end{aligned} \qquad \square$$

In order to further explore the structure of doubly stochastic matrices, in particular their zero patterns, we need the tool of the permanent, a mathematical concept whose definition is almost the same as that for the determinant of a square matrix.

Definition 3.2. The *permanent* of $A \in \mathbb{C}^{n \times n}$ is defined by

$$\mathrm{per}A = \sum_{\pi}\prod_{i=1}^{n} a_{i\pi(i)},$$

where the summation extends over all the $n!$ different permutations π of the index set $\{1, 2, \cdots, n\}$.

The products $\prod_{i=1}^{n} a_{i\pi(i)}$ in the above definition will be referred to as the *diagonal products* in what follows. The similarity between the definition of the determinant and that of the permanent for a given square matrix makes the following properties of the permanent easy to understand; their simple proof is left as Exercise 3.8.

Proposition 3.7. *Let $A \in \mathbb{C}^{n \times n}$. Then:*
   (i) *$\mathrm{per}A$ is a multi-linear functional of the row vectors of A.*
   (ii) *$\mathrm{per}A^T = \mathrm{per}A$.*
   (iii) *If P and Q are $n \times n$ permutation matrices, then*

$$\mathrm{per}(PAQ) = \mathrm{per}A.$$

   (iv) *If D and G are $n \times n$ diagonal matrices, then*

$$\mathrm{per}(DAG) = \mathrm{per}D \cdot \mathrm{per}A \cdot \mathrm{per}G.$$

In the following, we prove another property of the permanent, which is analogous to the Laplace expansion for determinants and will be used in the proof of Theorem 3.4. For $1 \leq k \leq n$ let $I_{k,n}$ denote the set of all sequences $\alpha = (\alpha_1, \cdots, \alpha_k)$ such that $1 \leq \alpha_1 < \cdots < \alpha_k \leq n$. For any α, $\beta \in I_{k,n}$ let $A[\alpha|\beta]$ be the $k \times k$ sub-matrix of A whose (i,j)-entry is a_{α_i, β_j} and let $A(\alpha|\beta)$ be the $(n-k) \times (n-k)$ sub-matrix of A obtained from A by deleting its rows indexed by α and columns indexed by β.

Proposition 3.8. *Let* $A \in \mathbb{C}^{n \times n}$ *and* $\alpha \in I_{k,n}$ *with* $k < n$. *Then*
$$\operatorname{per} A = \sum_{\beta \in I_{k,n}} \operatorname{per} A[\alpha|\beta] \cdot \operatorname{per} A(\alpha|\beta). \tag{3.1}$$
In particular, for any $i = 1, 2, \cdots, n$,
$$\operatorname{per} A = \sum_{j=1}^{n} a_{ij} \operatorname{per} A(i|j).$$

Proof Given $\beta \in I_{k,n}$, the permanent of $A[\alpha|\beta]$ is a sum of $k!$ diagonal products of its k entries, and the permanent of $A(\alpha|\beta)$ is a sum of $(n-k)!$ diagonal products of its $n-k$ entries. The product of a diagonal product of $A[\alpha|\beta]$ and a diagonal product of $A(\alpha|\beta)$ is a diagonal product of A. Thus, for a fixed β, $\operatorname{per} A[\alpha|\beta] \cdot \operatorname{per} A(\alpha|\beta)$ is a sum of $k!(n-r)!$ distinct diagonal products of A. Furthermore, for different β, different diagonal products of A are obtained. There are $\binom{n}{k}$ sequences in $I_{k,n}$, hence the right-hand side of (3.1) is the sum of
$$\binom{n}{k} k!(n-k)! = n!$$
such diagonal products, which is the sum of all the diagonal products of A, and thus is equal to $\operatorname{per} A$. $\square$

The following fundamental result on the zero patterns of nonnegative matrices in terms of the permanent is the so-called *Frobenius-König theorem*, which will be used in this section for the structure analysis of doubly stochastic matrices.

Theorem 3.4. (Frobenius-König) *Let* $A \in \mathbb{R}^{n \times n}$ *be nonnegative. Then* $\operatorname{per} A = 0$ *if and only if* A *contains an* $r \times t$ *zero sub-matrix with* $r + t = n + 1$.

Proof First suppose that there is an $r \times t$ sub-matrix of A with $r + t = n + 1$. If $r = n$, then $t = 1$, so A has a zero column. Hence $\operatorname{per} A = 0$. If $r < n$, then there is $\alpha \in I_{r,n}$ and $\beta \in I_{t,n}$ such that $A[\alpha|\beta] = 0$. Since the sub-matrix

$A[\alpha|(1,2,\cdots,n)]$ contains at most $n - t = r - 1$ nonzero columns, every $r \times r$ sub-matrix $A[\alpha|\gamma], \gamma \in I_{r,n}$, has a zero column, so $\mathrm{per} A[\alpha|\gamma] = 0$ for all $\gamma \in I_{r,n}$. It follows from Proposition 3.8 that

$$\mathrm{per} A = \sum_{\gamma \in I_{k,n}} \mathrm{per} A[\alpha|\gamma] \cdot \mathrm{per} A(\alpha|\gamma) = 0.$$

For the necessity part, suppose that $\mathrm{per} A = 0$. We use induction on n. If $n = 1$, then $A = 0$. Assume that the theorem is true for all square matrices of order less than $n > 1$. If $A = 0$, then we are done. Otherwise, A contains a nonzero entry, say a_{jk}. Since $0 = \mathrm{per} A \geq a_{jk} \mathrm{per} A(j|k)$, we have $\mathrm{per} A(j|k) = 0$. The induction hypothesis implies that $A(j|k)$ contains a $p \times q$ zero sub-matrix with $p + q = (n - 1) + 1 = n$. Let P and Q be permutation matrices such that

$$PAQ = \begin{bmatrix} X & Y \\ 0 & Z \end{bmatrix},$$

where X is $q \times q$ and Z is $p \times p$. Using Proposition 3.7 gives

$$0 = \mathrm{per} A = \mathrm{per}(PAQ) \geq \mathrm{per} X \cdot \mathrm{per} Z,$$

which implies that either $\mathrm{per} X = 0$ or $\mathrm{per} Z = 0$. In the first case, using the induction hypothesis again, we can see that the matrix X contains an $r \times s$ zero sub-matrix with $r + s = q + 1$. But then PAQ contains an $(r + p) \times s$ zero sub-matrix, hence

$$(r + p) + s = p + (r + s) = p + q + 1 = n + 1.$$

The second case can be proved similarly. $\qquad\square$

Theorem 3.5. (König) *Every doubly stochastic matrix $A = (a_{ij})$ has a positive diagonal entry.*

Proof If $a_{ii} = 0$ for all i, then $\mathrm{per} A = 0$. By Theorem 3.4, there are permutation matrices P and Q such that

$$PAQ = \begin{bmatrix} B & C \\ 0 & D \end{bmatrix},$$

where the zero block is $p \times q$ and $p + q = n + 1$. Then

$$n = \sum_{i,j=1}^{n} a_{ij} \geq \sum_{i,j=1}^{q} b_{ij} + \sum_{i,j=1}^{p} d_{ij} = q + p = n + 1.$$

This contradiction proves the theorem. $\qquad\square$

Corollary 3.5. *If A is doubly stochastic, then $\mathrm{per} A > 0$.*

Let $U = (u_{ij}) \in \mathbb{C}^{n \times n}$ be a unitary matrix. If we define $S = (s_{ij}) \in \mathbb{R}^{n \times n}$ by $s_{ij} = |u_{ij}|^2$, $\forall\ i,j = 1, 2, \cdots, n$, then the condition $U^H U = I$ implies that S is a doubly stochastic matrix. Such doubly stochastic matrices constructed from unitary matrices are called *Schur-stochastic*. If U is real and orthogonal, then the corresponding matrix $S = (s_{ij}) = (u_{ij}^2)$ is said to be *orthostochastic*. Schur-stochastic matrices can be used to prove the following result for normal matrices.

Proposition 3.9. *Suppose that* $A \in \mathbb{C}^{n \times n}$ *is a normal matrix with eigenvalues* $\lambda_1, \cdots, \lambda_n$, *counting algebraic multiplicity. Then there exists a Schur-stochastic matrix* S *such that* $d = Sv$, *where the vectors* $d = (a_{11}, \cdots, a_{nn})^T$ *and* $v = (\lambda_1, \cdots, \lambda_n)^T$.

Proof The equality (A.2) in Appendix A.2 ensures that there is a unitary matrix $U = (u_{ij})$ such that

$$A = U \operatorname{diag}(\lambda_1, \lambda_2, \cdots, \lambda_n) U^H.$$

The matrix $S = (|u_{ij}|^2)$ is Schur-stochastic, as mentioned above. Furthermore, for each $i = 1, 2, \cdots, n$,

$$a_{ii} = \sum_{j=1}^{n} u_{ij} \lambda_j \bar{u}_{ij} = \sum_{j=1}^{n} |u_{ij}|^2 \lambda_j = \sum_{j=1}^{n} s_{ij} \lambda_j. \qquad \square$$

Since a Hermitian matrix is normal with real eigenvalues, we obtain a result of Schur as a corollary of the above proposition:

Corollary 3.6. (Schur) *If* $A \in \mathbb{C}^{n \times n}$ *is a Hermitian matrix, then there is a real vector* v *and a Schur-stochastic matrix* S *such that* $d = Sv$, *where* $d = (a_{11}, a_{22}, \cdots, a_{nn})^T$.

Remark 3.2. Not every doubly stochastic matrix is Schur-stochastic. One simple example is

$$A = \frac{1}{2} \begin{bmatrix} 0 & 0 & 1 \\ 1 & 0 & 1 \\ 1 & 1 & 0 \end{bmatrix}.$$

The next theorem describes the structure of an irreducible doubly stochastic matrix after its entries' rearrangement.

Theorem 3.6. *Let* $A \in \mathbb{R}^{n \times n}$ *be doubly stochastic and irreducible with index of imprimitivity* $s \geq 2$. *Then* s *is a factor of* n, *and there exists an*

$n \times n$ *permutation matrix* P *such that*

$$P^T A P = \begin{bmatrix} 0 & A_{12} & 0 & \cdots & 0 & 0 \\ 0 & 0 & A_{23} & \cdots & 0 & 0 \\ \vdots & \vdots & \vdots & & \vdots & \vdots \\ 0 & 0 & 0 & \cdots & 0 & A_{s-1,s} \\ A_{s1} & 0 & 0 & \cdots & 0 & 0 \end{bmatrix}, \tag{3.2}$$

where all the blocks are square ones of order n/s.

Proof Formula (3.2) is guaranteed by the Frobenius standard form (2.15) in Theorem 2.20 for general irreducible nonnegative square matrices, where the zero blocks along the main diagonal are square. Since each of the blocks $A_{12}, \cdots, A_{s-1,s}, A_{s1}$ is doubly stochastic, they must be square. Consequently all the zero blocks along the main diagonal are of the same order. Hence s divides n. This proves the theorem. $\square$

A reducible doubly stochastic matrix also has a special structure, as shown by the following decomposition theorem.

Theorem 3.7. *Let* $A \in \mathbb{R}^{n \times n}$ *be doubly stochastic and reducible. Then there are* r *irreducible doubly stochastic matrices* $A_{11}, \cdots, A_{rr}$ *such that*

$$P^T A P = \mathrm{diag}(A_{11}, A_{22}, \cdots, A_{rr}) \tag{3.3}$$

is a block diagonal matrix, where P *is a permutation matrix.*

Proof By assumption, a permutation matrix P_1 exists such that

$$P_1^T A P_1 = \begin{bmatrix} B & C \\ 0 & D \end{bmatrix} \equiv \tilde{A},$$

where B is $k \times k$ and D is $(n-k) \times (n-k)$. Let s_X denote the sum of the entries of a matrix X. Since $\tilde{A}$ is also doubly stochastic, clearly $s_B = k$ and $s_D = n - k$. On the other hand,

$$n = s_{\tilde{A}} = s_B + s_C + s_D = k + s_C + (n-k) = n + s_C.$$

Thus $s_C = 0$, and so $C = 0$. Hence, $\tilde{A} = \mathrm{diag}(B, D)$. If B and D are already irreducible, then the theorem is proved. Otherwise, the above process can be continued till a matrix of the form (3.3) is obtained, in which all the diagonal blocks are irreducible. $\square$

Let $\mathbb{D}_n \subset \mathbb{S}_n$ denote the set of all $n \times n$ doubly stochastic matrices $A = (a_{ij})$. Then $\mathbb{D}_n$ is a polyhedron since it is the intersection of $\mathbb{S}_n$ with

n additional hyper-planes $\sum_{i=1}^{n} a_{ij} = 1$, $j = 1, 2, \cdots, n$. Under the natural order of the n^2 variables

$$a_{11}, a_{12}, \cdots, a_{1n}, a_{21}, a_{22}, \cdots, a_{2n}, \cdots, a_{n1}, a_{n2}, \cdots, a_{nn},$$

the $2n \times n^2$ coefficient matrix of the $2n$ equations

$$\sum_{i=1}^{n} a_{ij} = 1, \ j = 1, 2, \cdots, n, \quad \sum_{j=1}^{n} a_{ij} = 1, \ i = 1, 2, \cdots, n$$

has the $2 \times n$ block matrix form

$$\begin{bmatrix} I & \cdots & I \\ e_1 e^T & \cdots & e_n e^T \end{bmatrix}, \tag{3.4}$$

whose rank can be easily shown to be $2n - 1$ after performing elementary row operations, from which it follows that the dimension of the convex set $\mathbb{D}_n$ is $n^2 - (2n - 1) = (n - 1)^2$.

Since permutation matrices are doubly stochastic, so are all of their convex combinations. A permutation matrix is an extreme point of $\mathbb{S}_n$ by Proposition 3.1, therefore it is also an extreme point of $\mathbb{D}_n$. One of the fundamental results in the theory of doubly stochastic matrices is the following so-called *Birkhoff theorem* [Birkhoff (1938)] that answers the question whether the converse statement is also true or not.

Theorem 3.8. (Birkhoff) *The extreme points of $\mathbb{D}_n$ are exactly the $n \times n$ permutation matrices.*

Proof We prove the theorem by induction. The theorem is clearly true when $n = 1$. Suppose that the extreme points of $\mathbb{D}_{n-1}$ are precisely the $(n-1) \times (n-1)$ permutation matrices. Let $A = (a_{ij})$ be an extreme point of $\mathbb{D}_n$. Since $\mathbb{D}_n$ is the intersection of n^2 half-spaces defined by $a_{ij} \geq 0$, $i, j = 1, 2, \cdots, n$, inside the intersection of the $2n$ hyper-planes as defined above, and since dim $\mathbb{D}_n = (n-1)^2$, the extreme point A must lie on the boundary of at least $(n - 1)^2$ of these half-spaces since the single point set $\{A\}$ has dimension 0. This implies that A has at most $n^2 - (n-1)^2 = 2n-1$ positive entries, so not every row can contain more than one positive entry. This ensures that $a_{ij} = 1$ for at least one pair (i, j).

If we delete row i and column j from A to have an $(n - 1) \times (n - 1)$ doubly stochastic matrix $\tilde{A}$, then $\tilde{A}$ is an extreme point of $\mathbb{D}_{n-1}$ since A is an extreme point of $\mathbb{D}_n$. By the induction assumption, $\tilde{A}$ is a permutation matrix, therefore A is a permutation matrix. $\qquad\square$

Since every compact convex polyhedron is the convex hull of the set of its extreme points, Birkhoff's theorem implies that each $A \in \mathbb{D}_n$ can be expressed as a convex combination of permutation matrices. Actually we can prove a much stronger result:

Proposition 3.10. *If $A \in \mathbb{R}^{n \times n}$ is doubly stochastic, then there are $k \leq (n-1)^2 + 1$ permutation matrices $P_1, \cdots, P_k$ such that*

$$A = \sum_{i=1}^{k} c_i P_i, \quad c_i \geq 0, \ \forall \, i, \ \sum_{i=1}^{k} c_i = 1.$$

Proof Since $\dim \mathbb{D}_n = (n-1)^2$, each simplex contained in $\mathbb{D}_n$ has at most $(n-1)^2 + 1$ vertices. The result follows from the fact that each $A \in \mathbb{D}_n$ belongs to a simplex contained in $\mathbb{D}_n$. $\qquad\square$

If $A \in \mathbb{R}^{n \times n}$ is symmetric and *positive semi-definite*, that is, $x^T A x \geq 0$ for all $x \in \mathbb{R}^n$, then there is a unique symmetric positive semi-definite matrix $B \in \mathbb{R}^{n \times n}$ such that $B^2 = A$. The matrix B is denoted by $A^{1/2}$ and is called the *square root* of A. In general, the square root of a symmetric positive semi-definite doubly stochastic matrix is not necessarily doubly stochastic. The following result of Marcus and Minc [Marcus and Minc (1962)] characterizes square roots of doubly stochastic matrices.

Theorem 3.9. *The square root $A^{1/2}$ of a symmetric positive semi-definite doubly stochastic matrix $A = (a_{ij}) \in \mathbb{R}^{n \times n}$ is doubly quasi-stochastic. Moreover, If $a_{ii} \leq 1/(n-1)$ for all $i = 1, 2, \cdots, n$, then the matrix $A^{1/2}$ is also a doubly stochastic one.*

Proof It is well-known [Horn and Johnson (1985)] that the eigenvalues $1, \lambda_2, \cdots, \lambda_n$ of A are all nonnegative. Since A is a normal matrix, the equality (A.2) holds, where $U = (u_{ij}) \in \mathbb{R}^{n \times n}$ can be chosen to be an orthogonal matrix for which

$$U^T A U = \mathrm{diag}\,(1, \lambda_2, \cdots, \lambda_n) \equiv \Lambda$$

and whose first column is $e/\sqrt{n}$. Then the square root of A is

$$B = (b_{ij}) = U \mathrm{diag}\,(1, \sqrt{\lambda_2}, \cdots, \sqrt{\lambda_n}) U^T = U \Lambda^{\frac{1}{2}} U^T.$$

Since U is orthogonal with the first column $e/\sqrt{n}$,

$$Be = U \Lambda^{\frac{1}{2}} U^T e = U \Lambda^{\frac{1}{2}} \sqrt{n} e_1 = U \sqrt{n} e_1 = e.$$

This proves the first part of the theorem.

Suppose that $a_{ii} \leq 1/(n-1)$, $i = 1, 2, \cdots, n$. If B defined above is not nonnegative, then $b_{ij} < 0$ for some i and j. Since

$$\sum_{k \neq j} b_{ik} = \sum_{k=1}^{n} b_{ik} - b_{ij} = 1 - b_{ij} > 1,$$

we have, by the Cauchy-Schwarz inequality,

$$a_{ii} = \sum_{k=1}^{n} b_{ik}^2 > \sum_{k \neq j} b_{ik}^2 \geq \frac{1}{n-1} \left(\sum_{k \neq j} b_{ik} \right)^2 > \frac{1}{n-1}.$$

This contradicts the assumption that $a_{ii} \leq 1/(n-1)$. $\qquad\qquad\square$

3.3 Inequalities Related to Doubly Stochastic Matrices

Since rows and columns of doubly stochastic matrices are discrete probability distributions, they provide coefficients of convex combinations in some analytic problems. In this section we study a classic theorem by Hardy, Littlewood, and Pólya.

For any vector $x = (x_1, x_2, \cdots, x_n)^T \in \mathbb{R}^n$, we use the notation $x^d = (x_1^*, x_2^*, \cdots, x_n^*)^T$ to denote the vector x rearranged in non-increasing order, namely $x_1^* \geq x_2^* \geq \cdots \geq x_n^*$.

Definition 3.3. $x = (x_1, x_2, \cdots, x_n)^T \in \mathbb{R}_+^n$ is said to be *majorized* by $y = (y_1, y_2, \cdots, y_n)^T \in R_+^n$ if $x^T e = y^T e$ and

$$x_1^* + x_2^* + \cdots x_k^* \leq y_1^* + y_2^* + \cdots + y_k^*$$

for $k = 1, 2, \cdots, n-1$. This is denoted by $x \prec y$.

We need several preliminary results for the proof of Theorem 3.10 below on an equivalent condition for majorization.

Lemma 3.2. *Let $x, y \in \mathbb{R}^n$ be nonnegative. Then*

$$x^T y \leq (x^d)^T y^d. \tag{3.5}$$

Proof Without loss of generality, we assume that $x = x^d$. Suppose that $y_i < y_j$ for some $i < j$. Then

$$(x_i y_j + x_j y_i) - (x_i y_i + x_j y_j) = (x_i - x_j)(y_j - y_i) \geq 0.$$

In other words, if the two numbers y_i and y_j are exchanged, then the corresponding contribution to the left-hand side summation of (3.5) will not be decreased. Repeating the above argument finitely many times will produce the desired inequality (3.5). □

Lemma 3.3. *Let $1 \leq k \leq n$ be two integers. Suppose that $x, y \in \mathbb{R}_+^n$ be such that $x \leq e$, $e^T x = k$, and $y = y^d$. Then*

$$x^T y \leq \sum_{i=1}^{k} y_i.$$

Proof A direct computation gives

$$\sum_{i=1}^{k} y_i - x^T y = \sum_{i=1}^{k} y_i - \sum_{i=1}^{n} x_i y_i = \sum_{i=1}^{k} (1 - x_i) y_i - \sum_{i=k+1}^{n} x_i y_i$$

$$\geq \sum_{i=1}^{k} (1 - x_i) y_k - \sum_{i=k+1}^{n} x_i y_k$$

$$= y_k \left(k - \sum_{i=1}^{k} x_i \right) - y_k \sum_{i=k+1}^{n} x_i = 0. \qquad \square$$

The elementary but rather tedious proof of the following important result is adopted from [Minc (1988)].

Theorem 3.10. (Hardy-Littlewood-Pólya) *Let two vectors $x, y \in \mathbb{R}_+^n$. Then $x \prec y$ if and only if there exists an $n \times n$ doubly stochastic matrix A that satisfies the equality $x = Ay$.*

Proof Suppose that $A \in \mathbb{R}^{n \times n}$ is a doubly stochastic matrix, and x and y are nonnegative vectors such that $x = Ay$. Without loss of generality, we can assume that $x = x^d$. Let k be any integer with $1 \leq k < n$. Then

$$\sum_{i=1}^{k} x_i^* = \sum_{i=1}^{k} \sum_{j=1}^{n} a_{ij} y_j = \sum_{j=1}^{n} \left(\sum_{i=1}^{k} a_{ij} \right) y_j = \sum_{j=1}^{n} b_{kj} y_j,$$

where $b_{kj} \equiv \sum_{i=1}^{k} a_{ij} \leq 1$ for all j and

$$\sum_{j=1}^{n} b_{kj} = \sum_{j=1}^{n} \sum_{i=1}^{k} a_{ij} = \sum_{i=1}^{k} \sum_{j=1}^{n} a_{ij} = \sum_{i=1}^{k} 1 = k.$$

It follows from Lemmas 3.2 and 3.3 that

$$\sum_{j=1}^{n} b_{kj} y_j \leq \sum_{j=1}^{n} b_{kj}^* y_j^* \leq \sum_{j=1}^{k} y_j^*.$$

Hence,

$$\sum_{i=1}^{k} x_i^* \le \sum_{i=1}^{k} y_i^*, \quad k = 1, 2, \cdots, n-1.$$

Since A is doubly stochastic, $e^T x = e^T A y = e^T y$. Thus, $x \prec y$.

Conversely, let $x \prec y$. If $x^d = A y^d$ for a doubly stochastic matrix A, then $x = PAQy$, where P and Q are permutation matrices so that PAQ is still doubly stochastic. Thus, without loss of generality, we can assume that $x = x^d$, $y = y^d$, and $x \ne y$. Denote by $n(x, y)$ the number of nonzero coordinates of $y - x$. Then $n(x, y) \ge 2$. Since $e^T (x - y) = 0$, there is the least subscript t such that $x_t > y_t$ and there is the greatest subscript s, less than t, such that $x_s < y_s$. Thus, we have

$$x_s < y_s, \ x_{s+1} = y_{s+1}, \ \cdots, \ x_{t-1} = y_{t-1}, \ x_t > y_t.$$

Define a doubly stochastic matrix $A_1 \equiv A_1(\theta)$ by replacing the (s, s) and (t, t) entries by θ, and the (s, t) and the (t, s) entries by $1 - \theta$, of the $n \times n$ identity matrix I, where $\theta \in [0, 1]$. Then $(A_1 y)_s = \theta y_s + (1 - \theta) y_t$, $(A_1 y)_t = (1 - \theta) y_s + \theta y_t$, and $(A_1 y)_i = y_i$ for all indices $i \ne s, t$. If we choose

$$\theta = \theta_1 \equiv \frac{x_s - y_t}{y_s - y_t} \in (0, 1),$$

then

$$(A_1 y)_s = x_s \ \text{and} \ (A_1 y)_t = y_s - x_s + y_t.$$

If we let

$$\theta = \theta_2 \equiv \frac{y_s - x_t}{y_s - y_t} \in (0, 1),$$

then

$$(A_1 y)_s = y_s - x_t + y_t \ \text{and} \ (A_1 y)_t = x_t.$$

Thus, in either case $n(x, A_1 y) < n(x, y)$, provided $A_1 y = (A_1 y)^d$. This will be the case for $\theta = \theta_1$ if

$$y_{t-1} \ge y_s - x_s + y_t \ge y_{t+1}, \tag{3.6}$$

and for $\theta = \theta_2$ if

$$y_{s-1} \ge y_s - x_t + y_t \ge y_{s+1}. \tag{3.7}$$

Because of the inequalities $y_t + (y_s - x_s) > y_t \ge y_{t+1}$ and $y_s - (x_t - y_t) < y_s \le y_{s-1}$, the right-hand side inequality in (3.6) and the left-hand side inequality in (3.7) hold. Suppose that the left-hand side inequality in (3.6) fails to hold. Then we have

$$y_s - x_t + y_t > y_{t-1} + x_s - x_t = x_{t-1} + x_s - x_t \geq x_s \geq x_{s+1} = y_{s+1},$$

so (3.7) holds. Similarly, it can be verified that the violation of the right-hand side inequality in (3.7) implies (3.6). We thus conclude that with an appropriate choice, $\theta = \theta_1$ or θ_2,

$$n(x, A_1 y) < n(x, y), \quad x \prec A_1 y, \quad \text{and} \quad A_1 y = (A_1 y)^d.$$

Continuing in the same way, we can find a sequence of doubly stochastic matrices $A_1, A_2, \cdots, A_k$ with $n(x, A_k A_{k-1} \cdots A_1 y) = 0$. In other words, $x = A_k A_{k-1} \cdots A_1 y$. Define $A = A_k A_{k-1} \cdots A_1$. Then A is a doubly stochastic matrix as a product of doubly stochastic ones, and $x = Ay$. □

At the end of this section we give an application of the convexity of doubly stochastic matrices; the result is due to Schur.

Theorem 3.11. (Schur) *Let $A \in \mathbb{R}^{n \times n}$ be doubly stochastic, $x = (x_1, \cdots, x_n)^T \in \mathbb{R}_+^n$, and $y = (y_1, \cdots, y_n)^T = Ax$. Then,*

$$\prod_{i=1}^{n} y_i \geq \prod_{i=1}^{n} x_i.$$

Proof First assume that $x > 0$. Then $y > 0$. Since the function $\ln t$ is concave on $(0, \infty)$, for each $i = 1, 2, \cdots, n$,

$$\ln y_i = \ln \sum_{j=1}^{n} a_{ij} x_j \geq \sum_{j=1}^{n} a_{ij} \ln x_j.$$

Summing up the above n inequalities gives

$$\sum_{i=1}^{n} \ln y_i \geq \sum_{i=1}^{n} \sum_{j=1}^{n} a_{ij} \ln x_j = \sum_{j=1}^{n} \left(\sum_{i=1}^{n} a_{ij} \right) \ln x_j = \sum_{j=1}^{n} \ln x_j.$$

So $\prod_{i=1}^{n} y_i \geq \prod_{i=1}^{n} x_i$. The general case of $x \geq 0$ comes from the above conclusion and a limit process. □

Remark 3.3. By the same token, we can show that, if A is an $n \times n$ doubly stochastic matrix, then for any concave function $g : [0, \infty) \to \mathbb{R}$,

$$\sum_{i=1}^{n} g(y_i) \geq \sum_{i=1}^{n} g(x_i),$$

where $x \geq 0$ and $y = Ax$.

Motivated by Schur's theorem above, we give the following definition of Schur-convex (or Schur-concave) functions.

Definition 3.4. Let $n > 1$ be an integer and let a region $\Omega \subset \mathbb{R}^n_+$ be such that $A\Omega \subset \Omega$ for all matrices $A \in \mathbb{D}_n$. A real-valued function f defined on Ω is called *Schur-convex* if

$$f(Ax) \le f(x), \quad \forall\, x \in \Omega, \; A \in \mathbb{D}_n \tag{3.8}$$

and *strictly Schur-convex* if "$\le$" in (3.8) is replaced by "$<$" for $x \neq 0$. The function f is called *Schur-concave* (or *strictly Schur-concave*) if direction of the inequality in (3.8) is reversed.

It follows from Remark 3.3 that, given any convex (or concave) function $g : [0, \infty) \to \mathbb{R}$, the function f defined by $f(x) = \sum_{i=1}^{n} g(x_i)$ with $x = (x_1, \cdots, x_n)^T$ is Schur-convex (or Schur-concave) on $\mathbb{R}^n_+$.

3.4 Ergodic Theorems of Stochastic Matrices

The sequence of the powers of a stochastic matrix may not have a limit, even if the matrix is irreducible. One example is, as seen before, the 2×2 permutation matrix $A = [e_2, e_1]$. It was proved in Section 2.3 that any irreducible nonnegative matrix with spectral radius 1 is mean ergodic. However, the main result of this section is that any stochastic matrix, irreducible or not, is always mean ergodic. But first we give, as a direct consequence of Theorem 2.13, the following result about the mean ergodicity of irreducible stochastic matrices.

Theorem 3.12. *Let $A \in \mathbb{R}^{n \times n}$ be irreducible.*

(i) *If A is a stochastic matrix, then*

$$\lim_{k \to \infty} \frac{1}{k} \sum_{i=0}^{k-1} A^i = ey^T$$

with y the left Perron vector of A. If A is also primitive, then

$$\lim_{k \to \infty} A^k = ey^T.$$

(ii) *If A is a column stochastic matrix, then*

$$\lim_{k \to \infty} \frac{1}{k} \sum_{i=0}^{k-1} A^i = xe^T,$$

where x is the Perron vector of A. If A is also primitive, then

$$\lim_{k \to \infty} A^k = xe^T.$$

Remark 3.4. It can be shown that there is a constant C, which only depends on the matrix A, such that

$$\left\| \frac{1}{k} \sum_{i=0}^{k-1} A^i - E \right\|_\infty \leq \frac{C}{k}, \ \forall \, k \geq 1,$$

where $E = ey^T$ or xe^T, depending on whether A is a stochastic or column stochastic square matrix.

However, the assumption of irreducibility in the previous theorem is not necessary. Actually, Lemma 2.5 directly implies the mean ergodicity of arbitrary stochastic matrices.

Theorem 3.13. *Any stochastic matrix is mean ergodic.*

Proof Let $A \in \mathbb{R}^{n \times n}$ be a stochastic matrix. Then its powers A^k are also stochastic for all $k \geq 1$, which implies that they are uniformly bounded. So Lemma 2.5 is applicable. □

Remark 3.5. In this most general case, the limit matrix E may have a complicated expression (see Exercise 3.24).

As an application of Theorem 3.13 and Theorem 2.14, we have an interesting spectral characterization of stochastic lattice isomorphisms which also implies a spectral result of stochastic lattice homomorphisms.

Proposition 3.11. *Let $A \in \mathbb{R}^{n \times n}$ be a stochastic matrix. Then A is a permutation matrix if and only if $\sigma(A) \subset \partial \mathbb{D}^2$.*

Proof Suppose that A is a permutation matrix. Then A preserves the 2-norm of any vector in $\mathbb{C}^n$. If $Ax = \lambda x$ with $x \neq 0$, then $\|x\|_2 = \|Ax\|_2 = |\lambda| \|x\|_2$. Since $\|x\|_2 > 0$, we have $|\lambda| = 1$.

For the sufficiency part, suppose that $\sigma(A) \subset \partial \mathbb{D}^2$. Since $\sigma(A)$ is cyclic, there is $k \geq 1$ such that $\sigma(A^k) = \{1\}$. The stochastic matrix A^k is mean ergodic by Theorem 3.13, so Theorem 2.14 ensures that $\lambda = 1$ is a simple pole of the resolvent $R(\lambda, A^k)$. Since $\sigma(A^k) = \{1\}$, the minimal polynomial of A^k is $\lambda - 1$, so $A^k = I$. Thus $A^{-1} = A^{k-1} \geq 0$ and so A is a lattice isomorphism. By Corollary 3.3, A is a permutation matrix. □

Corollary 3.7. *If a stochastic matrix $A \in \mathbb{R}^{n \times n}$ defines a lattice homomorphism, then $\sigma(A) \subset \{0\} \cup \partial \mathbb{D}^2$.*

Proof Let M denote the ideal of all $x \in \mathbb{C}^n$ such that $A^k|x| = 0$ for some k. Then $AM \subset M$, and the induced operator $A_M : \mathbb{C}^n/M \to \mathbb{C}^n/M$ is still a lattice homomorphism. The matrix representation of A_M is also stochastic. Moreover, the definition of M ensures that A_M is invertible, so A_M is a permutation matrix by Corollary 3.3. Thus, from Proposition 3.11, $\sigma(A_M) \subset \partial \mathbb{D}^2$. Since $(A|_M)^l = 0$ for some l, it is true that $\sigma(A|_M) = \{0\}$. Therefore, from (A.1) we obtain

$$\sigma(A) = \sigma(A|_M) \cup \sigma(A_M) \subset \{0\} \cup \partial \mathbb{D}^2. \qquad \square$$

Proposition 3.11 combined with the Birkhoff theorem, Theorem 3.8 on the extreme points of $\mathbb{D}_n$, gives rise to another consequence concerning the spectrum of doubly stochastic matrices.

Corollary 3.8. *Let $A \in \mathbb{R}^{n \times n}$ be doubly stochastic. Then $\sigma(A) \subset C_n$, where C_n is the convex hull of the set of all rth roots of 1 with $1 \leq r \leq n$.*

Proof Let λ be any eigenvalue of $A \in \mathbb{D}_n$ and let x be a corresponding eigenvector with $\|x\|_2 = 1$. Since $A = \sum_{i=1}^k c_i P_i$ by Theorem 3.8, where each P_i is a permutation matrix, we have

$$\lambda = \lambda x^H x = x^H A x = \sum_{i=1}^k c_i x^H P_i x.$$

Since each P_i is a normal matrix, the range of the quadratic form $z^H P_i z$ with $z \in \mathbb{C}^n$ such that $\|z\|_2 = 1$ is the convex hull of $\sigma(P_i)$ [Horn and Johnson (1985)]. On the other hand, $\sigma(P_i) \subset \partial \mathbb{D}^2 \cap C_n$ from Proposition 3.11. It follows that $x^H P_i x \in C_n$ for $i = 1, 2, \cdots, k$, and so λ, a convex combination of $x^H P_i x$'s, also belongs to C_n. $\qquad \square$

3.5 Notes and Remarks

1. Proposition 3.4 was proved by [Fréchet (1933)], independent of the discovery of the Gersgörin disk theorem published two years earlier.

2. A comprehensive monograph on the theory of permanents is [Minc (1978)], in which its historical development and relationship to $(0,1)$-matrices and other topics are well addressed.

3. The concept of Schur-stochastic matrices has its origin in the paper [Schur (1923)], where Corollary 3.6 was established.

4. In relation to Birkhoff's theorem, let $\beta(A)$ be the smallest number of permutation matrices whose convex combination equals a doubly stochastic matrix $A \in \mathbb{D}_n$. Proposition 3.10, which is an improvement of the result that $\beta(A) < n(n-1)$ due to [Marcus and Newman (1959)], says that $\beta(A) \leq (n-1)^2 + 1$. It was further proved in [Marcus *et al.* (1961)] that $\beta(A) < s(n/s - 1)^2$ for $n \times n$ irreducible doubly stochastic matrices with the index of imprimitivity s; see also [Minc (1988)].

5. Theorem 3.10 extends the classic Muirhead theorem [Muirhead (1903)] in the area of majorization of nonnegative vectors: given n positive numbers $a_1, \cdots, a_n$ and two n-dimensional vectors x, y of nonnegative integers x_i and y_j, $x \prec y$ if and only if $\mathrm{per} A(x) \leq \mathrm{per} A(y)$, where the (i, j)-entry of $A(x)$ and $A(y)$ is $a_i^{x_j}$ and $a_i^{y_j}$, respectively.

Exercises

3.1 Let $a, b \in [0, 1]$ and consider the 2×2 stochastic matrix

$$A = \begin{bmatrix} 1-a & a \\ b & 1-b \end{bmatrix}.$$

(i) Find the values a and b such that $\lim_{k \to \infty} A^k$ exists.

(ii) For other values of a and b, what is $\lim_{k \to \infty} k^{-1} \sum_{j=0}^{k-1} A^j$?

3.2 Show that the dimension of $\mathbb{S}_n$ is $n^2 - n$.

3.3 Show that any 2×2 doubly stochastic matrix must be symmetric with same diagonal entries and orthostochastic.

3.4 Show that a doubly quasi-stochastic matrix must be square.

3.5 Find a stochastic matrix similar to $r(A)^{-1}A$, where

$$A = \begin{bmatrix} 3 & 2 & 0 \\ 1 & 1 & 1 \\ 0 & 1 & 3 \end{bmatrix}.$$

3.6 Find 3×3 doubly stochastic matrices A and X such that XAX^{-1} is not doubly stochastic.

3.7 Prove that if $A : \mathbb{C}^n \to \mathbb{C}$ is a lattice homomorphism, then $Ax = ax_j$ for some constant a and a fixed index j for all $x \in \mathbb{C}^n$.

3.8 Prove Proposition 3.7.

3.9 Let $A, B \in \mathbb{R}^{n \times n}$ be nonnegative. Show that

$$\mathrm{per} A \cdot \mathrm{per} B \leq \mathrm{per}(AB).$$

3.10 Let $A \in \mathbb{R}^{n \times n}$ be nonnegative. Show that
$$\operatorname{per} A \le (Ae)_1 (Ae)_2 \cdots (Ae)_n.$$

3.11 Let $A = (a_{ij}) \in \mathbb{R}^{n \times n}$ be a $(0,1)$-matrix such that $\operatorname{per} A > 0$. Define an $n \times n$ matrix $B = (b_{ij})$ by
$$b_{ij} = a_{ij} \frac{\operatorname{per} A(i|j)}{\operatorname{per} A}, \quad \forall\, i,\, j = 1, \cdots, n.$$
Show that B is doubly stochastic.

3.12 Show that the matrix (3.4) has rank $2n - 1$.

3.13 Show that the doubly stochastic matrix
$$A = \frac{1}{2} \begin{bmatrix} 0 & 1 & 1 \\ 1 & 0 & 1 \\ 1 & 1 & 0 \end{bmatrix}$$
is not Schur-stochastic.

3.14 Show that the doubly stochastic matrix
$$A = \frac{1}{3} \begin{bmatrix} 1 & 1 & 1 \\ 1 & 1 & 1 \\ 1 & 1 & 1 \end{bmatrix}$$
is Schur-stochastic but not orthostochastic.

3.15 Show that a stochastic matrix $A \in \mathbb{R}^{n \times n}$ is reducible if and only if there are disjoint sequences α and β with $\alpha \cup \beta = \{1, \cdots, n\}$ such that $A[\alpha|\beta]$ is a zero matrix.

3.16 Let $x = (9, 6, 5, 4, 4)^T$ and $y = (10, 10, 5, 2, 1)^T$. Show that $x \prec y$ and find a doubly stochastic matrix S such that $x = Ay$ by using the construction in the proof of Theorem 3.10.

3.17 Show that the condition $a_{ii} \le 1/(n-1)$ for all $i = 1, 2, \cdots, n$ in Theorem 3.9 is not necessary for the square root $A^{1/2}$ to be doubly stochastic, even if A is in addition a positive matrix.

3.18 Let $x = (x_1, \cdots, x_n)^T$, $a = (a_1, \cdots, a_n)^T \in \mathbb{R}_+^n$ be such that $e^T a = 1$. Prove the *arithmetic mean and geometric mean inequality*:
$$\prod_{i=1}^{n} x_i^{a_i} \le \sum_{i=1}^{n} a_i x_i,$$
where $0^0 = 1$ (so $0 \ln 0 = 0$) by convention. Moreover, equality holds if and only if all the x_i corresponding to positive a_i are equal.

3.19 Let $x = (x_1, \cdots, x_n)^T$, $y = (y_1, \cdots, y_n)^T \in \mathbb{R}_+^n$ be such that $e^T x = e^T y > 0$. Show that
$$\prod_{i=1}^{n} y_i^{x_i} \le \prod_{i=1}^{n} x_i^{x_i}$$

and equality holds if and only if $x = y$.

3.20 Prove the *Hadamard inequality*: If $A = (a_{ij}) \in \mathbb{R}^{n \times n}$ is symmetric and positive definite, then

$$\det A \leq a_{11} a_{22} \cdots a_{nn}.$$

Hint: Let $D = \mathrm{diag}(1/\sqrt{a_{11}}, \cdots, 1/\sqrt{a_{nn}})$ and $B = DAD$, and apply the arithmetic mean-geometric mean inequality to the positive eigenvalues $\lambda_1, \cdots, \lambda_n$ of B.

3.21 Let $A \in \mathbb{R}^{n \times n}$ be a nonnegative matrix with row sums at most 1 (an example of such matrices is the raw Google matrix to be defined in Section 4.2). Show that $y \in \mathbb{R}^n_+$ is a left Perron vector of some stochastic matrix $B \geq A$ if and only if $Ay \leq y$.

3.22 Let $A \in \mathbb{R}^{n \times n}$ be positive. Show that there are positive $x, y \in \mathbb{R}^n$ such that the matrix $(a_{ij} x_i y_j)$ is doubly stochastic.

Hint: Let K be the standard simplex of $\mathbb{R}^n$. Define $S : K \to K$ by $S(y) = z$, where $z_j = u_j / e^T u$ and

$$\frac{1}{u_j} = \sum_{i=1}^n \frac{a_{ij}}{\sum_{k=1}^n a_{ik} y_k}.$$

Then apply Brouwer's fixed point theorem.

3.23 Use a similar idea as for Exercise 3.22 to show that if A is in addition symmetric, then there is $z = (z_1, \cdots, z_n)^T > 0$ such that the matrix $(a_{ij} z_i z_j)$ is doubly stochastic.

3.24 The following 5×5 reducible matrix

$$A = \begin{bmatrix} 1 & 0 & 0 & 0 & 0 \\ 1-p & 0 & p & 0 & 0 \\ 0 & 1-p & 0 & p & 0 \\ 0 & 0 & 1-p & 0 & p \\ 0 & 0 & 0 & 0 & 1 \end{bmatrix}$$

is defined in [Wong and Zigarovich (2007)]. Show that

$$\lim_{n \to \infty} A^n = \begin{bmatrix} 1 & 0 & 0 & 0 & 0 \\ \frac{(1-p)(1-p+p^2)}{1-2p+2p^2} & 0 & 0 & 0 & \frac{p^3}{1-2p+2p^2} \\ \frac{(1-p)^2}{1-2p+2p^2} & 0 & 0 & 0 & \frac{p^2}{1-2p+2p^2} \\ \frac{(1-p)^3}{1-2p+2p^2} & 0 & 0 & 0 & \frac{p(1-p+p^2)}{1-2p+2p^2} \\ 0 & 0 & 0 & 0 & 1 \end{bmatrix}.$$

Hint: You may consult Exercises 2.27 and 2.28.

Chapter 4

Applications of Nonnegative Matrices

Nonnegative matrices arise in almost every branch of mathematical sciences with particular applications in the areas related to probability theory, stochastic analysis, and numerical analysis. In this chapter we give several classic and more recent applications of nonnegative matrices, in which the Perron-Frobenius theory of nonnegative matrices, in particular the theory of stochastic matrices, plays a key role in such areas.

Stochastic matrices are directly used in the theory of finite state stationary Markov chains, which constitutes Section 4.1. Stochastic matrices have found new applications in modern information retrieval methods such as the PageRank concept from the Google Web search engine. Section 4.2 is devoted to the presentation of this exciting application related to our daily life in the modern times of the information highway. In Section 4.3 we illustrate some interesting applications of stochastic matrices in the area of dynamical geometry of polygons, and Section 4.4 is devoted to the classic application of nonnegative matrices to the matrix iterative analysis for solving linear equations.

4.1 Finite State Stationary Markov Chains

Stochastic matrices have many applications in the problems related to probability theory and stochastic processes, particularly to the theory of Markov processes with discrete times and finite state spaces. In the first section, we give an introductory account and some basic features of the subject of finite state stationary Markov chains, in which the theory of stochastic matrices has direct applications. We assume that the reader is familiar with basic probability theory.

A *Markov chain* is a discrete collection of random variables

$$\Phi = \{\Phi_0, \Phi_1, \cdots, \Phi_k, \cdots\},$$

which is a particular type of *stochastic processes* taking, at times $k = 0, 1, 2, \cdots$, values Φ_k in a *state space* X. The state space X is called *countable* if X is finite or infinitely countable. In this case, the σ-algebra of X will be taken to be the family of all subsets of X. The key feature of a Markov chain is that its future trajectory depends on its present and its past states only through the current value.

We only consider a stationary Markov chain with finitely many states, which means that the state space X can be represented by the symbol set $\{1, 2, \cdots, n\}$ and the process moves from state i to state j with a fixed transition probability p_{ij} for all states $i, j = 1, 2, \cdots, n$. Using a more formal language, we say that a finite state Markov chain is *stationary* if the *conditional probabilities* satisfy

$$\text{prob}(\Phi_{k+1} = j | \Phi_k = i, \Phi_{k-1} = i_{k-1}, \cdots, \Phi_1 = i_1, \Phi_0 = i_0) = p_{ij}$$

for all states $i_0, i_1, \cdots, i_{k-1}, i, j$ and for all $k \geq 0$. Thus, together with an initial probability distribution Φ_0, the process is completely characterized by the n^2 *transition probabilities*

$$\text{prob}(\Phi_{k+1} = j | \Phi_k = i) = p_{ij}, \quad \forall \, i, j = 1, 2, \cdots, n; \; k \geq 0, \qquad (4.1)$$

which says that p_{ij} is the probability for the system to be in state j at time $k + 1$ if it was in state i at time k. Clearly, the sum $\sum_{j=1}^{n} p_{ij}$ equals 1 for each i, so the $n \times n$ matrix $P = (p_{ij})$, called the *transition matrix* of the Markov chain, is stochastic.

It is easy to see that the probability transition from state i to state j after two steps is given by

$$p_{ij}^{(2)} = \sum_{r=1}^{n} p_{ir} p_{rj},$$

since the transition can occur in n mutually exclusive ways $i \to r \to j$ for $r = 1, 2, \cdots, n$. Therefore, the matrix $(p_{ij}^{(2)})$ is just the square P^2 of P. In general, the probability transition from state i to state j in exactly k steps is nothing but the (i, j)-entry $p_{ij}^{(k)}$ of the matrix P^k. When $k = 1$, we write $p_{ij}^{(1)} = p_{ij}$. Since $P^{k+l} = P^k P^l$, there is the following *Chapman-Kolmogorov identity* for all positive integers k and l:

$$p_{ij}^{(k+l)} = \sum_{r=1}^{n} p_{ir}^{(k)} p_{rj}^{(l)}, \quad \forall \, i, j = 1, \cdots, n.$$

We present two examples of finite state stationary Markov chains before studying their general properties.

Example 4.1. (Markov chain interpretation of Ulam's method) Consider the unit closed interval $[0, 1]$ with the Lebesgue measure m. Given any positive integer n, the interval $[0, 1]$ is the union of the n disjoint subintervals $I_i = [(i-1)/n, i/n)$, $i = 1, \cdots, n-1$, and $I_n = [1 - 1/n, 1]$. For any Borel measurable transformation $S : [0, 1] \to [0, 1]$ set
$$p_{ij} = \frac{m(I_i \cap S^{-1}(I_j))}{m(I_i)}, \quad \forall \, i, j = 1, \cdots, n.$$
Let a sequence of the random variables $\{\Phi_k\}$ with state space $X = \{1, 2, \cdots, n\}$ be defined by letting $\Phi_k(x) = i$ if $S^k(x) \in I_i$, $\forall \, i = 1, \cdots, n$. Then the Markov property (4.1) is satisfied, so $\{\Phi_k\}$ is a finite state stationary Markov chain which is closely related to Ulam's method that will be studied in Chapter 10.

Example 4.2. (Random walks) A *random walk* is a Markov chain which is defined to be a random process in which there is a sequence of moves between n states such that the probability of going to state j in the next step depends only on the current state i and not on the previous history of the process and the time when it is observed. Suppose that there are n states and there are moves among these states. For the n states indexed as $1, \cdots, n$, let p_{ij} denote the observed fraction of all moves from a given state i to a state j. If a large number of separate moves are followed, the fraction p_{ij} represents the probability of a transition from i to j. In fact, this is nothing more than the common sense of probability as a frequency of occurrence of some event. The $n \times n$ transition matrix $P = (p_{ij})$ is obviously stochastic since each state i moves to a state j among $1, \cdots, n$.

We take a simple model of one dimensional random walks as an illustration. Suppose that a particle can move among the nodes represented by the integers $1, 2, \cdots, n$ on the number line by bouncing one step at a time either right or left. If the particle is at integer i, then it goes to $i + 1$ with probability p and to $i - 1$ with probability $q = 1 - p$, except when i is either 1 or n. At these boundary points, the particle stays there. The transition matrix for this problem is given by
$$\begin{cases} p_{i,i+1} = p \text{ and } p_{i,i-1} = q, \text{ for } i = 2, 3, \cdots, n-1, \\ p_{11} = p_{nn} = 1, \text{ and} \\ p_{ij} = 0, \text{ for all other } i, j. \end{cases}$$
The set of transitions from states $i = 1, \cdots, n$ to states $j = 1, \cdots, n$ is called a *random walk with absorbing barriers*. $\qquad \square$

Many problems in the study of finite state stationary Markov chains are related to the eventual behavior of the iterates P^k of the transition matrix P as $k \to \infty$, which can be investigated by means of the theory of stochastic matrices. For the notational convenience, in this section all the involved vectors $y \in \mathbb{R}^n$ are considered row vectors. If Φ_0 has the probability distribution $y^{(0)} = (y_1^{(0)}, y_2^{(0)}, \cdots, y_n^{(0)})$, then Φ_1 has the probability distribution $y^{(1)} = (y_1^{(1)}, y_2^{(1)}, \cdots, y_n^{(1)})$, where

$$y_j^{(1)} = \sum_{i=1}^{n} p_{ij} y_i^{(0)}, \quad j = 1, 2, \cdots, n.$$

In other words, $y^{(1)} = y^{(0)} P$. In general, Φ_k has the distribution

$$y^{(k)} = y^{(k-1)} P = y^{(0)} P^k.$$

It is clear that the asymptotic behavior of the Markov chain is thus reflected by the asymptotic behavior of the sequence of the powers P^k as k approaches infinity. In particular, if the limit $\lim_{k \to \infty} P^k = Q$ exists for some nonnegative matrix Q, then the vector $y^{(0)} Q$ will be an approximate distribution of Φ_k when k is large enough.

A probability distribution y is said to be *stationary* if $y = yP$. Hence, stationary distributions are exactly left eigenvectors of P corresponding to eigenvalue 1 such that $y \geq 0$ and $\|y\|_1 = 1$. Since the transition matrix P of the Markov chain is stochastic, $r(P) = 1$, and Theorem 2.5 (or Theorem 3.1) immediately gives rise to the existence of a stationary distribution of the Markov chain in the following theorem.

Theorem 4.1. *Every finite state stationary Markov chain has a stationary probability distribution.*

Although $\lim_{k \to \infty} P^k$ may not exist in general, the averages sequence of the powers does converge. In fact, Theorem 3.13 implies the following convergence result for Markov chains.

Proposition 4.1. *For any transition matrix $P \in \mathbb{S}_n$, the asymptotic mean transition probabilities*

$$q_{ij} = \lim_{k \to \infty} \frac{1}{k} \sum_{l=0}^{k-1} p_{ij}^{(l)}$$

always exist for all $i, j = 1, 2, \cdots, n$. Furthermore, the limit matrix $Q = (q_{ij})$ is a transition matrix.

For each initial distribution $y^{(0)}$, the *mean asymptotic distribution* is defined to be $y = y^{(0)}Q$, where Q is the above asymptotic mean transition matrix associated with the given Markov chain. An initial probability distribution $y^{(0)}$ of the Markov chain is said to be *stationary* if

$$y^{(0)} = y^{(k)} = y^{(0)}P^k, \quad k = 1, 2, \cdots.$$

It follows from the above analysis that $y^{(0)}$ is a stationary probability distribution if and only if it is a normalized left eigenvector of P associated with the maximal eigenvalue 1.

Although the kth probability distribution $y^{(0)}P^k$ may not converge to a stationary probability distribution of the Markov chain as k approaches infinity for arbitrary initial distributions in general, Theorem 3.13 guarantees that on the average, the kth distribution will converge to a stationary distribution. Therefore, we have

Theorem 4.2. *For any finite state stationary Markov chain with the transition matrix $P \in \mathbb{S}_n$, the probability distribution vector*

$$y = \lim_{k \to \infty} \frac{1}{k} \sum_{i=0}^{k-1} y^{(0)}P^i$$

associated with any given initial probability distribution $y^{(0)}$, is a stationary probability distribution of the chain.

Let i be a state of a finite state stationary Markov chain which has the transition matrix P. A state j is called a *consequent* of state i if there exists $k \geq 1$ such that $p_{ij}^{(k)} > 0$. In this case we also say that state i has *access* to state j, written $i \to j$. If $i \to j$ and $j \to i$, then states i and j are said to *communicate* and we write $i \leftrightarrow j$.

The communication relation is an equivalent one on the set of states, and it partitions $\{1, \cdots, n\}$ into equivalence classes.

Definition 4.1. The *classes* of a finite state stationary Markov chain are the equivalence classes which are induced by the communication relation on the set of the states. A class θ has *access* to a class κ, written as $\theta \to \kappa$, if $i \to j$ for some $i \in \theta$ and $j \in \kappa$. A class is said to be *final* if it has no access to other classes. If the final class contains only one state, then this state is said to be *absorbing*.

A state i is called *transient* if $i \to j$ for some j but $j \not\to i$. Otherwise, the state i is called *ergodic*. In other words, i is ergodic if and only if $i \to j$

implies $j \to i$. Thus, if a transient state i is an initial state, then it will pass, with a positive probability, through a state j from which it can never return to i. On the other hand, if i is ergodic then the set of all consequents of i forms a subset θ of $\{1, 2, \cdots, n\}$ which satisfies the following property: if the system is ever in the state $i \in \theta$, it will not leave θ thereafter with probability 1; moreover, θ is minimal with respect to this property.

It follows that if one state in a class θ of states associated with a Markov chain is transient or ergodic, then each state in θ is transient or ergodic. This motivates the following definition.

Definition 4.2. A class θ of states of a finite state stationary Markov chain induced by the communication relation is called *transient* if it contains a transient state, and is called *ergodic* otherwise.

Obviously, a class is ergodic if and only if it is final. A Markov chain is called *ergodic* if it consists of only one ergodic class, *regular* if at some fixed step k each i moves to every j with positive probability, and *periodic* if it is ergodic but not regular. Clearly a regular chain is ergodic, but not the converse. For example, a Markov chain whose transition matrix is the 2×2 permutation matrix $P = [e_2, e_1]$ is ergodic but not regular.

Now it is time to apply the theory of nonnegative matrices to study the classification problem of Markov chains.

Theorem 4.3. *Let $P \in \mathbb{S}_n$ be the transition matrix of a finite state stationary Markov chain. Then*
    (i) *the chain is ergodic if and only if P is irreducible;*
    (ii) *the chain is regular if and only if P is primitive;*
    (iii) *the chain is periodic if and only if P is irreducible and imprimitive.*

Proof (i) By Theorem 1.4, the matrix P is irreducible if and only if the associated directed graph $G(P)$ is strongly connected. But this is true if and only if all the states of the chain have access to each other, thus if and only if the chain is ergodic.

    (ii) The chain is regular if and only if $P^k > 0$ for some k, and so if and only if P is primitive from Theorem 2.18.

    (iii) It is obvious. □

Theorems 2.17 and 2.18 give rise to the characterizations of ergodicity and regularity of a Markov chain, respectively.

Theorem 4.4. *A finite state stationary Markov chain with the transition* *matrix* $P \in \mathbb{S}_n$ *is ergodic if and only if* 1 *is an algebraically simple eigenvalue of* P *with a corresponding positive left eigenvector.*

Corollary 4.1. *A finite state stationary Markov chain is ergodic if and only if it has a unique stationary probability distribution.*

Theorem 4.5. *A finite state stationary Markov chain with the transition matrix* $P \in \mathbb{S}_n$ *is regular if and only if* 1 *is an algebraically simple eigenvalue of* P *with a corresponding positive left eigenvector and*
  (i) *the limit* $\lim_{k \to \infty} P^k$ *exists, or*
  (ii) 1 *is the unique eigenvalue of* P *on the unit circle, or*
  (iii) *the limit* $\lim_{k \to \infty} y^{(0)} P^k$ *exists for all initial probability distribution vectors* $y^{(0)}$, *or*
  (iv) $P^k > 0$ *for some positive integer* k.

Theorem 4.6. *A finite state stationary Markov chain with the transition matrix* $P \in \mathbb{S}_n$ *is periodic if and only if* 1 *is an algebraically simple eigenvalue of* P *with a corresponding positive left eigenvector and the index of imprimitivity of* P *is greater than* 1.

The index of P is called the *period* of the periodic Markov chain. We refer the reader to the standard textbook [Chung (1967)] for a comprehensive study of Markov chains.

4.2 The Web Information Retrieval

We live in the information highway era, and Web surfing has become part of our life every day. The Web information retrieval is more challenging than the traditional information retrieval because of the huge number of Web pages (currently in the order of more than 10 billions). There are various information retrieval methods in use, one of which is the nonnegative matrix approach to vector space models.

In the Web information retrieval research, each Web page is represented as a vertex in a very large directed Web graph (see Definition 1.3). The directed edges connecting these vertices represent the *hyper-links* among the pages. Let E be the set of all directed edges in the Web graph and let e_{ij} represent the directed edge from vertex i to vertex j. This hyper-link structure is explored by such information retrieval methods as the HITS (hypertext induced topic search), the PageRank, and the SALSA (stochastic approach for link-structure analysis) [Langville and Meyer (2006)]. Although they use different concepts, their fundamental ideas are all based on linear algebra, particularly the theory of nonnegative matrices.

We briefly discuss the philosophy behind the HITS before we give a more detailed discussion of the PageRank.

In the HITS information retrieval method, the concepts of *authorities* related to in-links and *hubs* related to out-links are introduced, and the HITS assigns both an *authority score* x_i and a *hub score* y_i to each Web page represented by vertex i. Suppose that an initial authority score $x_i^{(0)}$ and a hub score $y_i^{(0)}$ are assigned for each vertex i. Then the iteration scheme defined recursively by

$$x_i^{(k)} = \sum_{j:e_{ji} \in E} y_j^{(k-1)} \quad \text{and} \quad y_i^{(k)} = \sum_{j:e_{ij} \in E} x_j^{(k-1)}, \quad k = 1, 2, \cdots \qquad (4.2)$$

refines these scores, that is, the updated authority score of page i is the contribution of the current hub scores of all pages j linked to i, and the updated hub score of page i is the contribution of the current authority scores of all pages j linked from page i. Let $L = (l_{ij})$ be the adjacency matrix of the directed Web graph as defined by Definition 1.4, that is,

$$l_{ij} = \begin{cases} 1, & \text{if } e_{ij} \in E, \\ 0, & \text{otherwise.} \end{cases}$$

Then the above iteration scheme (4.2) can be rewritten in the matrix multiplication form as

$$x^{(k)} = L^T y^{(k-1)} \quad \text{and} \quad y^{(k)} = L x^{(k-1)},$$

which leads to the following iteration scheme

$$x^{(k)} = L^T L x^{(k-1)} \quad \text{and} \quad y^{(k)} = L L^T y^{(k-1)}. \qquad (4.3)$$

These two new equations in (4.3) define the *power method* for computing the maximal eigenvector associated with the spectral radius for the matrices $L^T L$ and $L L^T$, respectively. Since the nonnegative matrix $L^T L$ determines the authority scores, it is called the *authority matrix*, and similarly $L L^T \geq 0$ is called the *hub matrix*. Since $L^T L$ and $L L^T$ are both symmetric and positive semi-definite, their distinctive eigenvalues $\{\lambda_1, \cdots, \lambda_r\}$ are necessarily real and nonnegative, and so can be arranged as $\lambda_1 \geq \lambda_2 \geq \cdots \geq \lambda_r \geq 0$, the location of which is closely related to the performance of the HITS.

In the remainder of the section we explore the concept of the PageRank which was introduced by Google founders Sergey Brin and Larry Page in 1998 and is the heart of Google's software.

The basic idea of the PageRank is, as described in [Langville and Meyer (2006)]: Links from important Web pages should carry more weight than those from less important ones, and the significance of a link from any source page should be scaled by the number of Wed sites that the source page is linking to. Let $r(p_i)$ represent the *rank* of any Web page p_i. Then the above idea of the PageRank can be expressed mathematically by the following equality for a given page p_i:

$$r(p_i) = \sum_{p_j \in I_i} \frac{r(p_j)}{\# \text{ of out-links from } p_j}, \tag{4.4}$$

where I_i is the set of all pages pointing to p_i.

The system of the above equations for all Web pages relating their ranks defines the PageRank implicitly. Introducing a suitable matrix, it can be re-formulated as a matrix fixed point problem. Suppose that there are totally n Web pages,

$$p_1, \ p_2, \ p_3, \ \cdots, \ p_n.$$

Let an $n \times n$ matrix $P = (p_{ij})$ be defined by

$$p_{ij} = \begin{cases} \dfrac{1}{\# \text{ of out-links from } p_i}, & \text{if } p_i \text{ links to } p_j, \\ 0, & \text{otherwise}, \end{cases}$$

for all $i, j = 1, 2, \cdots, n$, and denote

$$y = (r(p_1), \ r(p_2), \ \cdots, \ r(p_n))^T.$$

Then the system (4.4) can be written concisely as

$$y^T = y^T P. \tag{4.5}$$

The matrix P is nonnegative. Moreover, if a vertex p_i does have at least one out-link, the corresponding row of P sums up to be 1. This *raw Google matrix* P is not a stochastic matrix if there are Web pages without any out-links. Such pages are referred to as *dangling vertices*. In fact, most Web pages have this feature. So the matrix P may not have a left eigenvector y as a solution of (4.5). To solve this existence problem of the PageRank vector, we fill all zero rows of P, which correspond to dangling vertices, with the same probability row vector

$$v^T = (v_1, \ v_2, \ \cdots, \ v_n),$$

where each $v_i > 0$ and $\sum_{i=1}^n v_i = 1$. Then the resulting matrix

$$A = P + uv^T,$$

where the n-dimensional vector $u \in \mathbb{R}^n$ is chosen such that $u_i = 1$ if p_i is a dangling vertex and $u_i = 0$ otherwise, becomes stochastic. This guarantees the existence of a left eigenvector y^T corresponding to the maximal eigenvalue 1. However, since this rank-1 updated matrix A is not irreducible in general, so the uniqueness of the PageRank vector y is still not guaranteed. To make it irreducible, we fix a real number $a \in (0, 1)$ and define the following *Google matrix* G as a convex combination of the stochastic matrix S and the rank-one stochastic matrix ev^T:

$$G = aA + (1 - a)ev^T. \tag{4.6}$$

The Google matrix G is still a stochastic one since

$$Ge = aAe + (1 - a)e(v^T e) = ae + (1 - a)e = e.$$

Most importantly, G is irreducible (actually positive), so it has the left Perron vector y which is referred to as the *PageRank* of the Google matrix for the Web system.

The PageRank y can be calculated by the power method

$$y_k^T = y_{k-1}^T G, \quad \forall\, k = 1, 2, \cdots,$$

which is the only practical numerical method for computing the left maximal eigenvector of G because of the huge size of the matrix. Since G is a positive matrix and $r(G) = 1$, the power method converges by Theorem 2.3 for any initial vector $y_0 \in \mathbb{R}_+^n$. From numerical linear algebra, the convergence rate of this method depends on the absolute value of the second largest eigenvalue of G in modules. As a matter of fact, the spectrum of the Google matrix G can be completely determined by the following result, which is a direct consequence of Lemma 2.2.

Theorem 4.7. *The eigenvalues of the Google matrix G are*

$$1, a\lambda_2, \cdots, a\lambda_n,$$

counting algebraic multiplicity, where $1, \lambda_2, \cdots, \lambda_n$ are the eigenvalues of S, counting algebraic multiplicity.

Proof It is easy to see that the eigenvalues of the matrix aA are $a, a\lambda_2, \cdots, a\lambda_n$, and $(1 - a)u$ is an eigenvector of aA associated with the eigenvalue a. Therefore, from Lemma 2.2, the eigenvalues of G are $\mu, a\lambda_2, \cdots, a\lambda_n$, where μ is the eigenvalue of the 1×1 matrix

$$a + v^T(1 - a)e = a + 1 - a = 1. \square$$

More discussions on various issues related to the Google matrix can be found in the book [Langville and Meyer (2006)].

4.3 Discrete Dynamical Geometry

The subject of *dynamical geometry* studies geometric changes with respect to time. In *discrete dynamical geometry* one is interested in the eventual behavior of deterministic sequences of geometric figures generated by iteration. In this section we apply the Perron-Frobenius theory of nonnegative matrices and properties of stochastic matrices to several kinds of iterations of geometric objects in the Euclidean plane.

First we consider a dynamical system problem in Euclidean geometry. Take any acute triangle $\triangle A_0 B_0 C_0$ and construct the inscribed circle. The three points of tangency of the circle to the three sides of the triangle form a second triangle, $\triangle A_1 B_1 C_1$. Then construct the inscribed circle for $\triangle A_1 B_1 C_1$. The points of tangency on the three sides of $\triangle A_1 B_1 C_1$ form a third triangle $\triangle A_2 B_2 C_2$. Continuing this process, one gets a sequence $\{\triangle A_k B_k C_k\}_{k=0}^{\infty}$ of nested triangles. Although the size of the triangles will go to zero, we are interested in the eventual shape of $\triangle A_k B_k C_k$ as k is getting bigger and bigger.

To find the limiting shape of the iterated triangles $\triangle A_k B_k C_k$, we inscribe each of them into a fixed circle and use the three central angles, $\alpha_k, \beta_k, \gamma_k$, of $\triangle A_k B_k C_k$ to determine its shape. Then, from elementary geometry, we have

$$
\begin{bmatrix} \alpha_k \\ \beta_k \\ \gamma_k \end{bmatrix} = \begin{bmatrix} 0 & \frac{1}{2} & \frac{1}{2} \\ \frac{1}{2} & 0 & \frac{1}{2} \\ \frac{1}{2} & \frac{1}{2} & 0 \end{bmatrix} \begin{bmatrix} \alpha_{k-1} \\ \beta_{k-1} \\ \gamma_{k-1} \end{bmatrix}, \ \forall\, k \geq 1.
$$

Since the involved 3×3 matrix is not only doubly stochastic, but also primitive, the Perron-Frobenius theorem ensures that

$$
\lim_{k \to \infty} (\alpha_k, \beta_k, \gamma_k) = \left(\frac{2\pi}{3}, \frac{2\pi}{3}, \frac{2\pi}{3} \right).
$$

Consequently, no matter what the shape of the initial triangle $\triangle A_0 B_0 C_0$ is, the shape of the subsequent triangles $\triangle A_k B_k C_k$ will approach that of an equilateral triangle.

A variation of the above construction of $\{\triangle A_k B_k C_k\}$ is: Let $\triangle A_0 B_0 C_0$ be an acute triangle circumscribing a circle Γ_0 with center O_0. The line segments $A_0 O_0$, $B_0 O_0$, and $C_0 O_0$ intersect Γ_0 at points A_1, B_1 and C_1, forming a second triangle $\triangle A_1 B_1 C_1$ that circumscribes a circle Γ_1 with center O_1. Construct a third triangle $\triangle A_2 B_2 C_2$ from $\triangle A_1 B_1 C_1$ by the same way, and so on. We have a new sequence $\{\triangle A_k B_k C_k\}$ of triangles that are nested in a coherent manner. Now the iteration formula relating the central angles of $\triangle A_k B_k C_k$ and $\triangle A_{k-1} B_{k-1} C_{k-1}$ is

$$
\begin{bmatrix} \alpha_k \\ \beta_k \\ \gamma_k \end{bmatrix} = \begin{bmatrix} \frac{1}{2} & \frac{1}{4} & \frac{1}{4} \\ \frac{1}{4} & \frac{1}{2} & \frac{1}{4} \\ \frac{1}{4} & \frac{1}{4} & \frac{1}{2} \end{bmatrix} \begin{bmatrix} \alpha_{k-1} \\ \beta_{k-1} \\ \gamma_{k-1} \end{bmatrix}, \quad \forall \, k \geq 1,
$$

which implies that the triangles in this new sequence also approach an equilateral triangle in shape as n goes to infinity.

There are many other ways to construct nested triangles with their members being closer and closer to a fixed triangle in shape. For instance, the reader may consider iterated triangles formed by the intersection points of the three angle bisectors with the opposite sides of an acute triangle. No matter how to construct the new triangle from the old one, if the relation between $(\alpha, \beta, \gamma)_{\text{new}}$ and $(\alpha, \beta, \gamma)_{\text{old}}$ is given by a 3×3 primitive column stochastic matrix A, then the limiting shape of the sequence $\{\triangle A_k B_k C_k\}$ of the iterated triangles is determined by the Perron vector of A.

Since every triangle can be inscribed to a circle, the above idea for iterating triangles can be extended to cyclic polygons. Given an inscribed n-polygon on a unit circle with vertices $z_1, \cdots, z_n$. This cyclic polygon is determined by the corresponding vector $(\theta_1, \cdots, \theta_n)$ of the central angles subtended by each side of the polygon. If by some rule we choose a point on each side of the polygon and join the center of the circle to each chosen point and extend them to meet the circle at points $w_1, \cdots, w_n$, then we obtain a new cyclic polygon, determined by its n central angles $(\phi_1, \cdots, \phi_n)$. Repeating this process we have a sequence of cyclic n-polygons. What interests us is the asymptotic behavior of the sequence. Since the sum of the central angles of a cyclic n-polygon is the same constant 2π, it is easy to see that if the rule of obtaining the new polygon $(\phi_1, \cdots, \phi_n)$ from the old one $(\theta_1, \cdots, \theta_n)$ is linear, then it can be represented by a quasi-column or column stochastic matrix acting on $\mathbb{R}^n$. For example, the rule such that

$$
\begin{bmatrix} \phi_1 \\ \phi_2 \\ \vdots \\ \phi_{n-1} \\ \phi_n \end{bmatrix} = \begin{bmatrix} 1-\lambda_1 & \lambda_2 & 0 & \cdots & 0 \\ 0 & 1-\lambda_2 & \lambda_3 & \cdots & 0 \\ \vdots & \ddots & \ddots & \ddots & \vdots \\ 0 & 0 & \cdots & 1-\lambda_{n-1} & \lambda_n \\ \lambda_1 & 0 & \cdots & 0 & 1-\lambda_n \end{bmatrix} \begin{bmatrix} \theta_1 \\ \theta_2 \\ \vdots \\ \theta_{n-1} \\ \theta_n \end{bmatrix},
$$

where $\lambda_i \in (0,1)$ for each i, corresponds to a column stochastic matrix and is called a Λ-*transformation* in the dynamical geometry of polygons

[Hitt and Zhang (2001)]. In this case we could expect a regular eventual behavior for the dynamics since the involved column-stochastic iteration matrix is primitive. Thus, starting from any cyclic polygon, the sequence of the iterated polygons will converge to a polygon whose central angles are determined by the Perron vector of the matrix. In particular if $n = 3$ and $\lambda_1 = \lambda_2 = \lambda_3 = 1/2$, we obtain the sequence of the so-called *Kasner triangles* which converges to an equilateral triangle. See [Hitt and Zhang (2001); Ding *et al.* (2003a)] for more detailed discussions about the discrete dynamical geometry of cyclic polygons.

However, if we form sequences of pedal triangles, the situation is different. Let $\triangle A_0 B_0 C_0$ be any triangle with three inner angles A_0, B_0, and C_0 and three corresponding sides a_0, b_0, and c_0. The *pedal triangle* $\triangle A_1 B_1 C_1$ of $\triangle A_0 B_0 C_0$, with the inner angles A_1, B_1, and C_1 and the corresponding sides a_1, b_1, and c_1, is the triangle obtained by joining the three feet of altitudes of $\triangle A_0 B_0 C_0$. If $\triangle A_0 B_0 C_0$ is an acute triangle, then $\triangle A_1 B_1 C_1$ is inscribed inside $\triangle A_0 B_0 C_0$. If $\triangle A_0 B_0 C_0$ is an obtuse triangle, then the vertices of $\triangle A_1 B_1 C1$ may be on the extension of the sides of $\triangle A_0 B_0 C_0$. Furthermore, as a separation of the above two different cases, if $\triangle A_0 B_0 C_0$ is a right triangle, then its pedal triangle $\triangle A_1 B_1 C_1$ degenerates into a line segment which has area zero.

From the classic Euclidean geometry, the three smaller triangles $\triangle A_0 B_1 C_1, \triangle A_1 B_0 C_1$, and $\triangle A_0 B_0 C_1$ are all similar to their "parent" triangle $\triangle A_0 B_0 C_0$, from which it follows that

(i) If $\triangle A_0 B_0 C_0$ is an acute triangle, then

$$A_1 = \pi - 2A_0, \quad B_1 = \pi - 2B_0, \quad C_1 = \pi - 2C_0,$$

$$a_1 = a_0 \cos A_0, \quad b_1 = b_0 \cos B_0, \quad c_1 = c_0 \cos C_0.$$

(ii) If $\triangle A_0 B_0 C_0$ is obtuse, say, $A_0 > \pi/2$, then

$$A_1 = 2A_0 - \pi, \quad B_1 = 2B_0, \quad C_1 = 2C_0,$$

$$a_1 = -a_0 \cos A_0, \quad b_1 = b_0 \cos B_0, \quad c_1 = c_0 \cos C_0,$$

and similar formulas hold when $B_0 > \pi/2$ or $C_0 > \pi/2$.

Starting with an initial triangle $\triangle A_0 B_0 C_0$, we can construct a sequence $\{\triangle A_k B_k C_k\}$ of triangles such that $\triangle A_k B_k C_k$ is the pedal triangle of $\triangle A_{k-1} B_{k-1} C_{k-1}$ for all k. From (i) and (ii) above it is easy to see that the corresponding operator for the iteration of pedal triangles is no longer linear, but can be represented by four 3×3 quasi-column stochastic (but not always column-stochastic) matrices. As was shown in the papers [Kingston and Synge (1988); Lax (1990)], the iteration of the pedal triangles exhibits periodicity and a chaotic behavior.

We end this section with a class of fractals constructed via pedal triangles [Zhang *et al.* (2008)]. Given an arbitrary acute triangle $\triangle A_0 B_0 C_0$, first we draw the corresponding pedal triangle $\triangle A_1 B_1 C_1$. Then we delete the interior of the pedal triangle from the original triangle. The remaining part is the union of three smaller triangles $\triangle A_0 B_1 C_1$, $\triangle A_1 B_0 C_1$, and $\triangle A_1 B_1 C_0$, which are all similar to the original triangle $\triangle A_0 B_0 C_0$, as mentioned above. For each of the three triangles we repeat the same procedure, and the remaining set is the union of even smaller 3^2 similar triangles. In general, at the kth step, we have the union of 3^{k-1} similar triangles, and for each of them we delete the interior of the corresponding pedal triangle so that we obtain the union of 3^k similar smaller triangles. The limiting set of such nested subsets of $\triangle A_0 B_0 C_0$ as k goes to infinity is the *Sierpiński pedal triangle* associated with the initial triangle $\triangle A_0 B_0 C_0$. If the initial triangle $\triangle A_0 B_0 C_0$ is equilateral, then the three feet of the altitudes of $\triangle A_0 B_0 C_0$ coincide with the middle points of the three sides of $\triangle A_0 B_0 C_0$, hence the resulting Sierpiński pedal triangle is the same as the classic *Sierpiński triangle*, the most famous fractal in the theory of fractals [Barnsley (1988)].

It follows from the formulas in (i) above that the ratio of the areas of the pedal triangle $\triangle A_1 B_1 C_1$ and the original triangle $\triangle A_0 B_0 C_0$ is given by the formula

$$\frac{\text{area}\triangle A_1 B_1 C_1}{\text{area}\triangle A_0 B_0 C_0} = -2\cos A_0 \cos B_0 \cos(A_0 + B_0).$$

Let the *index domain* for the class of all acute triangles be

$$I = \left\{ (x, y) \in \mathbb{R}^2 : 0 < x, y < \frac{\pi}{2}, \ \frac{\pi}{2} < x + y < \pi \right\},$$

and define the area ratio function of the pedal triangle by

$$f(x, y) = -2\cos x \cdot \cos y \cdot \cos(x + y), \ \forall\, (x, y) \in I. \tag{4.7}$$

We need the following two lemmas for the proof of the next proposition; see [Zhang (1998); Zhang *et al.* (2008)].

Lemma 4.1. *Let* $f : I \to \mathbb{R}$ *be differentiable on its domain. Then* f *is strictly Schur-convex on* I *if*

$$(f_x(x, y) - f_y(x, y))(x - y) > 0, \ \forall\, (x, y) \in I, \ x \neq y,$$

and strictly Schur-concave if the above inequality is reversed.

Lemma 4.2. *If* $f : I \to \mathbb{R}$ *is Schur-convex (or Schur-concave) on* I, *then its minimal (or maximal) value of* f *over* I *is only obtained along the diagonal of* I, *that is, a minimizer (or maximizer) of* f *must have the same* x *and* y *coordinates. Moreover, if* f *is strictly Schur-convex (or Schur-concave), then the minimizer (or maximizer) is unique.*

Proposition 4.2. (Zhang) *The function f with the expression given by (4.7) is strictly Schur-concave on I. Furthermore*

$$\max_{(x,y)\in I} f(x,y) = f\left(\frac{\pi}{3},\frac{\pi}{3}\right) = \frac{1}{4}$$

and $f(x,y) \to 0$ as $(x,y) \to \partial I$.

Proof Since $f_x = 2\sin x \cos y \cos(x+y) + 2\cos x \cos y \sin(x+y)$ and $f_y = 2\cos x \sin y \cos(x+y) + 2\cos x \cos y \sin(x+y)$,

$$f_x - f_y = 2\cos(x+y) \cdot \sin(x-y).$$

Therefore for any $(x,y) \in I$ and $x \neq y$ we always have

$$(f_x - f_y)(x-y) = 2[\cos(x+y)\sin(x-y)] \cdot (x-y) < 0$$

since $\cos(x+y) < 0$ and $[\sin(x-y)] \cdot (x-y) > 0$ on I. Hence, by Lemma 4.1, f is a strictly Schur-concave function on I. From Lemma 4.2, f attains its maximal value on I only along the intersection of I with the main diagonal of the xy-plane. But along the diagonal f becomes

$$\hat{f}(x) = -2\cos^2 x \cos(2x) = 2\cos^2 x - 4\cos^4 x$$

which has only one critical point $x_0 = \pi/3$ in its domain, and $\hat{f}$ does attain the maximal value there.   □

Since Sierpiński pedal triangles are self-similar fractals, the fractal dimension $d = d(x,y)$ of a Sierpiński petal triangle corresponding to the initial triangle $\triangle A_0 B_0 C_0$ with $A_0 = x$ and $B_0 = y$ is determined by the so-called *Moran equation* [Barnsley (1988)]

$$\cos^d x + \cos^d y + \cos^d z = 1, \tag{4.8}$$

where $z = \pi - x - y < \pi/2$. In particular, in the special case of the Sierpiński triangle, namely $x = y = z = \pi/3$, the unique solution of the above equation (4.8) is $d = \ln 3/\ln 2$.

It has been proved [Ding and Li (2009)] that the dimension function $d = d(x,y)$ of Sierpiński pedal triangles has a strict global minimal value $\ln 3/\ln 2$ on I at $x = y = \pi/3$, which is the dimension of the Sierpiński triangle. It was conjectured in [Zhang *et al.* (2008)] that this dimension function d be a strictly Schur-convex function on its domain I.

4.4   Iterative Solutions of Linear Algebraic Equations

Almost all the scientific computing problems eventually lead to solving systems of linear algebraic equations. In numerical linear algebra, direct methods and iterative methods constitute the two main classes of computational schemes for solving the nonhomogeneous linear equation

$$Ax = b, \tag{4.9}$$

where $A = (a_{ij}) \in \mathbb{R}^{n \times n}$ is nonsingular and $b \in \mathbb{R}^n$. The direct method is based on the idea of Gaussian elimination and so it produces the solution $A^{-1}b$ after a finite number of arithmetic operations. On the other hand, the iterative method, which is usually more efficient for solving large-scale linear equations with sparse coefficient matrices, applies the direct iteration to a mathematically equivalent linear fixed point equation

$$x = Gx + c, \tag{4.10}$$

hence it is an infinite process like Newton's method for solving nonlinear equations. Therefore, the choice of the *iterative matrix* G in (4.10) determines not only whether the resulting fixed point iteration scheme converges, but also how fast it does.

In this section, we apply nonnegative matrices to construct the matrix G and give some convergence rate analysis of the iterative method for solving linear algebraic equations.

A simple way to write (4.9) in the form of (4.10) is to *split* A as the difference of two matrices B and C such that B is nonsingular and its inverse is easy to calculate. Then $G = B^{-1}C$ and $c = B^{-1}b$ in (4.10). For example, write $A = D - L - U$, where $D = \text{diag}(a_{11}, \cdots, a_{nn})$, and L and U are respectively strictly lower and strictly upper triangular matrices, then

(i) $B = D$ and $C = L + U$ give the *Jacobi iterative method*;

(ii) the choice of $B = D - L$ and $C = U$ leads to the *Gauss-Seidel iterative method*;

(iii) if $B = \omega^{-1}D - L$ and $C = U + (\omega^{-1} - 1)D$ with $\omega \in (0, 2)$, then one has the *successive over-relaxation iterative method*.

The decomposition $A = B - C$ is called a *regular splitting* of a matrix $A \in \mathbb{R}^{n \times n}$ if B is a nonsingular matrix with $B^{-1} \geq 0$ and $C \geq 0$, and is called a *weak regular splitting* of A if the condition $C \geq 0$ is replaced by the weaker ones $B^{-1}C \geq 0$ and $CB^{-1} \geq 0$. We only consider the regular splitting in the following discussion.

With respect to a regular splitting $A = B - C$, the original equation (4.9) can be written as

$$x = B^{-1}Cx + B^{-1}b = (A+C)^{-1}Cx + (A+C)^{-1}b$$
$$= (I + A^{-1}C)^{-1}A^{-1}Cx + (I + A^{-1}C)^{-1}A^{-1}b.$$

If we denote the matrix $A^{-1}C$ by F, then the corresponding fixed point equation (4.10) becomes

$$x = (I + F)^{-1}Fx + (I + F)^{-1}A^{-1}b.$$

It is a routine task to show that the eigenvalues λ of the matrix $G = B^{-1}C = (I + F)^{-1}F$ are related to the eigenvalues γ of the matrix F via the following explicit formula

$$\lambda = \frac{\gamma}{1 + \gamma}, \tag{4.11}$$

from which we can deduce a key convergence result of the corresponding iterative method as follows.

Theorem 4.8. *Let $A = B - C$ be a regular splitting of a matrix $A \in \mathbb{R}^{n \times n}$. Then A is nonsingular and $A^{-1} \geq 0$ if and only if the spectral radius $r(B^{-1}C) < 1$, in which case*

$$r(B^{-1}C) = \frac{r(A^{-1}C)}{1 + r(A^{-1}C)} \tag{4.12}$$

and the iterative scheme

$$x^{(k+1)} = B^{-1}Cx^{(k)} + B^{-1}b, \quad k = 0, 1, 2, \cdots \tag{4.13}$$

converges for any initial vector $x^{(0)} \in \mathbb{R}^n$.

Proof The assumption on A ensures that $B^{-1} \geq 0$ and $C \geq 0$, so $B^{-1}C \geq 0$. Assume that $r(B^{-1}C) < 1$. Then $B^{-1}A = I - B^{-1}C$ is invertible and its inverse matrix is nonnegative by Theorem 1.2, hence A is invertible and

$$A^{-1} = (I - B^{-1}C)^{-1}B^{-1} \geq 0.$$

Conversely, let A be nonsingular with $A^{-1} \geq 0$. Then, for the regular splitting $A = B - C$ we have

$$0 \leq B^{-1}C = (I + F)^{-1}F.$$

Since $r(B^{-1}C)$ is an eigenvalue of $B^{-1}C \geq 0$ and $r(F)$ is an eigenvalue of $F = A^{-1}C \geq 0$, and since the function $t/(1 + t)$ is strictly increasing on $[0, \infty)$, equality (4.12) is valid because of relation (4.11), so $r(B^{-1}C) < 1$. Finally, the convergence of the iterative method (4.13) follows from the Banach fixed point theorem [Ortega and Rheinboldt (1970)] since the mapping $x \to B^{-1}Cx + B^{-1}b$ is a contraction from $\mathbb{R}^n$ into itself.   □

Corollary 4.2. *Let $A \in \mathbb{R}^{n \times n}$ be an M-matrix, let $B \in \mathbb{R}^{n \times n}$ be obtained by making some off-diagonal entries of A to zero, and let $C = B - A$. Then $A = B - C$ is a regular splitting of A such that $r(B^{-1}C) < 1$.*

Proof By Theorem 1.10, B is an M-matrix. Clearly $C \geq 0$, so $A = B - C$ is a regular splitting of A. Since $A^{-1} \geq 0$ by the assumption, Theorem 4.8 gives the result immediately.   □

Corollary 4.3. *Let $A \in \mathbb{R}^{n \times n}$ be a Stieltjes matrix and let $A = B - C$ be a regular splitting of A with $C = C^T$. Then*

$$r(B^{-1}C) \leq \frac{r(A^{-1})r(C)}{1 + r(A^{-1})r(C)}. \tag{4.14}$$

Proof From Definition 1.9, A is a symmetric matrix with $A^{-1} \geq 0$. Hence the conclusion of Theorem 4.8 is valid. Since the spectral radius of a real symmetric matrix equals its matrix 2-norm, we have $r(A^{-1}) = \|A^{-1}\|_2$ and $r(C) = \|C\|_2$. It follows that

$$r(A^{-1}C) \leq \|A^{-1}C\|_2 \leq \|A^{-1}\|_2\|C\|_2 = r(A^{-1})r(C),$$

so the inequality (4.14) is satisfied from the equality (4.12) and the monotonicity of the function $t/(1 + t)$ for $t \geq 0$.   □

We finish this section by presenting a general comparison theorem for regular splittings of the same matrix.

Theorem 4.9. *Let $A = B_1 - C_1 = B_2 - C_2$ be two regular splittings of $A \in \mathbb{R}^{n \times n}$ with $A^{-1} \geq 0$. If $C_1 \leq C_2$, then*

$$r(B_1^{-1}C_1) \leq r(B_2^{-1}C_2). \tag{4.15}$$

Proof The given assumption implies that $0 \leq A^{-1}C_1 \leq A^{-1}C_2$, so $r(A^{-1}C_1) \leq r(A^{-1}C_2)$ by Corollary 1.1. Therefore, since $t \to t/(1 + t)$ is a monotonically increasing function on $[0, \infty)$, (4.15) follows directly from the conclusion of Theorem 4.8.   □

Remark 4.1. A strict inequality in (4.15) can be obtained if in addition $A^{-1}C_1$ is irreducible and $C_1 \neq C_2$, the proof of which is based on Corollary 2.7. However, the conclusion is still true even without the irreducibility assumption; see [Varga (2000)].

4.5 Notes and Remarks

1. Some concepts introduced in Section 4.1 are adopted from Section 1.9 of [Bapat and Raghavan (1997)], and the classic textbook [Chung (1967)] gives an excellent introduction to the rich subject of general Markov chains. The monograph [Meyn and Tweedie (1993)] contains more information on the topic of stochastic stability related to Markov chains.

2. There are several proofs of Theorem 4.7 that is a basis for the success of the power method used to compute the PageRank of the worldwide Web system, which is the left Perron vector of the Google matrix, the world's largest matrix. For example, we can find a proof using orthogonal transformations in the book [Langville and Meyer (2006)]. Because of constant creation and deletion of Web pages every day, the Google search engine recomputes the PageRank at every short time period (for example, weekly or biweekly) to reflect the ever changing Web page distributions. Developing faster and more efficient algorithms for the computation of the PageRank is an active research area; see [Langville and Meyer (2006)] and the references therein for the computational issues of the PageRank.

3. Discrete dynamical geometry studies iterations of geometric figures such as triangles and polygons. The Perron-Frobenius theorem is often used for the convergence analysis; see examples in [Ding and Fay (2005); Ding and Ye (2006)]. In this field, E. Kasner introduced *Kasner triangles* and his student J. Douglas, who was the first Fields medalist with L. Ahlfors in 1936, studied iterated polygons. J. Kingston and J. Synge investigated the periodicity of pedal triangle iterations [Kingston and Synge (1988)], and P. Lax [Lax (1990)] studied the ergodicity of the corresponding chaotic *pedal mapping* $p : D \subset \mathbb{R}^2 \to D$, defined by

$$
p(x, y) = \begin{cases} (\pi - 2x, \pi - 2y), & (x, y) \in D_0, \\ (2x, \pi - 2y), & (x, y) \in D_1, \\ (2x, 2y), & (x, y) \in D_2, \\ (\pi - 2x, 2y), & (x, y) \in D_3, \end{cases}
$$

where $D = \{(x, y) \in \mathbb{R}^2 : x \geq 0, y \geq 0, x + y \leq \pi\}$ represents the set of all triangles, $D_0 = \{(x, y) \in D : x \leq \pi/2, y \leq \pi/2, x + y \geq \pi/2\}$ corresponds to all acute triangles, and D_1, D_2, D_3 are associated with obtuse triangles $\triangle A_0 B_0 C_0$ with $A_0 > \pi/2, B_0 > \pi/2$, or $C_0 > \pi/2$, respectively.

Dynamical geometry also has a close relation to fractals which was introduced first in [Mandelbrot (1982)]. The book [Barnsley (1988)] contains many examples of fractals and practical methods to study them with ideas from iterated functions systems.

4. A major application of M-matrices is in the iterative solution of linear algebraic equations via the coefficient matrix splitting technique. A classic monograph on the matrix iterative analysis is [Varga (2000)].

Exercises

4.1 A particle at the origin starts to move on the x-axis and jumps every second. It moves either from i to $i+1$ with probability p or from i to $i-1$ with probability $q = 1 - p$. When it reaches the points -3 and 2, it is absorbed in these points permanently.

(i) Find the transition matrix of this random walk with finite states $-3, -2, -1, 0, 1, 2$.

(ii) Study the properties of this chain and classify states.

4.2 A particle moves as in Exercise 4.1, but now it is modified as follows: when it reaches -3, it jumps to -2 the next second and when it reaches 2, it jumps to 1.

(i) Find the transition matrix of this Markov chain.

(ii) Study the properties of this chain and classify states.

4.3 Prove or disprove the statements for a Markov chain:

(i) If state i is transient and $p_{ij}^{(k)} > 0$ for some positive integer k, then state j is transient.

(ii) If state j is transient and $p_{ij}^{(k)} > 0$ for some positive integer k, then state i is transient.

4.4 Classify states of the Markov chain with transition matrix

$$P = \begin{bmatrix} 0 & 0 & \frac{1}{2} & \frac{1}{2} \\ 0 & 1 & 0 & 0 \\ 0 & 0 & 0 & 1 \\ 1 & 0 & 0 & 0 \end{bmatrix}.$$

4.5 Prove: a class in a finite state stationary Markov chain is ergodic if and only if it is final.

4.6 Let $P = (p_{ij})$ be the transition matrix of a Markov chain. Prove the following statements:

(i) State j is transient if and only if for any $\epsilon > 0$, there is an integer $k_0 > 0$ such that $p_{ij}^{(k)} \leq \epsilon$ for all i and $k \geq k_0$.

(ii) If $\lim_{k \to \infty} P^k$ exists and is positive, then P is primitive.

4.7 Let $G = G(a)$ be the Google matrix as given by (4.6) for some parameter $a \in (0, 1)$. Show that the PageRank vector $y(a)$ can be expressed as

$$y(a) = \frac{1}{\sum_{i=1}^{n} D_i(a)} (D_1(a), D_2(a), \cdots, D_n(a))^T,$$

where $D_i(a)$ is the determinant of the ith $(n-1) \times (n-1)$ principal submatrix of $I - G(a)$ [Langville and Meyer (2006)].

4.8 [Langville and Meyer (2006)] Let the PageRank vector $y(a) = (y_1(a), y_2(a), \cdots, y_n(a))^T$. Show that $y(a)$ is differentiable. Moreover

$$\left| \frac{dy_i(a)}{da} \right| \leq \frac{1}{1-a}, \quad \forall\, i = 1, 2, \cdots, n$$

and

$$\left\| \frac{dy(a)}{da} \right\|_1 \leq \frac{2}{1-a}.$$

4.9 [Langville and Meyer (2006)] Let $y(a)$ be the PageRank vector associated with the Google matrix $G(a) = aS + (1-a)ev^T$. Show that

$$\frac{dy(a)}{da} = -(I - aS^T)^{-2}(I - S^T)v.$$

4.10 [Langville and Meyer (2006)] Let $G = aS + (1-a)ev^T$ and $\hat{G} = a\hat{S} + (1-a)ev^T$ be two Google matrices of the same size with their respective PageRank vectors y and $\hat{y}$, respectively. Show that

$$\|y - \hat{y}\|_1 \leq \frac{2a}{1-a} \sum_{j \in J} \pi_j,$$

where J is the set of the Web page indices j such that the jth rows of S and $\hat{S}$ are different.

4.11 Find the limiting shape of the sequence $\{\triangle A_n B_n C_n\}$ of the iterated triangles such that the triangle $\triangle A_n B_n C_n$ is formed by the intersection points of the three angle bisectors with the opposite sides of the triangle $\triangle A_{n-1} B_{n-1} C_{n-1}$ for all $n \geq 1$, starting with an acute triangle $\triangle A_0 B_0 C_0$.

4.12 Find the relation between the three central angles of the pedal triangle $\triangle A_1 B_1 C_1$ and the original triangle $\triangle A_0 B_0 C_0$ in matrix form for the cases

(i) $\triangle A_0 B_0 C_0$ is an acute triangle;

(ii) $\triangle A_0 B_0 C_0$ is an obtuse triangle with $A_0 > \pi/2$, $B_0 > \pi/2$, or $C_0 > \pi/2$, respectively.

4.13 Let $C = B - A$, where

$$A = \begin{bmatrix} 2 & -1 & 0 \\ 0 & 1 & -1 \\ -1 & 0 & 1 \end{bmatrix}, \quad B = \begin{bmatrix} 2 & -2 & 2 \\ 0 & 2 & -2 \\ -1 & 0 & 1 \end{bmatrix}.$$

Show that A is an M-matrix and that $A = B - C$ is a weak regular splitting of A but not a regular splitting.

4.14 Show that

$$\|x^{(k)} - x^*\| \le \|G\|^k \|x^{(0)} - x^*\|$$

and

$$\|x^{(k)} - x^*\| \le \frac{\|G\|^k}{1 - \|G\|} \|x^{(1)} - x^{(0)}\|$$

for all $k \ge 1$, where G is an $n \times n$ matrix with $\|G\| < 1$ and

$$x^{(k)} = Gx^{(k-1)} + c, \quad k = 1, 2, \cdots$$

with any initial vector $x^{(0)} \in \mathbb{R}^n$, and x^* is a unique solution of the fixed point equation (4.10).

4.15 Let $A = D - L - U$ be symmetric and positive definite and let $G = (D - L)^{-1}L^T$ be the Gauss-Seidel iterative matrix. Define $P = A - G^T AG$. Prove the following claims:

  (i) $G = I - Q$, where $Q = (D - L)^{-1}A$;
  (ii) $P = Q^T(AQ^{-1} - A + (Q^T)^{-1}A)Q$;
  (iii) $P = Q^T DQ$, so P is symmetric and positive definite;
  (iv) $r(G) < 1$;
  (v) the Gauss-Seidel iterative method is convergent.

4.16 Use the same method as for Exercise 4.15 to prove the convergence of the successive over-relaxation iterative method with $0 < \omega < 2$ when A is symmetric and positive definite.

4.17 Let $A = (a_{ij}) \in \mathbb{R}^{n \times n}$ with $a_{ii} \ne 0$ for all i and let $G_\omega = (D - \omega L)^{-1}[(1 - \omega)D + \omega U]$ be the matrix for the successive over-relaxation iterative method. Show that $r(G_\omega) \ge |\omega - 1|$.

4.18 Prove the convergence of the Jacobi iterative method and the Gauss-Seidel iterative method for solving (4.9) with an initial vector $x^{(0)}$ when $A \in \mathbb{R}^{n \times n}$ is strictly diagonally dominant.

4.19 Prove the *Stein-Rosenberg theorem*: Suppose that $A \in \mathbb{R}^{n \times n}$ is a Z-matrix with positive diagonal entries. Let G_j and G_g be the matrices of the Jacobi and Gauss-Seidel iterative method, respectively. Then exactly one of the following is true.

  (i) $0 \le r(G_g) < r(G_j) < 1$;
  (ii) $1 < r(G_j) < r(G_g)$;
  (iii) $r(G_j) = r(G_g) = 0$;
  (iv) $r(G_j) = r(G_g) = 1$.

Chapter 5

Fundamental Theory of Positive Operators

A linear operator between two ordered vector spaces that maps positive vectors onto positive vectors is referred to as a positive operator. Nonnegative matrices can be viewed as finite dimensional positive operators defined on Euclidean spaces $\mathbb{R}^n$ when the natural order for real vectors is imposed on $\mathbb{R}^n$. Like nonnegative matrices, positive operators are monotone with respect to the underlying order structure. The theory of general positive operators is rich and beautiful, and at the same time is more abstract and complicated than that of nonnegative matrices. From this chapter on we study positive operators and their finite dimensional approximations via nonnegative matrices. In the present chapter, we present the basic theory of abstract positive operators, in which many results are generalizations from nonnegative matrices to infinite dimensional positive operators.

We assume that the reader has taken a first course in functional analysis and is familiar with some basic concepts and facts for bounded linear operators between normed vector spaces and in particular bounded linear functionals on Banach spaces. Appendix B contains relevant results which will be needed for the study of positive operators between vector lattices.

In Section 5.1 we introduce various definitions and elementary results related to the concept of Banach lattices. Section 5.2 is devoted to the study of positive operators. Basic properties of ideals and projections will be given in Section 5.3. In Section 5.4 we focus on L-spaces and M-spaces, the two classes of abstract Banach lattices, in which we also introduce stochastic operators on L-spaces and abstract composition operators on M-spaces.

5.1 Banach Lattices

As the study of nonnegative matrices is related to the natural ordering in Euclidean spaces, the theory of positive operators is based on the lattice structure of ordered sets. Motivated by the partial ordering properties of $\mathbb{R}^n$, an *ordered set* is any set A endowed with an *ordering relation*, denoted by $\leq$, which is *reflexive, transitive,* and *anti-symmetric.* So if $(A, \leq)$ is an ordered set, then for all elements $x, y, z \in A$,

(i) $x \leq x$;

(ii) $x \leq y$ and $y \leq z$ imply $x \leq z$;

(iii) $x \leq y$ and $y \leq x$ imply $x = y$.

Let $(A, \leq)$ be an ordered set. The order relation $x \leq y$ can also be written as $y \geq x$. The relation $x < y$ means that $x \leq y$ and $x \neq y$ (note the difference of "<" here and "<" in previous chapters for real matrices). A subset B of A is called *majorized* (or *minorized*) if there exists $a \in A$ such that $b \leq a$ (or $b \geq a$) for all $b \in B$. In this case a is called an *upper bound* (or *lower bound*) of B in A. Let $x, \ y \in A$. The *interval* $[x, y]$ of A is the set $\{z \in A : x \leq z \leq y\}$. If a subset of A is contained in an interval of A, then it is said to be *order bounded.* If B has an upper bound a' that is majorized by all upper bounds of B in A, then a' must be unique and is called the *least upper bound* of B in A, and is denoted by $\sup B$. Similarly we can define the *greatest lower bound* of B in A and denote it by $\inf B$.

Definition 5.1. An ordered set $(A, \leq)$ is called a *lattice* if for any $x, y \in A$, the following elements exist in A:

$$x \vee y \equiv \sup\{x, y\} \quad \text{and} \quad x \wedge y \equiv \inf\{x, y\}.$$

A lattice A is called *distributive* if

$$(x \vee y) \wedge z = (x \wedge z) \vee (y \wedge z), \ \ \forall \, x, y, z \in A.$$

A distributive lattice A is called a *Boolean algebra* if A has a smallest element $0 \equiv \inf A$ and a largest element $1 \equiv \sup A$, and if for each $x \in A$, there exists a (unique) element x^c, called the *complement* of x in A, satisfying $x \wedge x^c = 0$ and $x \vee x^c = 1$. A subset of a lattice A is called a *sublattice* of A if it is a lattice under the ordering inherited from A.

Example 5.1. Let X be a topological space. The set of all open subsets of X is a distributive lattice under the set inclusion ordering $\subset$. The same is true for the set of all closed subsets of X.

Example 5.2. If G is a group, then the set of all subgroups of G is a lattice under the inclusion ordering but in general, not a sublattice of 2^G. Similarly, if V is a vector space, then the set of all vector subspaces of V is a lattice but in general, not a sublattice of 2^V.

Example 5.3. Let S be a nonempty set and let $(A, \leq)$ be a lattice. The set of all mappings $f : S \to A$ is a lattice under the natural *pointwise ordering* relation defined by "$f \leq g$ if and only if $f(s) \leq g(s)$, $\forall\, s \in S$".

Definition 5.2. A vector space X over the field of real numbers is called an *ordered vector space* if it is endowed with an order relation $\leq$ that satisfies the following two conditions:

(i) $x \leq y \Rightarrow x + z \leq y + z$ for all vectors $x, y, z \in X$;

(ii) $x \leq y \Rightarrow ax \leq ay$ for all vectors $x, y \in X$ and $a \in \mathbb{R}_+$.

If $(X, \leq)$ is an ordered vector space, the subset $X_+ = \{x \in X : x \geq 0\}$ is referred to as the *positive cone* of X, and every vector $x \in X_+$ is said to be a *positive* element.

Definition 5.3. An ordered vector space X is called a *vector lattice* if $x \vee y$ and $x \wedge y$ exist for all $x, y \in X$.

Most important vector lattices in applications are various function spaces X consisting of real-valued functions defined on a set Ω such that for any given functions $f, g \in X$, the two functions $\max\{f, g\}$ and $\min\{f, g\}$ defined by $\max\{f, g\}(x) = \max\{f(x), g(x)\}$ and $\min\{f, g\}(x) = \min\{f(x), g(x)\}$ for all $x \in \Omega$ both belong to X. Clearly, every such function space with the pointwise ordering, as was defined in Example 5.3, is a vector lattice which satisfies

$$f \vee g = \max\{f, g\} \quad \text{and} \quad f \wedge g = \min\{f, g\}.$$

Example 5.4. Such real function spaces are vector lattices:

(i) $\mathbb{R}^S$, the space of all functions from a set S into $\mathbb{R}$;

(ii) $l_\infty(S)$, the space of real-valued bounded functions on S;

(iii) $C(\Omega)$, the space of all real-valued continuous functions defined on a topological space Ω;

(iv) $C_b(\Omega)$, the space of all bounded real-valued continuous functions defined on a topological space Ω;

(v) the real Banach space $L^p(X, \Sigma, \mu)$ with $1 \leq p \leq \infty$.

In the following we assume that X is a general vector lattice.

Proposition 5.1. *For all vectors $x, y, z \in X$ and all numbers $a \in \mathbb{R}_+$, the following equalities are satisfied:*

(i) $a(x \vee y) = (ax) \vee (ay)$ *and* $a(x \wedge y) = (ax) \wedge (ay)$.

(ii) $-(x \vee y) = (-x) \wedge (-y)$ *and* $-(x \wedge y) = (-x) \vee (-y)$.

(iii) $x \vee y + x \wedge y = x + y$.

(iv) $(x + y) \vee (x + z) = x + (y \vee z)$ *and* $(x + y) \wedge (x + z) = x + (y \wedge z)$.

Proof (i) The case $a = 0$ is trivial. Suppose that $a > 0$. Then $(ax) \vee (ay) \leq a(x \vee y)$. If $ax \leq z$ and $ay \leq z$, then $x \leq a^{-1}z$ and $y \leq a^{-1}z$, hence $x \vee y \leq a^{-1}z$. Thus, $a(x \vee y) \leq z$. It follows that $a(x \vee y) = (ax) \vee (ay)$. The second equality of (i) can be shown similarly.

(ii) We prove the first equality only, since the second one follows by applying it to $-x$ and $-y$. Since $x \leq x \vee y$ and $y \leq x \vee y$, $-(x \vee y) \leq -x$ and $-(x \vee y) \leq -y$, so $-(x \vee y) \leq (-x) \wedge (-y)$. On the other hand, if $-x \geq z$ and $-y \geq z$, then $-z \geq x \vee y$, and thus $-(x \vee y) \geq z$. This shows that $-(x \vee y) = (-x) \wedge (-y)$.

(iii) $x \leq x + y - x \wedge y$ since $x \wedge y \leq y$, and $y \leq x + y - x \wedge y$ from $x \wedge y \leq x$. So $x \vee y \leq x + y - x \wedge y$. On the other hand, $x + y - x \vee y \leq x$ by $y \leq x \vee y$, and $x + y - x \vee y \leq y$ since $x \leq x \vee y$. Thus, $x + y - x \vee y \leq x \wedge y$. Therefore, we obtain $x \vee y + x \wedge y = x + y$.

(iv) $(x+y) \vee (x+z) \leq x + y \vee z$ from $x + y \leq x + y \vee z$ and $x + z \leq x + y \vee z$. Also $y \vee z \leq -x + (x+y) \vee (x+z)$ by $y = -x + x + y \leq -x + (x+y) \vee (x+z)$ and similarly $z \leq -x + (x+y) \vee (x+z)$. Therefore, $(x+y) \vee (x+z) \geq x + y \vee z$. This gives the first equality of (iv), and the second one can be proved with the same argument.   □

Properties (ii) and (iv) of the above proposition can be generalized to an arbitrary subset A of X as follows:

Proposition 5.2. *Let $A \subset X$. Then $\sup A$ exists in X if and only if $\inf(-A)$ exists in X. Moreover,*

$$\sup A = -\inf(-A),$$

and for every vector $x \in X$,

$$\sup(x + A) = x + \sup A \quad and \quad \inf(x + A) = x + \inf A.$$

Definition 5.4. Let $x \in X$. We define $x^+ = x \vee 0$, $x^- = -(x \wedge 0)$, and $|x| = x \vee (-x)$, which are called the *positive part*, the *negative part*, and the *absolute value* of x, respectively.

Similar to real numbers, the four vectors x, x^+, x^-, and $|x|$ satisfy the following important relations.

Proposition 5.3. *Let* $x \in X$. *Then*
(i) $x = x^+ - x^-$;
(ii) $|x| = x^+ + x^-$;
(iii) $x^+ \wedge x^- = 0$;
(iv) *if* $x = y - z$ *and* $y \wedge z = 0$, *then* $y = x^+$ *and* $z = x^-$.

Proof (i) From Proposition 5.1(iii),

$$x = x + 0 = x \vee 0 + x \wedge 0 = x^+ - x^-.$$

(ii) Proposition 5.1(iv) and (i), and (i) here imply

$$|x| = x \vee (-x) = (2x) \vee 0 - x = 2(x \vee 0) - x$$
$$= 2x^+ - x = 2x^+ - (x^+ - x^-) = x^+ + x^-.$$

(iii) By Proposition 5.1(iv), and (i) here,

$$x^+ \wedge x^- = (x^+ - x^-) \wedge 0 + x^- = x \wedge 0 + x^- = -x^- + x^- = 0.$$

(iv) Suppose that $x = y - z$ is such that $y \wedge z = 0$. Then Proposition 5.1(iv) and (iii) imply that

$$x^+ = (y - z) \vee 0 = y \vee z - z = y + z - y \wedge z - z = y.$$

Similarly, we have $x^- = z$. □

Remark 5.1. Proposition 5.3(i) and (ii) imply that

$$x^+ = \frac{|x| + x}{2} \quad \text{and} \quad x^- = \frac{|x| - x}{2}.$$

The following proposition lists more properties of the absolute value.

Proposition 5.4. *Let* $x, y \in X$. *Then*
(i) $|x| = 0$ *if and only if* $x = 0$;
(ii) $|ax| = |a||x|$ *for all numbers* $a \in \mathbb{R}$;
(iii) $x \vee y = 2^{-1}(x + y + |x - y|)$ *and* $x \wedge y = 2^{-1}(x + y - |x - y|)$;
(iv) $|x - y| = x \vee y - x \wedge y$;
(v) $|x| \vee |y| = 2^{-1}(|x + y| + |x - y|)$;
(vi) $|x| \wedge |y| = 2^{-1}(|x + y| - |x - y|)$.

Proof (i) Let $x = 0$. Then $x^+ = 0$ and $x^- = 0$, so $|x| = x^+ + x^- = 0$. Conversely, assume $|x| = 0$. Then $0 \leq x^+ \leq |x| = 0$, hence $x^+ = 0$. Similarly, $x^- = 0$. Thus $x = x^+ - x^- = 0$.

(ii) The equality is obvious for $a \geq 0$ by Proposition 5.1(i). If $a < 0$, then by Proposition 5.1(ii), $(ax)^+ = (-|a|x) \vee 0 = -(|a|x \wedge 0) = (|a|x)^- = |a|x^-$. Similarly $(ax)^- = |a|x^+$. Thus,

$$|ax| = (ax)^+ + (ax)^- = |a|x^- + |a|x^+ = |a||x|.$$

(iii) By Proposition 5.3(ii) and Proposition 5.1(iv),

$$
\begin{aligned}
x + y + |x - y| &= x + y + (x - y) \vee (y - x) \\
&= [(x + y) + (x - y)] \vee [(x + y) + (y - x)] \\
&= (2x) \vee (2y) = 2(x \vee y),
\end{aligned}
$$

so the first equality of (iii) holds. The second one follows similarly.

(iv) Subtracting the two identities of (iii) gives (iv).

(v) From Proposition 5.1(iv) and (iii) here,

$$
\begin{aligned}
|x + y| + |x - y| &= (x + y) \vee (-x - y) + |x - y| \\
&= (x + y + |x - y|) \vee (-x - y + |x - y|) \\
&= 2\{(x \vee y) \vee [(-x) \vee (-y)]\} \\
&= 2\{[x \vee (-x)] \vee [y \vee (-y)]\} = 2(|x| \vee |y|).
\end{aligned}
$$

(vi) It follows from (iv), (v) and Proposition 5.11(iii) that

$$
\begin{aligned}
\big|\, |x + y| - |x - y| \,\big| &= 2(|x + y| \vee |x - y|) - (|x + y| + |x - y|) \\
&= 2(|x| + |y|) - 2(|x| \vee |y|) = 2(|x| \wedge |y|). \square
\end{aligned}
$$

Remark 5.2. Proposition 5.4(iii) implies that an ordered vector space X is a vector lattice if $|x| = x \vee (-x)$ exists, $\forall\, x \in X$.

We introduce the concept of "disjointness" of two vectors in a vector lattice, which is so similar in its properties to that of "orthogonality" of two vectors in a Hilbert space that it deserves the same notation "$\perp$" to be used. Two vectors $x, y \in X$ are called *lattice disjoint* if $|x| \wedge |y| = 0$, denoted $x \perp y$. Proposition 5.4(vi) shows that $x \perp y$ if and only if $|x + y| = |x - y|$. Two subsets A, B of X are called *lattice disjoint* if $x \perp y$, $\forall\, x \in A$, $y \in B$, and in particular, if $A = \{x\}$, then x is said to be *lattice disjoint* to B. The *order complement*, denoted $A^\perp$, of A in X is defined to be the set of all $x \in X$ that are lattice disjoint to A. It is clear to see that $A \cap A^\perp = \{0\}$.

Some elementary and useful inequalities which involve various lattice operations are presented by the next proposition.

Proposition 5.5. *Let $x, y, z, w \in X$. Then*

(i) $(x+y)^+ \leq x^+ + y^+$ *and* $(x+y)^- \leq x^- + y^-$;

(ii) $|x+y| \leq |x| + |y|$ *(triangle inequality)*;

(iii) $|\, |x| - |y| \,| \leq |x-y|$;

(iv) $|x \vee y - z \vee w| \leq |x-z| + |y-w|$ *and* $|x \wedge y - z \wedge w| \leq |x-z| + |y-w|$;

(v) $|x \vee y - z \vee y| \leq |x-z|$ *and* $|x \wedge y - z \wedge y| \leq |x-z|$;

(vi) $|x^+ - z^+| \leq |x-z|$ *and* $|x^- - z^-| \leq |x-z|$.

Proof (i) Since $x = x^+ - x^- \leq x^+$ and $y = y^+ - y^- \leq y^+$, we have $x + y \leq x^+ + y^+$, and so $(x+y)^+ = (x+y) \vee 0 \leq x^+ + y^+$. The other inequality $(x+y)^- \leq x^- + y^-$ can be proved the same way.

(ii) From (i), $|x+y| = (x+y)^+ + (x+y)^- \leq x^+ + y^+ + x^- + y^- = x^+ + x^- + y^+ + y^- = |x| + |y|$.

(iii) By (ii), $|x| = |(x-y) + y| \leq |x-y| + |y|$, so $|x| - |y| \leq |x-y|$. Similarly, $|y| - |x| \leq |x-y|$. This gives (iii).

(iv) Using (i) and Proposition 5.1(iv), we see that

$$x \vee y - z \vee w = [(x-y) \vee 0 + y] - [(z-w) \vee 0 + w]$$
$$= (x-y)^+ + y - (z-w)^+ - w$$
$$= (x-z+z-y)^+ - [z-y-(w-y)]^+ + y - w$$
$$\leq (x-z)^+ + (z-y)^+ - [(z-y)^+ - (w-y)^+] + y - w$$
$$= (x-z)^+ + (w-y)^+ - (w-y)$$
$$= (x-z)^+ + (w-y)^- \leq |x-z| + |y-w|.$$

Similarly, $z \vee w - x \vee y \leq |x-z| + |y-w|$. This proves the first inequality of (iv). The other one can be proved similarly.

(v) It is the special case of (iv) with $w = y$.

(vi) This comes from (v) by taking $y = 0$. $\qquad\qquad\square$

The following lemma is a key to proving the *Riesz decomposition property* which has many applications in lattice operations.

Lemma 5.1. *Let $|x| \leq |\sum_{k=1}^{n} y_k|$ in X. Then there exist $x_1, \cdots, x_n \in X$ with $x = \sum_{k=1}^{n} x_k$ and $|x_k| \leq |y_k|$ for each k. If in addition $x \geq 0$, then each x_k can be chosen to be positive.*

Proof Assume $n = 2$. Then the assumption and the triangle inequality imply that $-|y_1| - |y_2| \leq x \leq |y_1| + |y_2|$. Define $x_1 = [x \vee (-|y_1|)] \wedge |y_1|$. Then $|x_1| \leq |y_1|$. Let $x_2 = x - x_1$. By Proposition 5.1(iv), we have

$$x_2 = x - [x \vee (-|y_1|)] \wedge |y_1| = [0 \wedge (x + |y_1|)] \vee (x - |y_1|).$$

It follows that

$$-|y_2| = (-|y_2|) \wedge 0 \leq (x + |y_1|) \wedge 0 \leq x_2 \leq 0 \vee (x - |y_1|) \leq |y_2|,$$

from which we have $|x_2| \leq |y_2|$. If in addition $x \geq 0$, then clearly $0 \leq x_1 \leq x$, and therefore $x_2 = x - x_1 \geq 0$.

The general case can be proved by induction on n. □

Theorem 5.1. (Riesz) *Let* $x_1, \cdots, x_n, \ y_1, \cdots, y_m$ *be positive vectors of* X. *If* $\sum_{k=1}^{n} x_k = \sum_{j=1}^{m} y_j$, *then there exist* mn *positive vectors* z_{kj}, $k = 1, \cdots, n, j = 1, \cdots, m$, *such that*

$$x_k = \sum_{j=1}^{m} z_{kj}, \ \forall \, k = 1, \cdots, n$$

and

$$y_j = \sum_{k=1}^{n} z_{kj}, \ \forall \, j = 1, \cdots, m.$$

Proof We shall use mathematical induction on m. The case $m = 1$ is trivial since we just let $z_{k1} = x_k$, $k = 1, \cdots, n$. Assume that the theorem is true for some $m \geq 2$ and all $n \geq 1$. Let

$$\sum_{k=1}^{n} x_k = \sum_{j=1}^{m+1} y_j,$$

where all x_k and y_j are positive. Since $\sum_{j=1}^{m} y_j \leq \sum_{k=1}^{n} x_k$, Lemma 5.1 ensures that there exist elements $u_1, \cdots, u_n$ such that $\sum_{k=1}^{n} u_k = \sum_{j=1}^{m} y_j$ and $0 \leq u_k \leq x_k$ for all $k = 1, \cdots, n$. By the induction hypothesis, there exist positive vectors z_{kj}, $k = 1, \cdots, n, j = 1, \cdots, m$, such that

$$u_k = \sum_{j=1}^{m} z_{kj}, \ k = 1, \cdots, n; \ \ y_j = \sum_{k=1}^{n} z_{kj}, \ j = 1, \cdots, m.$$

Define $z_{k,m+1} = x_k - u_k \geq 0$ for $k = 1, \cdots, n$. Then the positive vectors z_{kj}, $k = 1, \cdots, n, j = 1, \cdots, m + 1$, satisfy

$$x_k = \sum_{j=1}^{m+1} z_{kj}, \ k = 1, \cdots, n; \ \ y_j = \sum_{k=1}^{n} z_{kj}, \ j = 1, \cdots, m + 1.□$$

A sequence $\{x_n\}$ in a vector lattice is said to be *increasing* or *decreasing* if $m \geq n$ implies $x_m \geq x_n$ or $x_m \leq x_n$ respectively, and is denoted by $x_n \uparrow$ or $x_n \downarrow$. Increasing or decreasing sequences are called *monotonic*. The notion $x_n \uparrow x$ means that $x_n \uparrow$ and $\sup\{x_n : n \geq 1\} = x$. The meaning of

$x_n \downarrow x$ is defined analogously. A vector lattice X is called *Archimedean* if the sequence $\{n^{-1}x\}$ converges to 0 monotonically for each $x \in X_+$. It is easy to see that X is Archimedean if and only if $x, y \in X$ and $nx \leq y$ for all n imply $x \leq 0$. Many classic real vector spaces in functional analysis, in particular the function spaces used in this book, are Archimedean. Unless otherwise stated, all vector lattices will be assumed to be Archimedean.

Definition 5.5. A vector lattice X is called *Dedekind complete* if every nonempty subset of X that is bounded from above has a least upper bound, *σ-Dedekind complete* if each countable subset of X which is bounded from above has a least upper bound, and *complete* if every subset of X possesses a least upper bound and a greatest lower bound.

It is obvious that X is σ-Dedekind complete if and only if $0 \leq x_n \leq x$ for all positive integers n imply the existence of $\sup\{x_n : n \geq 1\}$ for any increasing sequence $\{x_n\}$ in X.

A sequence $\{x_n\}$ in a vector lattice X is called *order convergent* to $x \in X$, written $x_n \xrightarrow{o} x$, if there is a sequence $\{y_n\}$ in X such that $y_n \downarrow 0$ and $|x_k - x| \leq y_k$ for all $k \geq n$ and $n \geq 1$.

Definition 5.6. Let X be a vector lattice. A norm $\| \ \|$ on X is called a *lattice norm* if $|x| \leq |y|$ implies $\|x\| \leq \|y\|$ for all $x, y \in X$. If $\| \ \|$ is a lattice norm on X, then the pair $(X, \| \ \|)$ is called a *normed lattice*. If, in addition, $(X, \| \ \|)$ is a Banach space, then it is called a *Banach lattice*.

It can be shown that in a normed lattice X, $\|x\| = \| \ |x| \ \|$ for all $x \in X$. Also, Proposition 5.5(vi) implies the inequalities

$$\|x^+ - y^+\| \leq \|x - y\| \quad \text{and} \quad \|x^- - y^-\| \leq \|x - y\|.$$

Furthermore, it follows from Proposition 5.5(iii) that

$$\| \ |x| - |y| \ \| \leq \|x - y\|.$$

Proposition 5.6. *Let X be a normed lattice. Then the three mappings $x \to x^+$, $x \to x^-$, and $x \to |x|$ are all uniformly continuous from X into X, and the two mappings $(x, y) \to x \vee y$, $(x, y) \to x \wedge y$ are both uniformly continuous from $X \times X$ into X.*

Proof The uniform continuity of the mappings $(x, y) \to x \vee y$ and $(x, y) \to x \wedge y$ results directly from Proposition 5.5(iv) by virtue of the monotonicity of the norm of X, and the uniform continuity of the mappings $x \to x^+$ and $x \to x^-$, and therefore the uniform continuity of the mapping $x \to |x| = x^+ + x^-$, come from Proposition 5.5(vi). $\qquad \square$

Corollary 5.1. *Let X be a normed lattice. Then the positive cone X_+ is a closed subset of X. In particular, X is Archimedean.*

Proof $X_+ = \{x \in X : x^- = 0\}$ is closed as the inverse image of $\{0\}$ under the continuous mapping $x \to x^-$. If $x, y \in X$ and $nx \leq y$ for all n, then $x - n^{-1}y \in -X_+$ for all n. Since $-X_+$ is closed, $x \in -X_+$, namely $x \leq 0$. Hence, X is Archimedean. $\qquad\qquad\qquad\qquad\qquad\qquad\qquad\qquad\qquad\square$

Definition 5.7. A lattice norm $\| \ \|$ on a vector lattice X is said to be *σ-order continuous* if $x_n \downarrow 0$ in X implies $\|x_n\| \downarrow 0$ in $\mathbb{R}$.

The $L^p(\mu)$ spaces with $1 \leq p < \infty$ are examples of Banach lattices with σ-order continuous norms. More generally, reflexive Banach lattices have σ-order continuous norms. If each interval of a Banach lattice X is weakly compact, then X has a σ-order continuous norm. See [Schaefer (1974); Aliprantis and Burkinshaw (1985)] for more on order continuous norms.

5.2   Positive Operators

We turn to the study of positive operators between vector lattices. In this book, the term "positive operator" always means a positive linear operator. Throughout the section, unless otherwise stated we always assume that X and Y are vector lattices.

Definition 5.8. Let $T : X \to Y$ be a linear operator.

(i) T is called a *positive operator*, denoted $T \geq 0$, if $Tx \geq 0$ for all $x \geq 0$. If $T \geq 0$ and $T \neq 0$, then we write $T > 0$. T is called *strictly positive*, written $T \gg 0$, if $Tx > 0$ for all $x > 0$.

(ii) T is called a *lattice homomorphism* between vector lattices if $T(x \vee y) = Tx \vee Ty$ and $T(x \wedge y) = Tx \wedge Ty$ for all $x, y \in X$.

(iii) T is said to be *σ-order continuous* if for any sequence $\{x_n\}$ in X, $x_n \xrightarrow{o} 0$ in X implies $Tx_n \xrightarrow{o} 0$ in Y.

A linear operator $T : X \to Y$ between two vector lattices is positive if and only if $TX_+ \subset Y_+$, and also if and only if $x \leq y$ implies $Tx \leq Ty$ for all $x, y \in X$. The set $\mathcal{C} \equiv \mathcal{C}(X, Y)$ of all positive operators $T : X \to Y$ satisfies $\mathcal{C} + \mathcal{C} \subset \mathcal{C}$ and $a\mathcal{C} \subset \mathcal{C}$ for all real numbers $a \geq 0$, thus it is an additive cone. We first present a simple property of positive operators.

Proposition 5.7. *If $T : X \to Y$ is a positive operator, then*
$$(Tx)^+ \leq Tx^+ \quad and \quad (Tx)^- \leq Tx^-$$

for all $x \in X$. Consequently,

$$|Tx| \leq T|x|, \quad \forall\, x \in X.$$

Proof $Tx \leq Tx^+$ since $x \leq x^+$. Since $Tx^+ \geq 0$,

$$(Tx)^+ = (Tx) \vee 0 \leq Tx^+.$$

Similarly, $(Tx)^- \leq Tx^-$. The last conclusion comes from

$$|Tx| = (Tx)^+ + (Tx)^- \leq Tx^+ + Tx^- = T(x^+ + x^-) = T|x|.\square$$

Remark 5.3. The property that $|Tx| \leq T|x|$ for a positive operator T also follows from the fact that $-|x| \leq x \leq |x|$ implies $-T|x| \leq Tx \leq T|x|$.

Among the most beautiful properties of positive operators between Banach lattices is the following continuity property.

Theorem 5.2. *Let X be a Banach lattice and let Y be a normed lattice. Then every positive operator $T : X \to Y$ is continuous.*

Proof If $T : X \to Y$ is not continuous, then there exists a sequence $\{x_n\}$ of positive vectors such that $\|x_n\| = 1$ and $\|Tx_n\| \geq n^3$ for all n. On the other hand, since X is a Banach space and X_+ is closed, the series $\sum_{n=1}^{\infty} n^{-2} x_n$ converges to some $\hat{x} \in X_+$. From $\hat{x} \geq n^{-2} x_n$ we have $T\hat{x} \geq n^{-2} T x_n$. Hence $\|T\hat{x}\| \geq n^{-2} \|T x_n\| \geq n$ for all positive integers n, which is impossible since $\|T\hat{x}\|$ is a finite number. $\square$

An immediate but at the same time surprising consequence of the preceding result is that a vector lattice admits essentially at most one lattice norm under which it is a Banach lattice.

Theorem 5.3. (Goffman) *All lattice norms that make a vector lattice a Banach lattice are equivalent.*

Proof Let X be a Banach lattice under norms $\|\ \|$ and $\|\ \|'$. Then, by Theorem 5.2, the identity operator $I : (X, \|\ \|) \to (X, \|\ \|')$ is homeomorphic. Therefore, $\|\ \|$ and $\|\ \|'$ are equivalent. $\square$

It is clear that every lattice homomorphism T from a vector lattice X to a vector lattice Y must be positive since if $x \geq 0$, then $Tx = T(x \vee 0) = Tx \vee T0 = Tx \vee 0 \geq 0$. The following result gives several equivalent conditions for a lattice homomorphism.

Theorem 5.4. *Let $T : X \to Y$ be a linear operator. Then the following statements are equivalent:*

(i) *T is a lattice homomorphism.*

(ii) *$Tx^+ \wedge Tx^- = 0$ for all $x \in X$.*

(iii) *$(Tx)^+ = Tx^+$ and $(Tx)^- = Tx^-$ for all $x \in X$.*

(iv) *$|Tx| = T|x|$ for all $x \in X$.*

Proof (i) $\Rightarrow$ (ii): Since $x^+ \wedge x^- = 0$ for all $x \in X$, (i) implies that $Tx^+ \wedge Tx^- = T(x^+ \wedge x^-) = T0 = 0$.

(ii) $\Rightarrow$ (iii). By Proposition 5.3(i), $Tx = Tx^+ - Tx^-$ and $Tx = (Tx)^+ - (Tx)^-$. Since $Tx^+ \wedge Tx^- = 0$, Proposition 5.3(iv) guarantees the equalities $(Tx)^+ = Tx^+$ and $(Tx)^- = Tx^-$.

(iii) $\Rightarrow$ (iv). $|Tx| = (Tx)^+ + (Tx)^- = Tx^+ + Tx^- = T|x|$.

(iv) $\Rightarrow$ (i). If $x \geq 0$, then $x = |x|$, so $Tx = T|x| = |Tx| \geq 0$. Thus, T is positive, which implies that $(Tx)^+ \leq Tx^+$ and $(Tx)^- \leq Tx^-$ by Proposition 5.7. Since $(Tx)^+ + (Tx)^- = Tx^+ + Tx^-$ from the hypothesis, $(Tx)^+ = Tx^+$ and $(Tx)^- = Tx^-$ for all $x \in X$. Therefore for any $x, y \in X$, since $x \vee y = x + (y - x)^+$ from Remark 5.1 and Proposition 5.4(iii),

$$T(x \vee y) = Tx + T(y - x)^+ = Tx + (T(y - x))^+ = Tx \vee Ty.$$

By the same token, $T(x \wedge y) = Tx \wedge Ty$. □

The following fundamental extension theorem due to Kantorovich says that a necessary and sufficient condition for a given mapping $T : X_+ \to Y_+$ to be the restriction of a positive operator from X into Y is that T is *additive* on X_+, namely $T(x + y) = Tx + Ty$ for all $x, y \in X_+$.

Theorem 5.5. (Kantorovich) *Let $T : X_+ \to Y_+$ be additive. Then T can be extended uniquely to a positive operator $T : X \to Y$. Moreover, the unique extension T has the expression*

$$Tx = Tx^+ - Tx^-, \ \forall \, x \in X.$$

Proof The uniqueness of the extension follows from Proposition 5.3(i), so we only need to show its existence. Let $S : X \to Y$ be defined by $Sx = Tx^+ - Tx^-$. We show that S is linear.

Let $x, y \in X$ and write $w = x + y$. Then $w^+ + x^- + y^- = x^+ + y^+ + w^-$, hence the additivity assumption of T on X_+ implies that $Tw^+ + T(x^- + y^-) = T(x^+ + y^+) + Tw^-$. Thus

$$S(x + y) = Sw = Tw^+ - Tw^- = T(x^+ + y^+) - T(x^- + y^-)$$
$$= (Tx^+ - Tx^-) + (Ty^+ - Ty^-) = Sx + Sy.$$

This proves the additivity of S, which also implies that $S(ax) = aSx$ for any rational number $a \geq 0$.

For the homogeneity of S, we first assume that $x \in X_+$ and $a > 0$. Let $\{r_n\}$ and $\{s_n\}$ be two sequences of rational numbers such that $0 < r_n < a < s_n$ and $\lim_{n \to \infty} r_n = \lim_{n \to \infty} s_n = a$. By the additivity of T on X_+, if $0 \leq y \leq z$, then $Ty \leq Tz$, so

$$r_n Tx = T(r_n x) \leq T(ax) \leq T(t_n x) = t_n Tx,$$

which implies that, using the fact that X is Archimedean, $T(ax) = aTx$. Now, let $x \in X$ and $a \in \mathbb{R}$. If $a \geq 0$, then

$$S(ax) = T(ax^+) - T(ax^-) = aTx^+ - aTx^- = aSx,$$

and if $a < 0$, then

$$S(ax) = -S(-ax) = -(-a)Sx = aSx. \square$$

The above Kantorovich theorem says that a positive operator $T : X \to Y$ is determined completely by its restriction to X_+. Thus, we can use the phrase "*the mapping* $T : X_+ \to Y_+$ *defines a positive operator*" to mean that T is additive on X_+.

The vector space of all linear operators from X into Y will be denoted by $L(X,Y)$. A partial ordering on $L(X,Y)$ is defined by $T \leq S$ whenever $S - T$ is a positive operator. $L(X,Y)$ is an ordered vector space, but not a vector lattice in general.

Definition 5.9. Let $T \in L(X,Y)$. We say that T possesses a *modulus* $|T|$ if $|T| \equiv T \vee (-T)$ exists in $L(X,Y)$.

Kantorovich's theorem implies the following important sufficient condition for the existence of the modulus of $T \in L(X,Y)$.

Theorem 5.6. *Let* $T \in L(X,Y)$. *If* $\sup\{|Ty| : |y| \leq x\}$ *exists in* Y *for every* $x \in X_+$, *then* $|T|$ *exists, and*

$$|T|x = \sup\{|Ty| : |y| \leq x\}, \quad \forall\, x \in X_+.$$

Proof Define $S : X_+ \to Y_+$ by

$$Sx = \sup\{|Ty| : |y| \leq x\} = \sup\{Ty : |y| \leq x\}, \quad \forall\, x \in X_+.$$

We show that S defines a positive operator.

Let $x, y \in X_+$. If $|z| \leq x$ and $|w| \leq y$, then $|z + w| \leq |z| + |w| \leq x + y$, so $Tz + Tw \leq S(x + y)$. Thus $Sx + Sy \leq S(x + y)$. Now, if $|z| \leq x + y$, then by Lemma 5.1 there exist z_1 and z_2 such that $|z_1| \leq x$, $|z_2| \leq y$, and $z = z_1 + z_2$. Hence $Tz = Tz_1 + Tz_2 \leq Sx + Sy$, from which we have $S(x+y) \leq Sx + Sy$. Thus, $S(x+y) = Sx + Sy$, and Theorem 5.5 ensures that the mapping S can be extended to a positive operator from X into Y.

It remains to show the equality $S = T \vee (-T)$. The inequalities $T \leq S$ and $-T \leq S$ in $L(X,Y)$ are obvious from the definition of S. If $T \leq \hat{T}$ and $-T \leq \hat{T}$ for some operator $\hat{T} \in L(X,Y)$, then clearly $\hat{T} \geq 0$. Let $x \in X_+$. If $|y| \leq x$, then we have

$$Ty = Ty^+ - Ty^- \leq \hat{T}y^+ + \hat{T}y^- = \hat{T}|y| \leq \hat{T}x.$$

It follows that $Sx \leq \hat{T}x$ for all $x \in X_+$, namely $S \leq \hat{T}$. Therefore, $S = T \vee (-T)$. This completes the proof. □

Remark 5.4. Theorem 5.6 also implies that if $|T|$ exists, then

$$|Tx| \leq |T||x|, \quad \forall \; x \in X.$$

A linear operator $T : X \to Y$ is called *order bounded* if it maps order bounded subsets of X onto order bounded subsets of Y. Positive operators are clearly order bounded. The vector subspace of $L(X,Y)$ consisting of all order bounded linear operators from X into Y is denoted by $L_b(X,Y)$. When Y is Dedekind complete, $L_b(X,Y)$ is actually a Dedekind complete vector lattice, as the following important theorem shows.

Theorem 5.7. (Riesz-Kantorovich) *If Y is Dedekind complete, then $L_b(X,Y)$ is a Dedekind complete vector lattice, and its lattice operations for all $S, T \in L_b(X,Y)$ satisfy*

$$(S \vee T)x = \sup\{Sy + Tz : y, z \in X_+, \; y + z = x\}, \quad \forall \; x \in X_+$$

and

$$(S \wedge T)x = \inf\{Sy + Tz : y, z \in X_+, \; y + z = x\}, \quad \forall \; x \in X_+.$$

Proof We only show that $L_b(X,Y)$ is a vector lattice and verify the expressions for $S \vee T$ and $S \wedge T$. For the proof of Dedekind completeness of $L_b(X,Y)$ we refer to [Schaefer (1974)] or [Aliprantis and Burkinshaw (1985)]. Since Y is Dedekind complete and $T \in L_b(X,Y)$ is order bounded,

$$\sup\{|Ty| : |y| \leq x\} = \sup\{Ty : |y| \leq x\} = \sup T[-x, x]$$

exists for each $x \in X_+$. By Theorem 5.6, $|T|$ exists and $|T|x = \sup\{Ty : |y| \leq x\}$. From Remark 5.2, $L_b(X,Y)$ is a vector lattice.

Let $S, T \in L_b(X, Y)$ and $x \in X_+$. Note that $y + z = x$ with $y, z \in X_+$ if and only if there is $|u| \leq x$ such that $y = 2^{-1}(x + u)$ and $z = 2^{-1}(x - u)$. It follows from Proposition 5.4(iii) that

$$
\begin{aligned}
(S \vee T)x &= \frac{1}{2}(Sx + Tx + |S - T|x) \\
&= \frac{1}{2}[Sx + Tx + \sup\{(S - T)u : |u| \leq x\}] \\
&= \frac{1}{2}\sup\{Sx + Su + Tx - Tu : |u| \leq x\} \\
&= \sup\left\{S\left(\frac{x + u}{2}\right) + T\left(\frac{x - u}{2}\right) : |u| \leq x\right\} \\
&= \sup\{Sy + Tz : y, z \in X_+, \ y + z = x\}.
\end{aligned}
$$

The formula for $S \wedge T$ can be proved similarly. $\qquad\square$

From the above Riesz-Kantorovich theorem, if Y is Dedekind complete, then each order bounded operator $T : X \to Y$ satisfies
(i) $|T|x = \sup\{Ty : |y| \leq x\}, \ \forall\, x \in X_+$;
(ii) $T^+x = \sup\{Ty : 0 \leq y \leq x\}, \ \forall\, x \in X_+$;
(iii) $T^-x = \sup\{-Ty : 0 \leq y \leq x\}, \ \forall\, x \in X_+$.
A linear operator $T : X \to Y$ is said to be *regular* if it can be written as a difference of two positive operators. T is regular if and only if there is a positive operator $S : X \to Y$ such that $T \leq S$. Regular operators are order bounded since positive operators are order bounded. Conversely, order bounded operators T are also regular from the decomposition $T = T^+ - T^-$.

Definition 5.10. Let X and Y be Banach lattices. If $T \in L(X, Y)$ has the modulus $|T|$, then the *regular norm* of T, which is abbreviated as the *r-norm* of T, is defined to be

$$
\|T\|_r = \| \, |T| \, \| = \sup\{\| \, |T|x \, \| : \|x\| \leq 1\}.
$$

Theorem 5.8. *Let X and Y be Banach lattices and assume that Y is Dedekind complete. Then $L_b(X, Y)$ is a Dedekind complete Banach lattice under the regular norm $\| \ \|_r$.*

Proof Let $\{T_n\}$ be a $\| \ \|_r$-Cauchy sequence of $L_b(X, Y)$. Then it is also a Cauchy sequence of $B(X, Y)$ since $\|T_n - T_m\| \leq \|T_n - T_m\|_r$, so there exists $T \in B(X, Y)$ such that $\lim_{n \to \infty} \|T_n - T\| = 0$. Without loss of generality, we may assume that

$$
\|T_{n+1} - T_n\|_r = \| \, |T_{n+1} - T_n| \, \| < \frac{1}{2^n}, \ \forall\, n.
$$

Let $x \in X_+$. Then, for each $y \in X$ with $|y| \leq x$, we have

$$(T_n - T)y = \sum_{k=n}^{\infty}(T_{k+1} - T_k)y \leq \sum_{k=n}^{\infty}|T_{k+1} - T_k|x.$$

Thus, the modulus of $T - T_n$ exists and satisfies

$$|T - T_n|x = \sup\{(T - T_n)y : |y| \leq x\}$$
$$\leq \sum_{k=n}^{\infty}|T_{k+1} - T_k|x, \quad \forall\, x \in X_+.$$

In particular, $T - T_n \in L_b(X,Y)$, which implies that $T \in L_b(X,Y)$. It follows from the above inequality that

$$\|T - T_n\|_r \leq \sum_{k=n}^{\infty}\|T_{k+1} - T_k\|_r \leq \frac{1}{2^{n-1}} \to 0$$

as $n \to \infty$. Therefore, $(L_b(X,Y), \|\ \|_r)$ is a Banach lattice. $\qquad\square$

A linear functional $\xi : X \to \mathbb{R}$ on a vector lattice X is called a *positive functional* if $x \geq 0$ implies $\xi(x) \geq 0$, and is said to be *order bounded* if ξ maps order bounded subsets of X onto bounded subsets of $\mathbb{R}$. The corresponding vector space $L_b(X,\mathbb{R})$ is called the *order dual space* of X, denoted as X^*. Since $\mathbb{R}$ is Dedekind complete, X^* is a Dedekind complete vector lattice by Theorem 5.7. One can show [Aliprantis and Burkinshaw (1985)] that the dual space X' of a normed lattice X is a Banach lattice.

Based on the above results we can characterize the σ-order continuity for order bounded linear operators.

Theorem 5.9. *Let Y be σ-Dedekind complete and let $T \in L_b(X,Y)$. Then the following statements are equivalent:*
(i) *T is σ-order continuous.*
(ii) *$x_n \downarrow 0$ in X implies $Tx_n \xrightarrow{o} 0$ in Y.*
(iii) *$x_n \downarrow 0$ in X implies $\inf\{|Tx_n| : n \geq 1\} = 0$ in Y.*
(iv) *T^+ and T^- are both σ-order continuous.*
(v) *$|T|$ is σ-order continuous.*

Proof It is clear that (i) $\Rightarrow$ (ii) $\Rightarrow$ (iii).

(iii) $\Rightarrow$ (iv). Let $x_n \downarrow 0$. Then $T^+x_n \downarrow y$ for some $y \in Y_+$. Given x_{n_0}. Then for any $z \in [0, x_{n_0}]$ and any $n \geq n_0$, we have

$$0 \leq z - z \wedge x_n = z \wedge x_{n_0} - z \wedge x_n \leq x_{n_0} - x_n,$$

which implies

$$Tz - T(z \wedge x_n) = T(z - z \wedge x_n)$$
$$\leq T^+(x_{n_0} - x_n) = T^+x_{n_0} - T^+x_n.$$

It follows that

$$0 \leq y \leq T^+x_n \leq T^+x_{n_0} + |T(z \wedge x_n)| - Tz. \tag{5.1}$$

Since $z \wedge x_n \downarrow_{n \geq n_0} 0$, the hypothesis ensures that

$$\inf_{n \geq n_0} \{|T(z \wedge x_n)|\} = 0,$$

so from (5.1) we see that $0 \leq y \leq T^+x_{n_0} - Tz$ for all $z \in [0, x_{n_0}]$, which implies $y = 0$. Thus T^+ is σ-order continuous, and the same argument gives the σ-order continuity of T^-.

(iv) $\Rightarrow$ (v). It follows from the equality $|T| = T^+ + T^-$.

(v) $\Rightarrow$ (i). It is obvious from Remark 5.3. $\qquad\qquad \square$

We give a local approximation result for positive operators.

Theorem 5.10. *Let Y be σ-Dedekind complete, $T \in L(X, Y)_+$, and $x \in X_+$. Then there is $S \in L(X, Y)$ such that*

(i) $0 \leq S \leq T$;

(ii) $Sx = Tx$;

(iii) $Sy = 0$ *for all $y \perp x$.*

Proof For the given $x \in X_+$ define $S : X_+ \to Y_+$ by

$$Sy = \sup\{T(y \wedge nx) : n = 1, 2, \cdots\}.$$

Let $y, z \in X_+$. Since $(y + z) \wedge nx \leq y \wedge nx + z \wedge nx$,

$$T((y + z) \wedge nx) \leq T(y \wedge nx) + T(z \wedge nx) \leq Sy + Sz,$$

so $S(y + z) \leq Sy + Sz$. On the other hand, for any m and n we have $y \wedge mx + z \wedge nx \leq (y + z) \wedge (m + n)x$, thus

$$T(y \wedge mx) + T(z \wedge nx) \leq T((y + z) \wedge (m + n)x) \leq S(y + z).$$

Therefore, $Sy + Sz \leq S(y + z)$, so $S(y + z) = Sy + Sz$.

By Theorem 5.2, the mapping S defines a positive operator on X for which (i), (ii), and (iii) are easily verified. $\qquad\qquad \square$

An interesting application of the above result gives the expressions of Tx^+, Tx^-, and $T|x|$ for a positive operator T.

Theorem 5.11. *Let Y be σ-Dedekind complete. If $T \in L(X,Y)$ is positive, then for each $x \in X$,*

    (i) $Tx^+ = \max\{Sx : S \in L(X,Y),\ 0 \leq S \leq T\}$;

    (ii) $Tx^- = \max\{-Sx : S \in L(X,Y),\ 0 \leq S \leq T\}$;

    (iii) $T|x| = \max\{Sx : S \in L(X,Y),\ -T \leq S \leq T\}$.

Proof Let $x \in X$. By Theorem 5.10, there exists $S_0 \in L(X,Y)$ such that $0 \leq S_0 \leq T$, $S_0 x^+ = Tx^+$, and $S_0 x^- = 0$ since $x^- \perp x^+$. So $Tx^+ = S_0 x$. Given $S \in L(X,Y)$ with $0 \leq S \leq T$,

$$Sx = Sx^+ - Sx^- \leq Sx^+ \leq Tx^+.$$

Hence we obtain (i). Since $x^- = (-x)^+$, applying (i) to $-x$ gives (ii). To verify (iii), note that if $-T \leq S \leq T$, then

$$Sx = Sx^+ - Sx^- \leq Tx^+ + Tx^- = T|x|.$$

On the other hand, Theorem 5.10 ensures that there exist $S_1, S_2 \in L(X,Y)$ with $0 \leq S_1,\ S_2 \leq T$ such that

$$S_1 x^+ = Tx^+,\ \ S_1 x^- = 0; \quad S_2 x^- = Tx^-,\ S_2 x^+ = 0.$$

If we let $S_0 = S_1 - S_2$, then $-T \leq S_0 \leq T$ and $T|x| = Sx$.   □

5.3   Ideals and Projections

We now introduce the notion of ideals, a key concept for studying positive operators. Let X and Y be vector lattices.

Definition 5.11. $A \subset X$ is *solid* if $x \in A, y \in X$, and $|y| \leq |x|$ imply $y \in A$. A solid vector subspace of X is called an *ideal*.

The union of solid sets is solid, and $A \subset X$ is a solid set if and only if $A = \bigcup_{x \in A}[-|x|, |x|]$. If A and B are solid subsets of X, then so is their sum $A + B$. In particular, the sum of two ideals is also an ideal. Every ideal I is a vector sublattice of X by Proposition 5.4(iii). A vector sublattice A is an ideal if the interval $[x,y] \subset A$ whenever $x, y \in A$. The family of all ideals of X is denoted by $\mathbf{I}(X)$. Given a subset A of X, the smallest solid subset of X that contains A is called the *solid hull* of A and denoted by so$A$, and the smallest ideal of X containing A is called the *ideal generated by A* and written as $I(A)$. The ideal generated by a singleton set $\{u\}$ is called a *principal ideal*, which is denoted by X_u. It can be shown that

$$X_u = \{y \in X : |y| \leq a|u| \text{ for some } a > 0\}. \tag{5.2}$$

An element $e \in X_+$ is called an *order unit* of X if $X = X_e$. From (5.2), $e > 0$ is an order unit of X if and only if for each $x \in X$, there is a number $a > 0$ such that $|x| \le ae$.

Example 5.5. If $x \in X_+$, then the interval $[-x, x]$ is solid, and the principal ideal X_x is the vector subspace $\bigcup_{n=1}^{\infty} n[-x, x]$.

Proposition 5.8. Let $A \subset X$. Then so$A = \{y \in X : |y| \le |x| \text{ for some } x \in A\}$. If A is solid, then so is its convex hull co$A$. If $A_1, \cdots, A_k$ are solid, then so is their sum $\sum_{i=1}^{k} A_i$.

Proof Let $B = \{y \in X : |y| \le |x| \text{ for some } x \in A\}$. Then B is solid. If $A \subset W$ and W is solid, then $B \subset W$. Thus so$A = B$.

Let A be a solid subset of X. If $y \in X$ and $|y| \le |x|$ for some $x = \sum_{i=1}^{n} a_i x_i \in \text{co}A$, where $x_i \in A$, $a_i \ge 0$, $\forall i$, and $\sum_{i=1}^{n} a_i = 1$, then $y^+ \le \sum_{i=1}^{n} a_i |x_i|$ and $y^- \le \sum_{i=1}^{n} a_i |x_i|$. Theorem 5.1 implies that there exist $y_i, z_i \in [0, |x_i|]$, $i = 1, \cdots, n$, such that $y^+ = \sum_{i=1}^{n} a_i y_i$ and $y^- = \sum_{i=1}^{n} a_i z_i$. Since A is solid, $y_i - z_i \in [-|x_i|, |x_i|] \subset A$ for all i. So $y = y^+ - y^- = \sum_{i=1}^{n} a_i (y_i - z_i) \in \text{co}A$. Hence, co$A$ is solid.

The last conclusion follows from Theorem 5.1. $\square$

Now we explore the structure of $\mathbf{I}(X)$. If $I, J \in \mathbf{I}(X)$, then clearly $I \cap J \in \mathbf{I}(X)$, and $I + J \in \mathbf{I}(X)$ by Proposition 5.8. So, under set inclusion $\mathbf{I}(X)$ becomes a lattice with largest element X and smallest element $\{0\}$. Furthermore, we have

Proposition 5.9. The lattice $\mathbf{I}(X)$ is distributive.

Proof We need to show that $(I + J) \cap K = I \cap K + J \cap K$, $\forall I, J, K \in \mathbf{I}(X)$. It is obvious that $(I + J) \cap K \supset I \cap K + J \cap K$. If $z \in (I + J) \cap K$, then $z = x + y$ with $x \in I$, $y \in J$. So $|z| \le |x| + |y|$. By Lemma 5.1, $|z| = u + v$, where $u \in [0, |x|]$ and $v \in [0, |y|]$. Clearly, $u \in I \cap K$ and $v \in J \cap K$, and it follows that $|z| \in I \cap K + J \cap K$. Therefore, $z \in I \cap K + J \cap K$. $\square$

Remark 5.5. In general $\mathbf{I}(X)$ is not a Boolean algebra.

If $T : X \to Y$ is a positive operator, then the set $\{x \in X : T|x| = 0\}$ is an ideal of X, called the *null ideal* of T. For a lattice homomorphism T, its null space $N(T)$ equals its null ideal, and $T^{-1}(B)$ is solid for any solid subset B of Y. Moreover, a positive operator T with $TX_+ = Y_+$ is a lattice homomorphism if $N(T)$ is an ideal of X [Schaefer (1974)].

The next result concerns mutually complementary ideals.

Proposition 5.10. *Let I, J be ideals of X such that $I \oplus J = X$. Then the projection $P : X \to X$ with $R(P) = I$ and $N(P) = J$ is positive and $I = J^{\perp}$. Thus, $X = I \oplus I^{\perp}$. Moreover, $I = I^{\perp\perp}$.*

Proof For any $x \in X_{+}$ let $x = y + z = y^{+} - y^{-} + z^{+} - z^{-}$ be such that $y \in I$ and $z \in J$. Then $0 \le y^{-} + z^{-} \le y^{+} + z^{+}$. Since $y^{-} \wedge z^{+} \in I \cap J = \{0\}$ and $y^{-} \wedge y^{+} = 0$, we have $y^{-} \perp y^{+} + z^{+}$, which implies $y^{-} = 0$. Thus, $y = Px \ge 0$, so the projection P is a positive operator. By symmetry, the projection $x \to z$ is also positive.

Clearly $I \subset J^{\perp}$ since $I \cap J = \{0\}$. Let $x \in J^{\perp}$. Then $|x| \in J^{\perp}$ and $|x| = u + v$ with $u \in I_{+}$ and $v \in J_{+}$ from the previous paragraph. This implies that $v \le |x| \in J^{\perp}$, hence $v \in J \cap J^{\perp} = \{0\}$. Thus $|x| = u \in I$ and therefore $x \in I$.   □

Definition 5.12. An ideal I of X is called a *band* if $A \subset I$ and $\sup A \in X$ imply that $\sup A \in I$. A complemented ideal I of X is referred to as a *projection band* and the associated projection $P : X \to I$ with $N(P) = I^{\perp}$ is said to be a *band projection*.

The smallest band containing a subset A of X is called the *band generated by A*. Clearly, the band generated by A is the same as that generated by the ideal generated by A. The band generated by a singleton set $\{u\}$ is called a *principal band*, written X^{u}. An element $e \in X_{+}$ is called a *weak order unit* of X if $X = X^{e}$. Since $A^{\perp}$ is solid for any $A \subset X$, it follows that $A^{\perp}$ is a band. Since X is Archimedean, if B is a band, then $B = B^{\perp\perp}$.

Theorem 5.12. *Let $\mathbf{B}(X)$ and $\mathbf{P}(X)$ be the set of all projection bands and band projections of X, respectively. Then*

(i) *$\mathbf{B}(X)$ is a Boolean algebra and a sublattice of $\mathbf{I}(X)$;*

(ii) *a positive operator $P : X \to X$ is a band projection if and only if $P^{2} = P$ and $Px \le x$ for all $x \in X_{+}$, and every band projection is a lattice homomorphism;*

(iii) *$\mathbf{P}(X)$ is a Boolean algebra under the lattice operation*
$$P \vee Q := P + Q - PQ \quad \text{and} \quad P \wedge Q = PQ,$$
and $\mathbf{P}(X)$ is isomorphic with $\mathbf{B}(X)$ under the mapping $P \to R(P)$. In particular, every pair of band projections commute.

Proof (i) Let $A, B \in \mathbf{B}(X)$. Then $X = A + A^{\perp} = B + B^{\perp}$. Since $\mathbf{I}(X)$ is distributive by Proposition 5.9,
$$
\begin{aligned}
X &= (A + A^{\perp}) \cap (B + B^{\perp}) \\
&= A \cap B + (A \cap B^{\perp} + A^{\perp} \cap B + A^{\perp} \cap B^{\perp}).
\end{aligned}
$$

The expression in the last parentheses above gives an ideal which is lattice disjoint to $A \cap B$, therefore $A \cap B \in \mathbf{B}(X)$. On the other hand, writing the above decomposition as

$$X = (A \cap B + A \cap B^\perp + A^\perp \cap B) + A^\perp \cap B^\perp$$

and noting the two equalities $A = A \cap (B + B^\perp) = A \cap B + A \cap B^\perp$ and $B = (A + A^\perp) \cap B = A \cap B + A^\perp \cap B$, we find that the expression in the parentheses is the ideal $A + B$ which is clearly lattice disjoint to $A^\perp \cap B^\perp$. So, $A + B \in \mathbf{B}(X)$. This proves that $\mathbf{B}(X)$ is a sublattice of $\mathbf{I}(X)$. it is easy to see that $\mathbf{B}(X)$ is a Boolean algebra.

(ii) Let P_A be the band projection with range $A \in \mathbf{B}(X)$. Then P_A is a positive operator and $P_A x \le x$ for all $x \in X_+$ by Proposition 5.10. Conversely, if a positive operator $P \in L(X)$ satisfies $P^2 = P$ and $Px \le x$ for all $x \in X_+$, then for each $x \in X$ we have $0 \le Px^+ \le x^+$ and $0 \le Px^- \le x^-$, so $0 \le Px^+ \wedge Px^- \le x^+ \wedge x^- = 0$, which implies that P is a lattice homomorphism by Theorem 5.4. Thus $R(P)$ and $N(P)$ are ideals, so $P \in \mathbf{P}(X)$ by Proposition 5.10.

(iii) The correspondence $P \to R(P)$ is one-to-one and onto between $\mathbf{P}(X)$ and $\mathbf{B}(X)$. It remains to show that if P_A and P_B are band projections with ranges $A, B \in \mathbf{B}(X)$ respectively, then $P_A \vee P_B$ and $P_A \wedge P_B$ are band projections with ranges $A + B$ and $A \cap B$, respectively. Since $P_A P_B$ vanishes on the ideal $A \cap B^\perp + A^\perp \cap B + A^\perp \cap B^\perp$ and leaves each element of $A \cap B$ fixed, $P_A P_B$ is the band projection with range $A \cap B$. This proves that $P_A \wedge P_B$ is the band projection with range $A \cap B$. By symmetry, $P_B P_A$ is the band projection with range $B \cap A$, so $P_A P_B = P_B P_A$. Therefore $P_A \vee P_B$ is a projection. Since $P_A \vee P_B$ leaves each element of $A + B = A \cap B + A \cap B^\perp + A^\perp \cap B$ fixed and vanishes on $A^\perp \cap B^\perp$, it follows that $P_A \vee P_B$ is the band projection with range $A + B$. □

The proof of the following fundamental result of Riesz is referred to [Aliprantis and Burkinshaw (1985)].

Theorem 5.13. (Riesz) *If X is Dedekind complete, then $X = A \oplus A^\perp$ for every band A of X.*

We give an expression of band projections to end this section. The band projection P_A associated with a band A is defined by

$$P_A x = y, \quad \forall\, x \in X,$$

where $x = y + z$ with $y \in A$ and $z \in A^\perp$.

Proposition 5.11. P_A *has the expression*

$$P_A(x) = \sup \left(A \cap [0, x] \right), \ \forall \, x \in X_+.$$

Proof Fix $x \in X_+$. Let $x = y + z$ with $0 \leq y \in A$ and $0 \leq z \in A^\perp$. Since $y = x - z \leq x$, we have $y \in A \cap [0, x]$. Let $u \in A \cap [0, x]$. Then $(u - y)^+ = 0$ from $0 \leq (u - y)^+ \leq z \in A^\perp$ and $(u - y)^+ \in A$. So $u \leq y$. Thus, $y = \sup(A \cap [0, x])$.   □

5.4 *L*-spaces and *M*-spaces

Two important classes of Banach lattices in applications are L-spaces and M-spaces, which are abstract versions of $L^1(\mu)$ and $C(K)$, respectively, where K is a compact space. Let X be a Banach lattice throughout.

Definition 5.13. Let $1 \leq p < \infty$. X is called an *abstract L^p-space*, if its norm is *p-additive*, that is,

$$\|x + y\|^p = \|x\|^p + \|y\|^p$$

for all $x, y \in X$ with $x \wedge y = 0$. An abstract L^1-space is called an *L-space*. X is called an *M-space*, if

$$\|x \vee y\| = \max\{\|x\|, \|y\|\}$$

for all $x, y \in X$ with $x \wedge y = 0$.

The concept of L-spaces was introduced by Birkhoff in 1938 [Birkhoff (1938)], and L-spaces and M-spaces were investigated by Kakutani in 1941 [Kakutani (1941a,b)]. We first give an important duality relation between L-spaces and M-spaces with the help of an approximation lemma.

Lemma 5.2. *Let $\xi, \eta \in X'$ and $\epsilon > 0$. If $\xi \wedge \eta = 0$ in X', then there exist two unit vectors $x, y \in X_+$ with $x \wedge y = 0$ such that*

$$\|\xi\| < \xi(x) + \epsilon \quad \text{and} \quad \|\eta\| < \eta(y) + \epsilon.$$

Proof Pick $u, v \in X_+$ with $\|u\| = \|v\| = 1$ such that $\|\xi\| < \xi(u) + \epsilon/3$ and $\|\eta\| < \eta(v) + \epsilon/3$. Since $\xi \wedge \eta(u) = 0$, by Theorem 5.7, there exist $u_1, u_2 \in X_+$ with $u = u_1 + u_2$ such that $\xi(u_1) + \eta(u_2) < \epsilon/3$. Similarly, there are $v_1, v_2 \in X_+$ with $v = v_1 + v_2$ such that $\xi(v_1) + \eta(v_2) < \epsilon/3$. Let

$$x = u_2 - v_1 \wedge u_2 \quad \text{and} \quad y = v_1 - v_1 \wedge u_2,$$

and note that $\|x\| = \|y\| = 1$ and $x \wedge y = 0$. Now

$$\xi(x) = \xi(u_2) - \xi(v_1 \wedge u_2) > \xi(u_2) - \frac{\epsilon}{3} = \xi(u) - \xi(u_1) - \frac{\epsilon}{3}$$

$$> \left(\|\xi\| - \frac{\epsilon}{3}\right) - \frac{\epsilon}{3} - \frac{\epsilon}{3} = \|\xi\| - \epsilon,$$

and similarly, $\eta(y) > \|\eta\| - \epsilon$ from the same argument. $\qquad\square$

Theorem 5.14. *A Banach lattice X is an L-space (or M-space) if and only if X' is an M-space (or L-space).*

Proof Suppose that X is an L-space. Let $\xi \wedge \eta = 0$ in X'. Then $\max\{\|\xi\|, \|\eta\|\} \leq \|\xi + \eta\|$. For any $\epsilon > 0$ choose $x \in X_+$ with $\|x\| = 1$ and $\|\xi + \eta\| < (\xi + \eta)(x) + \epsilon/3$. Since $\xi \wedge \eta(x) = 0$, Theorem 5.7 ensures that there are $u, v \in X_+$ with $u + v = x$ and $\xi(u) + \eta(v) < \epsilon/3$. Since X is an L-space, from $(v - v \wedge u) \wedge (u - v \wedge u) = 0$ and $0 \leq u + v - 2(u \wedge v) \leq x$,

$$\|v - v \wedge u\| + \|u - v \wedge u\| = \|u + v - 2(v \wedge u)\| \leq \|x\| = 1.$$

It follows that

$$\|\xi + \eta\| < \xi(x) + \eta(x) + \frac{\epsilon}{3}$$

$$= \xi(v) + \eta(u) + \xi(u) + \eta(v) + \frac{\epsilon}{3} < \xi(v) + \eta(u) + \frac{2\epsilon}{3}$$

$$= \xi(v - v \wedge u) + \eta(u - v \wedge u) + (\xi + \eta)(v \wedge u) + \frac{2\epsilon}{3}$$

$$\leq \|\xi\|\|v - v \wedge u\| + \|\eta\|\|u - v \wedge u\| + \xi(u) + \eta(v) + \frac{2\epsilon}{3}$$

$$\leq \max\{\|\xi\|, \|\eta\|\} + \epsilon \leq \|\xi + \eta\| + \epsilon.$$

Since $\epsilon > 0$ is an arbitrary number, $\|\xi \vee \eta\| = \|\xi + \eta\| = \max\{\|\xi\|, \|\eta\|\}$. In other words, X' is an M-space.

Assume now that X is an M-space. Let $\xi \wedge \eta = 0$ in X' and let $\epsilon > 0$ be given. From Lemma 5.2, there exist two vectors x and y in X such that $\|x\| = \|y\| = 1$ and

$$x \wedge y = 0, \quad \|\xi\| < \xi(x) + \epsilon, \quad \text{and} \quad \|\eta\| < \eta(y) + \epsilon.$$

Since X is an M-space, $\|x + y\| = \max\{\|x\|, \|y\|\} = 1$, so

$$\|\xi\| + \|\eta\| < \xi(x + y) + \eta(x + y) + 2\epsilon$$

$$\leq \|\xi\|\|x + y\| + \|\eta\|\|x + y\| + 2\epsilon$$

$$= \|\xi\| + \|\eta\| + 2\epsilon.$$

Therefore, $\|\xi + \eta\| = \|\xi\| + \|\eta\|$. Thus X' is an L-space.

Finally, if X' is an L-space, then X'' is an M-space, hence the closed vector sublattice X of X'' is also an M-space. By the same token, if X' is an M-space, then X is an L-space.   □

Because of Theorem 5.14, we shall mainly study M-spaces.

Theorem 5.15. *Let $u \in X$. Then the mapping*

$$p_u(x) = \inf\{a > 0 : |x| \le a|u|\}$$

defines a lattice norm $\| \ \|_\infty$ on the ideal X_u. Under this norm X_u is an M-space, whose closed unit ball is the interval $[-|u|, |u|]$.

Proof It is not difficult to verify that $\|x\|_\infty = p_u(x)$ defines a lattice norm on X_u with $[-|u|, |u|]$ as its closed unit ball.

Let $\{x_n\}$ be a $\| \ \|_\infty$-Cauchy sequence in X_u. By passing to a subsequence, we can assume that

$$|x_{n+k} - x_n| \le \frac{|u|}{2^n} \tag{5.3}$$

for all n and k. Hence, $\{x_n\}$ is also a $\| \ \|$-Cauchy sequence of the Banach lattice X, so $\lim_{n\to\infty} x_n = x \in X$. Letting $k \to \infty$ in (5.3), we have the inequality $|x - x_n| \le 2^{-n}|u|$ for all n. This shows that $x \in X_u$ and $\lim_{n\to\infty} \|x - x_n\|_\infty = 0$. Therefore, $(X_u, \| \ \|_\infty)$ is a Banach lattice.

We now show that $(X_u, \| \ \|_\infty)$ is an M-space. Let $x \wedge y = 0$ in X_u. Then $\max\{\|x\|_\infty, \|y\|_\infty\} \le \|x + y\|_\infty = \|x \vee y\|_\infty$. From

$$0 \le x + y \le (\|x\|_\infty|u|) \vee (\|y\|_\infty|u|) \le \max\{\|x\|_\infty, \|y\|_\infty\}|u|,$$

we have $\|x + y\|_\infty \le \max\{\|x\|_\infty, \|y\|_\infty\}$. So

$$\|x + y\|_\infty = \max\{\|x\|_\infty, \|y\|_\infty\}.\square$$

If a Banach lattice X has an order unit e, then the norm

$$\|x\|_\infty = \inf\{a > 0 : |x| \le ae\}$$

is equivalent to the original norm of X, according to Goffman's theorem. Therefore, Theorem 5.15 ensures that if a Banach lattice X has an order unit e, then X can be re-normed so that it becomes an M-space with the interval $[-e, e]$ as its closed unit ball.

Example 5.6. The real number axis $\mathbb{R}$ under the natural ordering and the absolute value norm is an M-space with order unit 1. Let X_n be an M-space with order unit e_n for each n. The Banach lattice X of all bounded sequences $x = \{x_n\}$ with $x_n \in X_n$, endowed with the norm $\|x\| =$

$\sup\{\|x_n\| : n \geq 1\}$, is an M-space with order unit $e = \{e_n\}$ under its coordinate-wise ordering. If $X_n = \mathbb{R}$ for all n, then $X = l^\infty$, which is not separable. Its closed vector sublattice of all convergent real-valued sequences $\{x_n\}$ is a separable M-space with order unit $e = \{1, 1, \cdots\}$, and its closed vector sublattice of all sequences $\{x_n\}$ such that $\lim_{n\to\infty} x_n = 0$ is an M-space without order unit.

Example 5.7. Let Ω be a topological space. Then $C_b(\Omega)$ introduced in Example 5.4 is an M-space with order unit 1, In particular, if K is a compact topological space, then the Banach lattice $C(K)$ with the max-norm is an M-space with order unit 1.

Example 5.8. Let (X, Σ, μ) be a measure space. The Banach space $L^\infty(\mu)$ is an M-space with order unit 1.

Example 5.7 provides a concrete example of Banach lattices which are M-spaces. However, the next theorem, which will be proved by means of a classic approximation result of M. H. Stone as the following lemma, says that every M-space is isomorphic to the Banach lattice $C(K)$ with an appropriate compact topological space K.

Lemma 5.3. (The Stone-Weierstrass theorem) *Let K be a compact space, and let F be a vector sublattice of $C(K)$ containing 1. If F separates points of K, namely there is $f \in F$ such that $f(s) \neq f(t)$ for any two distinct points s and t in K, then F is dense in $C(K)$.*

Proof Let $h \in C(K)$ and $\epsilon > 0$. Fix $s \in K$. For each $t \in K$, there is $f_t \in F$ such that $f_t(s) = h(s)$ and $f_t(t) = h(t)$. The set $U_t = \{r \in K : f_t(r) > h(r) - \epsilon\}$ is open and contains t, so $K = \bigcup_{t \in K} U_t$ and the compactness of K implies the existence of a finite set $\{t_1, \cdots, t_n\}$ such that $K = \bigcup_{k=1}^n U_{t_k}$. Since F is a lattice, the function $g_s = \max\{f_{t_k} : k = 1, \cdots, n\} \in F$. It is clear that $g_s(t) > h(t) - \epsilon$ for all $t \in K$, and $g_s(s) = h(s)$.

Now for each $s \in K$, the set $V_s = \{r \in K : g_s(r) < h(r) + \epsilon\}$ is open and contains s. Since $K = \bigcup_{s \in K} V_s$, there is a finite set $\{s_1, \cdots, s_m\}$ such that $K = \bigcup_{j=1}^m V_{s_j}$. Using the lattice property of F again we see that $g = \inf\{g_{s_j} : j = 1, \cdots, m\} \in F$. It is clear that $|g(r) - h(r)| < \epsilon$, from which we have $\|g - h\| < \epsilon$. $\square$

In the following statement of Lemma 5.4 and proof to Theorem 5.16, we shall employ the notation $\overline{B'}_+$ to denote the intersection of the closed unit ball of the dual space X' of X and the positive cone of X'.

Lemma 5.4. *Let X be an M-space, and let $0 \leq \xi \in X'$ satisfy the condition $\|\xi\| = 1$. Then ξ is an extreme point of $\overline{B'}_+$ if and only if ξ is a lattice homomorphism from X onto $\mathbb{R}$.*

Proof Let ξ be an extreme point of $\overline{B'}_+$. If $0 < \eta < \xi$, then

$$\xi = \|\eta\| \frac{\eta}{\|\eta\|} + \|\xi - \eta\| \frac{\xi - \eta}{\|\xi - \eta\|}.$$

Since X' is an L-space, $\|\eta\| + \|\xi - \eta\| = \|\xi\| = 1$, so $\eta = \|\eta\|\xi$. Thus, if $|\eta| \leq \xi$, then there exists a number a with $|a| \leq 1$ such that $\eta = a\xi$. Given $x \in X$, by Theorem 5.6, there exists an $\eta \in X'$ such that $|\eta| \leq \xi$ and $\xi(|x|) = |\eta(x)|$, so it follows from

$$|\xi(x)| \leq \xi(|x|) = |\eta(x)| = |a\xi(x)| \leq |\xi(x)|$$

that $|\xi(x)| = \xi(|x|)$. Thus, ξ is a lattice homomorphism.

Let now ξ be a lattice homomorphism. $\xi(x) = 0$ if and only if $\xi(|x|) = 0$. So if $0 \leq \eta \leq \xi$, then the null space of η contains that of ξ, thus $\eta = a\xi$ for some $0 \leq a \leq 1$. Let $\xi = a\eta + (1 - a)\zeta$ with $\eta, \zeta \in \overline{B'}_+$ and $0 < a < 1$. Clearly, $\|\eta\| = \|\zeta\| = 1$. On the other hand, from $0 \leq a\eta \leq \xi$ and $0 \leq (1 - a)\zeta \leq \xi$, there exist two numbers $b, c > 0$ such that $\eta = b\xi$ and $\zeta = c\xi$. Consequently, $b = c = 1$, and ξ is an extreme point of $\overline{B'}_+$. □

Theorem 5.16. (Kakutani-Bohnenblust and Krein-Krein) *A Banach lattice X is an M-space with order unit e if and only if it is lattice isometric to some $C(K)$, where K is a compact Hausdorff space.*

Proof Only the necessity part of the theorem needs a proof. Let X be an M-space with order unit e. Define

$$K = \{\xi \in \overline{B'}_+ : \xi \text{ is an extreme point of } \overline{B'}_+, \ \|\xi\| = \xi(e) = 1\}.$$

By Lemma 5.4, K is the set of lattice homomorphisms of norm 1 from X onto $\mathbb{R}$. By Alaoglu's theorem [Dunford and Schwartz (1957)], K is w'-compact. Thus, (K, w') is a compact Hausdorff space. Define the evaluation mapping $T : X \to C(K)$ by

$$Tx(\xi) = \xi(x), \ \forall \, x \in X, \ \xi \in K.$$

Since $\|x\| = \| \, |x| \, \| = \sup\{\xi(|x|) : \xi \in K\} = \sup\{T|x|(\xi) : \xi \in K\} = \sup\{|Tx(\xi)| : \xi \in K\}$ for each $x \in X$, it follows that T is a lattice isometry from X into $C(K)$. Also, $Te(\xi) = \xi(e) = 1$ for all $\xi \in K$. Since K separates points of X, so $T(X)$ separates points of K. It follows from Lemma 5.3 that T is onto, so X is lattice isometric to $C(K)$. □

Corollary 5.2. *A Banach lattice X is an M-space if and only if it is lattice isometric to a closed vector sublattice of $C(K)$, where K is some suitable compact topological space.*

Proof By Theorem 5.16, if X is an M-space, then X' is an L-space and X'' is an M-space with order unit. Since X is a closed vector sublattice of X'', the conclusion follows. □

The proof of the next result is referred to [Schaefer (1974)].

Proposition 5.12. *Let K be a compact topological space. The Banach space $C(K)$ is separable if and only if K is metrizable.*

The representation of an M-space as a closed subspace of $C(K)$ leads to the following property of the lattice operations for M-spaces. Denote $|A| = \{|x| : x \in A\}$ for $A \subset X$.

Proposition 5.13. *Let X be an M-space and let $A \subset X$ be norm bounded. If A is pre-compact, so is $|A|$, and $\sup A$ and $\inf A$ exist in X. If A is weakly pre-compact, so is $|A|$. Moreover, the mappings $x \to x^+, x \to x^-$, and $x \to |x|$ are weakly sequentially continuous.*

Proof It is enough to let $X = C(K)$, where K is compact. The conclusion is from the facts that a bounded subset A of $C(K)$ is pre-compact if and only if A is equi-continuous and that A is weakly pre-compact if and only if every sequence in A contains a subsequence which converges pointwisely to a function in $C(K)$. □

We omit the proof of the following theorem since it is too technical; see [Schaefer (1974); Aliprantis and Burkinshaw (1985)].

Theorem 5.17. (Kakutani) *X is an L-space if and only if X is lattice isometric to some $L^1(A, \Sigma, \mu)$, where A is a locally compact space and μ is a strictly positive Radon measure.*

Finally, we introduce two special kinds of positive operators defined on L-spaces and M-spaces. They are abstract versions of Markov operators defined on $L^1(X)$, which will be the focus of Chapter 7, and composition operators defined on $C(K)$, which will be the main theme of Chapter 9.

Definition 5.14. A positive operator T on an L-space is said to be a *stochastic operator* if $\|Tx\| = \|x\|$ for all vectors $x \geq 0$. As a dual concept, a positive operator T on an M-space with order unit e is called an *abstract composition operator* if $Te = e$.

Remark 5.6. Stochastic operators are also called *abstract Markov operators*, and abstract composition operators are sometimes referred to as Markov operators in the literature. However, we reserve the term of "Markov operator" for a class of positive operators on L^1 spaces which will be defined in Chapter 7.

Example 5.9. Any $n \times n$ column stochastic matrix A defines a stochastic operator on $\mathbb{R}^n$ and if A is an $n \times n$ stochastic matrix, then it defines an abstract composition operator on $\mathbb{R}^n$.

5.5   Notes and Remarks

1. Vector lattices are also called *Riesz spaces* in the literature. It is F. Riesz's address, "On the Decomposition of Linear Functionals," at the International Congress of Mathematicians in Bologna in 1928, that initiated the research of vector lattices. H. Freudenthal and L. V. Kantorovich made first contributions to this area in the 1930s.

2. The most general definitions on the order structure of normed lattices in, for example, [Schaefer (1974); Aliprantis and Burkinshaw (1985)], such as the order continuity of norms or operators, are expressed in terms of the concept of *nets*. We only introduced simplified versions of such concepts by using only sequences instead of nets. Therefore, we only defined σ-order continuous lattice norms in Definition 5.7 and σ-order continuous linear operators in Definition 5.8, rather than order continuous ones. Generalizations of some results, such as Theorem 5.9, can be found in the above mentioned books on which much of the presentation of this chapter is based.

3. The usefulness of Theorem 5.5 from [Kantorovich (1937)] is that any additive mapping $T : X_+ \to Y_+$ can be extended uniquely to a positive operator from X into Y, which can be used for the definition of Frobenius-Perron operators in Chapter 8.

4. [Kantorovich *et al.* (1950); Birkhoff (1967)] are two classic books on vector lattices and positive operators. An excellent recent book on vector lattices and operator theory at the introductory level is [Zaanen (1997)] for readers with minimal functional analysis preparation.

Exercises

5.1 Let G be a group. Show that the set of all subgroups of G is a lattice under inclusion but in general, not a sublattice of 2^G.

5.2 Let V is a vector space. Show that the set of all vector subspaces of V is a lattice but in general, not a sublattice of 2^V.

5.3 Prove the conclusions of Example 5.3.

5.4 Prove Proposition 5.2.

5.5 Let X be a vector lattice and $x, y \in X$. Show that

(i) $x \le y$ if and only if $x^+ \le y^+$ and $x^- \ge y^-$;

(ii) $x \perp y$ if and only if $|x| \vee |y| = |x| + |y|$;

(iii) $x \perp y$ implies $(x+y)^+ = x^+ + y^+$ and $|x+y| = |x| + |y|$.

5.6 Suppose that $\{x_a\}_{a\in\Lambda}$ and $\{y_a\}_{a\in\Lambda}$ be two families of vectors in a vector lattice X. Let $y \in X$. Show that

$$\left(\sup_{a\in\Lambda} x_a\right) \wedge y = \sup_{a\in\Lambda}(x_a \wedge y), \quad \left(\inf_{a\in\Lambda} x_a\right) \vee y = \inf_{a\in\Lambda}(x_a \vee y).$$

5.7 Let $A \subset X$. Show that $A^\perp$ is a vector subspace of X.

5.8 Let $x, y \in X_+$. Show that $[0, x+y] = [0, x] + [0, y]$.

5.9 Let X be a vector lattice and $x, y, z \in X_+$. Show that

$$(x+y) \wedge z \le x \wedge z + y \wedge z.$$

5.10 Show that in a normed lattice X, $\| \, |x| \, \| = \|x\|$ for $x \in X$.

5.11 Show that $L^p(X)$ is a Banach lattice for any $p \in [1, \infty]$.

5.12 Let $x_n \xrightarrow{o} x$ and $y_n \xrightarrow{o} y$ in a vector lattice X. Show that

(i) $ax_n + by_n \xrightarrow{o} ax + by$ for all $a, b \in \mathbb{R}$;

(ii) $|x_n| \xrightarrow{o} |x|$ and $(x_n - y_n)^+ \xrightarrow{o} (x-y)^+$;

(iii) $x_n \vee y_n \xrightarrow{o} x \vee y$ and $x_n \wedge y_n \xrightarrow{o} x \wedge y$.

5.13 Let X be an Archimedean vector lattice. Show that $e > 0$ is a weak order unit of X if and only if $x \perp e$ implies $x = 0$.

5.14 Show that the norm of the Banach lattice $L^p(X)$ with $1 \le p < \infty$ is σ-order continuous.

5.15 Show that if each interval of a Banach lattice X is weakly compact, then X has a σ-order continuous norm.

5.16 Show directly that $C[0, 1]$ is a Banach lattice with order unit in the max-norm. Is the norm σ-order continuous?

5.17 Investigate whether the L^∞-norm of the Banach lattice $L^\infty(0, 1)$ is a σ-order continuous norm.

5.18 Show that $L(X, Y)$ is an ordered vector space in the partial ordering defined by: $T \le S$ if and only if $S - T \ge 0$, $\forall \, T, S \in L(X, Y)$.

5.19 Define $T : C[0,1] \to C[0,1]$ by
$$Tf(x) = f(\sin x) - f(\cos x), \quad \forall\, x \in [0,1].$$
Show that T^+ and T^- both exist.

5.20 Let $h(x) = x$ on $[0,1/2]$ and $h(x) = 1/2$ on $(1/2,1]$. Define $T : C[0,1] \to C[0,1]$ by
$$Tf(x) = f(h(x)) - f(1/2), \quad \forall\, x \in [0,1].$$
Show that T is a regular operator without a modulus.

5.21 Find the modulus $|A|$ of $A \in \mathbb{R}^{m \times n}$ via Definition 5.9.

5.22 Show that the sum of two solid sets of a vector lattice is solid and the sum of two ideals is an ideal.

5.23 Let A and B be projection bands. Show that $A \subset B$ if and only if $P_A \le P_B$ and if and only if $P_A P_B = P_B P_A = P_A$.

5.24 Let A and B be projection bands. Show that
  (i) $A^\perp$, $A \cap B$, $A + B$ are projection bands;
  (ii) $P_{A \cap B} = P_A P_B = P_B P_A$, and $P_{A+B} = P_A + P_B - P_A P_B$.

5.25 Let P_A be a band projection in a vector lattice X. Show that
$$|P_A x| = P_A |x|, \quad \forall\, x \in X.$$

5.26 Show that $T : L^1(0,1) \to L^1(0,1)$ defined by
$$Tf(x) = \int_0^1 f(t)dt, \quad \forall\, x \in [0,1]$$
is a projection, but not a band projection.

5.27 Let T be as in Exercise 5.26. Show that $I \wedge T = 0$.

5.28 Show that $\|x\|_\infty = p_u(x)$ defines a lattice norm on X_u, with $[-|u|, |u|]$ as its closed unit ball in Theorem 5.15.

5.29 Let X be a vector lattice and let $\| \ \|$ be a norm on X. Show that there exists a lattice norm on X which is equivalent to $\| \ \|$ if and only if the set of all vectors $y \in X$ such that $|y| \le |x|$ for some $x \in X$ with $\|x\| \le 1$ is bounded in $(X, \| \ \|)$.

5.30 Prove the claims of Examples 5.4-5.8.

5.31 Show that $I \cap J \in \mathbf{I}(X)$, and $I + J \in \mathbf{I}(X)$, $\forall\, I, J \in \mathbf{I}(X)$.

5.32 Consider the vector lattice $\mathbb{R}^S$ of all real-valued functions on $S \ne \emptyset$. For nonempty $S_0 \subset S$, show that the *restriction map* $f \to f|_{S_0}$ is a lattice homomorphism from $\mathbb{R}^S$ onto $\mathbb{R}^{S_0}$.

5.33 Let $a \in (0,1]$. Show that the set
$$B_a = \{f \in C[0,1] : f(t) = 0, \ \forall\, t \ge a\}$$
is a band of $C[0,1]$, but not a projection band.

5.34 Let X be a Banach lattice and Y a normed vector lattice. If $T \in L(X,Y)$ satisfies $|Tx| \le Sx$ for all $x \in X_+$, where $S : X \to Y$ is a positive operator, show that $T \in B(X,Y)$.

Chapter 6

The Spectral Theory of Positive Operators

We continue the study of positive operators in this chapter, but our emphasis will be on their spectral properties. Much of the spectral theory of positive operators is largely based on the spectral theory for general bounded linear operators which will be presented in Section 6.1 first as an introduction, and then we apply it to positive operators in the next section. Section 6.3 concerns the ergodic theory of bounded linear operators, and its applications to positive operators will be covered in Section 6.4.

6.1 The Spectral Theory of Bounded Linear Operators

In this section we present a general spectral theory for bounded linear operators which will be applied to the spectral analysis of positive operators in Section 6.2. Throughout the section we assume that $X \neq \{0\}$ is a *complex* Banach space and $T : X \to X$ is a bounded linear operator. The Banach space of all bounded linear operators on X is denoted by $B(X)$. See Appendix B for the basic theory of bounded linear operators.

Let f be a function from an open domain Ω in the complex plane $\mathbb{C}$ into a Banach space X. If for $\lambda_0 \in \Omega$ the limit

$$\lim_{\lambda \to \lambda_0} \frac{f(\lambda) - f(\lambda_0)}{\lambda - \lambda_0}$$

exists in X, then f is said to be *differentiable* at the point $\lambda = \lambda_0$, and the limit value, which is denoted as $f'(\lambda_0)$, is called the *derivative* of f at λ_0. If f is differentiable at every point of Ω, it is said to be *analytic* in Ω. If f is analytic in an open disk centered at λ_0, then f is said to be *analytic at* λ_0. The *Riemann integral* $\int_\Gamma f(\lambda) d\lambda$ of a continuous function f on a Jordan curve Γ in Ω can be defined in the same way as in complex analysis, in

which the classic Cauchy's integral theorem and Cauchy's integral formula
are still valid in this more general setting.

Definition 6.1. Let $T \in B(X)$. The *resolvent set* $\rho(T)$ of T is the set of
all complex numbers λ for which the inverse $(\lambda I - T)^{-1}$ exists and belongs
to $B(X)$, where I is the identity operator. The *spectrum* $\sigma(T)$ of T is
defined to be the complement of $\rho(T)$ in $\mathbb{C}$. The operator-valued function
$R(\lambda) \equiv R(\lambda, T) = (\lambda I - T)^{-1}$ of the complex variable λ, defined on $\rho(T)$,
is called the *resolvent* of T.

The spectrum of T is the disjoint union of $\sigma_a(T)$, the *approximate point
spectrum*, which is defined to be the set of all complex numbers λ for
which there is a sequence $\{x_n\}$ of vectors in X such that $\|x_n\| \equiv 1$ and
$\lim_{n \to \infty} \|(\lambda I - T)x_n\| = 0$, and $\sigma_r(T)$, the *residual spectrum*, which is the
set of all λ such that $\overline{R(\lambda I - T)} \neq X$. The *point spectrum* $\sigma_p(T)$ is the
set of all λ such that $Tx = \lambda x$ for some nonzero vector $x \in X$ called an
eigenvector associated with the *eigenvalue* $\lambda \in \sigma_p(T)$.

Some properties of the spectrum of matrices are still satisfied by general
bounded linear operators. We list them in the following theorem. But first
we give a useful lemma.

Lemma 6.1. *The resolvent set $\rho(T)$ is an open subset of $\mathbb{C}$. Let $d(\lambda)$ be
the distance from λ to the spectrum $\sigma(T)$. Then*

$$\|R(\lambda, T)\| \geq \frac{1}{d(\lambda)}, \quad \forall \, \lambda \in \rho(T).$$

Therefore $\lim_{d(\lambda) \to 0} \|R(\lambda, T)\| = \infty$.

Proof Banach's lemma implies that if $\lambda_0 \in \rho(T)$, then $\lambda \in \rho(T)$ for all
numbers $\lambda \in \mathbb{C}$ such that $|\lambda - \lambda_0| < \|R(\lambda_0, T)\|^{-1}$. This proves the first
conclusion. The above also ensures that $d(\lambda_0) \geq \|R(\lambda_0, T)\|^{-1}$, from which
the second statement follows. $\square$

Let $|\sigma| = \{|\lambda| : \lambda \in \sigma\}$ for any subset σ of the complex plane.

Theorem 6.1. *The resolvent $R(\lambda, T)$ is an analytic function in $\rho(T)$, and
the spectrum $\sigma(T)$ of T is a nonempty and compact subset of $\mathbb{C}$. Moreover,*

$$\sup |\sigma(T)| = \lim_{n \to \infty} \|T^n\|^{\frac{1}{n}} \leq \|T\|$$

and the Laurent series

$$R(\lambda, T) = \sum_{n=0}^{\infty} \frac{T^n}{\lambda^{n+1}} \tag{6.1}$$

converges in the norm of $B(X)$ for all λ with $|\lambda| > \sup |\sigma(T)|$.

Proof Since the resolvent set $\rho(T)$ is open by Lemma 6.1, the spectrum $\sigma(T)$ is closed. To prove the analyticity of $R(\lambda, T)$, we note the equality

$$T_2^{-1} - T_1^{-1} = T_1^{-1}(T_1 - T_2)T_2^{-1}$$

for any two invertible operators T_1, T_2. So the *resolvent equation*

$$R(\lambda, T) - R(\lambda_0, T) = R(\lambda_0, T)(\lambda_0 - \lambda)R(\lambda, T) \tag{6.2}$$

holds for any $\lambda_0, \lambda \in \rho(T)$, from which we get

$$\frac{R(\lambda, T) - R(\lambda_0, T)}{\lambda - \lambda_0} = -R(\lambda_0, T)R(\lambda, T).$$

Letting $\lambda \to \lambda_0$, we see that the derivative of $R(\lambda, T)$ exists at $\lambda_0 \in \rho(T)$ with the expression

$$R'(\lambda_0, T) = -R(\lambda_0, T)^2.$$

We now show that $\sigma(T)$ is a bounded subset of $\mathbb{C}$. Since the scalar Laurent series $\sum_{n=0}^{\infty} \lambda^{-(n+1)} \|T^n\|$ converges absolutely to a scalar-valued function in the region $\Omega = \{\lambda : |\lambda| > \limsup_{n\to\infty} \|T^n\|^{1/n}\}$, the operator Laurent series $\sum_{n=0}^{\infty} \lambda^{-(n+1)}T^n$ converges to an operator-valued function $f(\lambda)$ under the operator norm of $B(X)$ in Ω. It is easy to see that

$$f(\lambda)(\lambda I - T) = (\lambda I - T)f(\lambda) = I$$

for $\lambda \in \Omega$, so $\Omega \subset \rho(T)$ and thus $\sigma(T)$ is bounded.

If $\sigma(T) = \emptyset$, then $R(\lambda, T)$ is an entire function which is also bounded from (6.1). It follows from Liouville's theorem [Rudin (1986)] that $R(\lambda, T)$ is identically the same operator for all $\lambda \in \mathbb{C}$. Hence, the coefficient of λ^{-1} in the Laurent series of $R(\lambda, T)$ vanishes. This means that $I = 0$, which contradicts the assumption $X \neq \{0\}$. Thus $\sigma(T)$ is a nonempty compact subset of the complex plane.

Since $\rho(T)$ is the natural domain of analyticity for $R(\lambda, T)$, the Laurent series (6.1) of the resolvent $R(\lambda, T)$ will have the domain of convergence $\{\lambda : |\lambda| > \sup |\sigma(T)|\}$. Thus we must have $\sup |\sigma(T)| = \limsup_{n\to\infty} \|T^n\|^{1/n}$. For any n, the factorization

$$\lambda^n I - T^n = (\lambda I - T)p_n(T) = p_n(T)(\lambda I - T)$$

for some polynomial p_n shows that the operator $\lambda^n I - T^n$ has a bounded inverse only if the operator $\lambda I - T$ has a bounded inverse. It follows that $|\lambda|^n \le \|T^n\|$, $\forall \lambda \in \sigma(T)$, so

$$\sup |\sigma(T)| \le \|T^n\|^{\frac{1}{n}} \le \|T\|, \ \forall n.$$

Therefore, we have by taking the lower limit that

$$\sup |\sigma(T)| \leq \liminf_{n \to \infty} \|T^n\|^{1/n} \leq \|T\|. \square$$

Definition 6.2. The nonnegative number

$$r(T) \equiv \sup |\sigma(T)| = \lim_{n \to \infty} \|T^n\|^{\frac{1}{n}}$$

is called the *spectral radius* of T.

It follows from Definition B.13 that $\sigma(T') = \sigma(T)$ for any $T \in B(X)$. Further, $R(\lambda, T') = R(\lambda, T)'$ for $\lambda \in \rho(T') = \rho(T)$.

If $\dim X = \infty$, then T may not satisfy any nonzero polynomial equation $p(T) = 0$, so the class of polynomials is not enough for further exploring the spectral properties of T.

Definition 6.3. By $\mathcal{H}(T)$, we denote the set of all complex functions f that are analytic in some neighborhood, which depends on f and may be disconnected, of $\sigma(T)$ in $\mathbb{C}$.

We introduce the concept of *Dunford integrals* of bounded linear operators for $f \in \mathcal{H}(T)$. As in complex analysis, we assume that the contour of integration, which is the boundary of an open set, is regular enough, for example, a finite union of smooth curves.

Definition 6.4. Let $f \in \mathcal{H}(T)$, and let $\Omega \subset \mathbb{C}$ be open such that $\Omega \supset \sigma(T)$ and $\overline{\Omega} = \Omega \cup \partial\Omega$ is contained in the domain of analyticity of f. Then the operator $f(T)$ is defined by

$$f(T) = \frac{1}{2\pi i} \int_{\partial\Omega} f(\lambda) R(\lambda, T) d\lambda.$$

It follows from Cauchy's integral theorem in complex analysis that the operator $f(T)$ depends only on the function f, but not on the domain Ω specifically chosen in the above definition.

Some fundamental properties of the Dunford integral of bounded linear operators are listed in the following proposition.

Proposition 6.1. *Let* $f, g \in \mathcal{H}(T)$ *and let* $a, b \in \mathbb{C}$. *Then*
   (i) $af + bg \in \mathcal{H}(T)$ *and* $(af + bg)(T) = af(T) + bg(T)$;
   (ii) $fg \in \mathcal{H}(T)$ *and* $(fg)(T) = f(T)g(T)$;
   (iii) *If* $f(\lambda) = \sum_{k=0}^{\infty} a_k \lambda^k$ *in an open disk that contains* $\sigma(T)$, *then* $f(T) = \sum_{k=0}^{\infty} a_k T^k$ *under the operator norm of* $B(X)$.
   (iv) $f \in \mathcal{H}(T')$ *and* $f(T') = f(T)'$.

Proof Properties (i) and (iv) are obvious. It is clear that $fg \in \mathcal{H}(T)$. Let Ω_1 and Ω_2 be open neighborhoods of $\sigma(T)$ such that $\overline{\Omega_1} \subset \Omega_2$ and $\overline{\Omega_2}$ is contained in the domain of analyticity of fg. Then, using (6.2) and the Cauchy integral formula,

$$
\begin{aligned}
f(T)g(T) &= \frac{1}{2\pi i} \int_{\partial \Omega_1} f(\lambda) R(\lambda, T) d\lambda \cdot \frac{1}{2\pi i} \int_{\partial \Omega_2} g(\gamma) R(\gamma, T) d\gamma \\
&= -\frac{1}{4\pi^2} \int_{\partial \Omega_1} \int_{\partial \Omega_2} f(\lambda) g(\gamma) R(\lambda, T) R(\gamma, T) d\gamma d\lambda \\
&= -\frac{1}{4\pi^2} \int_{\partial \Omega_1} \int_{\partial \Omega_2} \frac{f(\lambda) g(\gamma)[R(\lambda, T) - R(\gamma, T)]}{\gamma - \lambda} d\gamma d\lambda \\
&= -\frac{1}{4\pi^2} \int_{\partial \Omega_1} f(\lambda) R(\lambda, T) \int_{\partial \Omega_2} \frac{g(\gamma)}{\gamma - \lambda} d\gamma d\lambda \\
&\quad + \frac{1}{4\pi^2} \int_{\partial \Omega_2} g(\gamma) R(\gamma, T) \int_{\partial \Omega_1} \frac{f(\lambda)}{\gamma - \lambda} d\lambda d\gamma \\
&= \frac{1}{2\pi i} \int_{\partial \Omega_1} f(\lambda) g(\mu) R(\lambda, T) d\lambda = (fg)(T).
\end{aligned}
$$

This proves (ii). To prove (iii), note that the power series $\sum_{k=0}^{\infty} a_k \lambda^k$ converges uniformly on the circle $\Gamma = \{\lambda : |\lambda| = r(T) + \epsilon\}$ for some $\epsilon > 0$. By Theorem 6.1 and the Cauchy integral formula,

$$
\begin{aligned}
f(T) &= \frac{1}{2\pi i} \int_{\Gamma} \left(\sum_{k=0}^{\infty} a_k \lambda^k \right) R(\lambda, T) d\lambda \\
&= \frac{1}{2\pi i} \sum_{k=0}^{\infty} a_k \int_{\Gamma} \lambda^k R(\lambda, T) d\lambda \\
&= \frac{1}{2\pi i} \sum_{k=0}^{\infty} a_k \int_{\Gamma} \lambda^k \left(\sum_{j=0}^{\infty} \frac{T^j}{\lambda^{j+1}} \right) d\lambda = \sum_{k=0}^{\infty} a_k T^k. \square
\end{aligned}
$$

Theorem 6.2. (The spectral mapping theorem) *For $f \in \mathcal{H}(T)$,*
$$
f(\sigma(T)) = \sigma(f(T)).
$$

Proof Let $\lambda_0 \in \sigma(T)$. Define a function g by
$$
g(\lambda) = \frac{f(\lambda) - f(\lambda_0)}{\lambda - \lambda_0}, \quad \lambda \neq \lambda_0,
$$
and $g(\lambda_0) = f'(\lambda_0)$. Then $g \in \mathcal{H}(T)$. By Proposition 6.1(ii), $f(\lambda_0)I - f(T) = (\lambda_0 I - T)g(T)$. So if $f(\lambda_0) \in \rho(f(T))$, then $g(T)(f(\lambda_0)I - f(T))^{-1}$ is the bounded inverse of $\lambda_0 I - T$, which leads to a contradiction to the assumption. Thus, $f(\lambda_0) \in \sigma(f(T))$.

Conversely, let $\gamma \in \sigma(f(T))$, and suppose that $\gamma \notin f(\sigma(T))$. Then the function h defined by $h(\lambda) = (f(\lambda) - \gamma)^{-1}$ belongs to $\mathcal{H}(T)$. From Proposition 6.1(ii), we have the equality $h(T)(f(T) - \gamma I) = I$, which contradicts the assumption that $\gamma \in \sigma(f(T))$. $\qquad \square$

Theorem 6.3. *Let $f \in \mathcal{H}(T)$ and $g \in \mathcal{H}(f(T))$. Then $g \circ f \in \mathcal{H}(T)$ and $(g \circ f)(T) = g(f(T))$.*

Proof That the composition function $g \circ f \in \mathcal{H}(T)$ comes from Theorem 6.2. Let Ω be a neighborhood of $\sigma(f(T))$ such that $\overline{\Omega}$ is contained in the domain of analyticity of g, and let U be a neighborhood of $\sigma(T)$ such that $\overline{U}$ is contained in the domain of analyticity of f and $f(\overline{U}) \subset \Omega$. By Proposition 6.1(ii), for $\lambda \in \rho(f(T))$, the operator

$$A(\lambda) = \frac{1}{2\pi i} \int_{\partial U} \frac{R(\gamma, T)}{\lambda - f(\gamma)} d\gamma$$

satisfies $(\lambda I - f(T))A(\lambda) = A(\lambda)(\lambda I - f(T)) = I$. Thus $A(\lambda) = R(\lambda, f(T))$. Consequently, by Cauchy's integral theorem,

$$
\begin{aligned}
g(f(T)) &= \frac{1}{2\pi i} \int_{\partial\Omega} g(\lambda) R(\lambda, f(T)) d\lambda \\
&= -\frac{1}{4\pi^2} \int_{\partial\Omega} \int_{\partial U} \frac{g(\lambda) R(\gamma, T)}{\lambda - f(\gamma)} d\gamma d\lambda \\
&= \frac{1}{2\pi i} \int_{\partial U} g(f(\gamma)) R(\gamma, T) d\gamma = (g \circ f)(T). \square
\end{aligned}
$$

With the help of the above algebraic rules of the Dunford integral, we can present some important analytic properties of the resolvent $R(\lambda, T)$ of T in the remaining part of this section.

Definition 6.5. $\lambda_0 \in \sigma(T)$ is said to be an *isolated point* of $\sigma(T)$, if there is a neighborhood U of λ_0 such that $\sigma(T) \cap U = \{\lambda_0\}$. An isolated point λ_0 of $\sigma(T)$ is called a *pole* of T, if $R(\lambda, T)$ has a pole at λ_0. By the *order* $\nu = \nu(\lambda_0) \equiv \nu_T(\lambda_0)$ of λ_0 is meant the order of λ_0 as a pole of $R(\lambda, T)$.

The following result gives an equivalent condition for the operator equality $f(T) = g(T)$ with given $f, g \in \mathcal{H}(T)$.

Theorem 6.4. (The minimal equation) *Let $f, g \in \mathcal{H}(T)$. Then $f(T) = g(T)$ if and only if $f(\lambda) = g(\lambda)$ identically in a neighborhood of $\sigma(T)$ except for finitely many poles $\lambda_1, \cdots, \lambda_k$ of T, and $f - g$ has a zero of order at least $\nu(\lambda_j)$ at λ_j for each $j = 1, \cdots, k$.*

Proof Without loss of generality, we assume that $g = 0$. Let $f \in \mathcal{H}(T)$ be such that $f(\lambda) = 0$ for all $\lambda \neq \lambda_1, \cdots, \lambda_k$, where λ_j, $j = 1, \cdots, k$, give all the poles of T in $\sigma(T)$. Then

$$f(T) = \frac{1}{2\pi i} \int_{\partial \Omega} f(\lambda) R(\lambda, T) d\lambda = \sum_{j=1}^{k} \frac{1}{2\pi i} \int_{\Gamma_j} f(\lambda) R(\lambda, T) d\lambda,$$

in which Γ_j is a small circle centered at λ_j for each j. If $f(\lambda)$ has a zero of order at least $\nu(\lambda_j)$ at $\lambda = \lambda_j$, then, since $R(\lambda, T)$ has a pole of order $\nu(\lambda_j)$ at $\lambda = \lambda_j$, $f(\lambda) R(\lambda, T)$ is analytic inside Γ_j. By Cauchy's integral formula, each term in the above summation for $f(T)$ equals 0, so $f(T) = 0$.

Conversely, let $f(T) = 0$ for some f which is analytic in a neighborhood Ω of $\sigma(T)$. Then $f(\sigma(T)) = 0$ by Theorem 6.2. For each $\lambda \in \sigma(T)$, there is an open disk $D_\lambda \subset \Omega$ centered at λ. Then the family $\{D_\lambda : \lambda \in \sigma(T)\}$ gives an open covering of $\sigma(T)$. Since $\sigma(T)$ is compact, there are finitely many such disks, denoted $D_1, \cdots, D_n$, which cover $\sigma(T)$. If some D_j contains infinitely many points of $\sigma(T)$, then f is identically zero in D_j, from complex analysis. Thus, if Ω_1 is the union of those disks D_j which contain an infinite number of points of $\sigma(T)$, then $f(\lambda) \equiv 0$ in Ω_1. Hence, Ω_1 contains all but a finite number of isolated points, $\lambda_1, \cdots, \lambda_r$, of $\sigma(T)$.

Suppose that f does not vanish identically in any neighborhood of λ_1. Then, since $f(\sigma(T)) = 0$, f has a zero of finite order n at λ_1. It follows that the function $g(\lambda) = (\lambda_1 - \lambda)^n / f(\lambda)$ is analytic in a neighborhood of λ_1. Let $h \in \mathcal{H}(T)$ be identically 1 in a neighborhood of λ_1 and identically 0 in a neighborhood of every other point of $\sigma(T)$, and define $g_1 = gh$. Then $(\lambda_1 I - T)^n h(T) = f(T) g_1(T) = 0$. Let the Laurent series of $R(\lambda, T)$ at the point $\lambda = \lambda_1$ be given by

$$R(\lambda, T) = \sum_{k=-\infty}^{\infty} A_k (\lambda_1 - \lambda)^k,$$

where, with Γ_1 being a small circle centered at λ_1 and $k \geq 1$,

$$A_{-k} = -\frac{1}{2\pi i} \int_{\Gamma_1} (\lambda_1 - \lambda)^{k-1} R(\lambda, T) d\lambda = -(\lambda_1 I - T)^{k-1} h(T).$$

Then $A_{-(k+1)} = -(\lambda_1 I - T)^k h(T) = 0$ for $k \geq n$, and therefore λ_1 is a pole of order $\leq n$. The same is true for $\lambda_2, \cdots, \lambda_r$. $\qquad \square$

Definition 6.6. A subset σ_0 of $\sigma(T)$, which is both relatively open and relatively closed in $\sigma(T)$, is called a *spectral set* of T.

It is clear that an isolated point of $\sigma(T)$ is a spectral set. If $\sigma_0 \subset \sigma(T)$ is a spectral set, then there is $h \in \mathcal{H}(T)$ which is identically 1 on σ_0 and identically 0 on $\sigma(T) \setminus \sigma_0$. We put $E(\sigma_0) \equiv E(\sigma_0, T) = h(T)$ for this special h. By Theorem 6.4, $E(\sigma_0)$ depends only on σ_0, not on a particular choice of $h \in \mathcal{H}(T)$ for its definition. In particular, if the spectral set σ_0 consists of only one point λ_0, an extremely important case for our future use, then we write $E(\{\lambda_0\}, T)$ as $E(\lambda_0, T)$ or just $E(\lambda_0)$.

Definition 6.7. Let $T \in B(X)$ and let $\lambda \in \mathbb{C}$. The *index* of λ with respect to T is the smallest integer $\nu \geq 0$ such that

$$N((\lambda I - T)^{\nu}) = N((\lambda I - T)^{\nu+1}).$$

Theorem 6.5. *An isolated point λ_0 of the spectrum $\sigma(T)$ is a pole of $T \in B(X)$ with order ν if and only if*

$$(\lambda_0 I - T)^{\nu} E(\lambda_0, T) = 0 \ \ and \ \ (\lambda_0 I - T)^{\nu-1} E(\lambda_0, T) \neq 0. \qquad (6.3)$$

Moreover, If λ_0 is a pole of T with order ν, then λ_0 has index ν.

Proof For the given $\lambda_0 \in \sigma(T)$, the Laurent series expansion of the resolvent $R(\lambda, T)$ at the point $\lambda = \lambda_0$ is

$$R(\lambda, T) = \sum_{k=-\infty}^{\infty} A_k (\lambda_0 - \lambda)^k$$

with the coefficients

$$A_{-(k+1)} = -(\lambda_0 I - T)^k E(\lambda_0, T), \ \ \forall\, k \geq 0.$$

Thus λ_0 is a pole of T with order ν if and only if (6.3) is satisfied. For the last part of the theorem, assume the point λ_0 is a pole of T with order ν. Then there is $x \in X$ such that

$$(\lambda_0 I - T)^{\nu} x = 0, \ \ (\lambda_0 I - T)^{\nu-1} x \neq 0,$$

which shows that the index n of λ_0 is at least ν. By Definition 6.7 for the index of λ_0, there is $y \in X$ such that

$$(\lambda_0 I - T)^n y = 0 \ \ and \ \ (\lambda_0 I - T)^{n-1} y \neq 0.$$

Since

$$R(\lambda, T) = -\sum_{j=0}^{\infty} \frac{(\lambda_0 I - T)^j}{(\lambda_0 - \lambda)^{j+1}}, \ \ \forall\, \lambda \in \mathbb{C} \ \text{with} \ |\lambda - \lambda_0| > \|\lambda_0 I - T\|,$$

the vector-valued function of λ,

$$R(\lambda, T)y = -\sum_{j=0}^{n-1} \frac{(\lambda_0 I - T)^j y}{(\lambda_0 - \lambda)^{j+1}},$$

is actually a rational one, so it is analytic over the entire plane, except possibly at the point $\lambda = \lambda_0$. Hence, if Ω is any neighborhood of $\sigma(T)$ and Γ is a small circle centered at λ_0, then

$$y = \frac{1}{2\pi i} \int_{\partial \Omega} R(\lambda, T)y d\lambda = \frac{1}{2\pi i} \int_{\Gamma} R(\lambda, T)y d\lambda = E(\lambda_0, T)x.$$

Thus $(\lambda_0 I - T)^\nu y = (\lambda_0 I - T)^\nu E(\lambda_0, T)y = 0$, and therefore $n \leq \nu$. This proves the assertion that $n = \nu$. □

Let $X_{\sigma_0} = R(E(\sigma_0, T))$ for any spectral set σ_0 of T. Then $TX_{\sigma_0} \subset X_{\sigma_0}$, and the restriction of any $S \in B(X)$ with $SX_{\sigma_0} \subset X_{\sigma_0}$ to X_{σ_0} will be denoted by S_{σ_0}. The following theorem gives the relation between the spectra of $T \in B(X)$ and $T_{\sigma_0} \in B(X_{\sigma_0})$.

Theorem 6.6. $\sigma(T_{\sigma_0}) = \sigma_0$ *for any spectral set σ_0 of T. Moreover,* $R(\lambda, T_{\sigma_0}) = R(\lambda, T)_{\sigma_0}$ *for $\lambda \notin \sigma_0$. An isolated point $\lambda_0 \in \sigma_0$ is a pole of T with order ν if and only if it is a pole of T_{σ_0} with order ν.*

Proof Suppose that $\lambda_0 \notin \sigma(T_{\sigma_0})$. Then there is $S \in B(X_{\sigma_0})$ such that $(\lambda_0 I - T)Sx = S(\lambda I - T)x = x$ for all $x \in X_{\sigma_0}$. Choose $g \in \mathcal{H}(T)$ such that $g(\lambda) = 0$ for λ in a neighborhood of σ_0 and $g(\lambda) = (\lambda_0 - \lambda)^{-1}$ for λ in a neighborhood of $\sigma(T) \setminus \sigma_0$. Then $g(T)(\lambda I - T) = (\lambda I - T)g(T) = I - E(\sigma_0)$. Let $S_0 = SE(\sigma_0)$. Then $S_0 \in B(X)$, and

$$(\lambda_0 I - T)(S_0 + g(T)) = (S_0 + g(T))(\lambda_0 I - T) = I,$$

from which $\lambda_0 \in \rho(T)$. This shows that $\sigma_0 \subset \sigma(T_{\sigma_0})$.

Suppose, conversely, that $\lambda_0 \notin \sigma_0$. Then there is $h \in \mathcal{H}(T)$ such that $h(\lambda) = (\lambda_0 - \lambda)^{-1}$ for λ in a neighborhood of σ_0 not containing λ_0, and $h(\lambda) \equiv 0$ in a neighborhood of $\sigma(T) \setminus \sigma_0$. It follows that $h(T)(\lambda_0 I - T) = (\lambda_0 I - T)h(T) = E(\sigma_0)$. Consequently, the restriction $h(T)_{\sigma_0}$ of $h(T)$ to X_{σ_0} satisfies $h(T)_{\sigma_0}(\lambda_0 I - T_{\sigma_0}) = (\lambda_0 I - T_{\sigma_0})h(T)_{\sigma_0} = I$, thus $\lambda_0 \in \rho(T_{\sigma_0})$. This proves that $\sigma(T_{\sigma_0}) \subset \sigma_0$, and that $R(\lambda_0, T_{\sigma_0}) = R(\lambda_0, T)_{\sigma_0}$.

Finally, let $\lambda \in \sigma_0$. Then $E(\lambda)E(\sigma_0) = E(\lambda)$, and so

$$(\lambda I - T)^k E(\lambda) = (\lambda I - T_{\sigma_0})^k E(\lambda), \quad \forall k = 1, 2, \cdots.$$

Theorem 6.5 tells us that λ is a pole of T with order ν if and only if (6.3) is satisfied. Therefore, λ is a pole of T with order ν if and only if it is a pole of T_{σ_0} with order ν. □

Theorem 6.7. *Let $\sigma_0 = \{\lambda_1, \cdots, \lambda_k\}$ be a set of poles of T with respective orders $\nu_1, \cdots, \nu_k$. Then, for any $f \in \mathcal{H}(T)$,*

$$f(T)E(\sigma_0) = \sum_{j=1}^{k} \sum_{r=0}^{\nu_j - 1} \frac{f^{(r)}(\lambda_j)}{r!} (T - \lambda_j I)^r E(\lambda_j). \tag{6.4}$$

Proof Let

$$g(\lambda) = \sum_{j=1}^{k} \sum_{r=0}^{\nu_j - 1} \frac{f^{(r)}(\lambda_j)}{r!} (\lambda - \lambda_j)^r.$$

Then the right-hand side of (6.4) is the expression of $g(E)E(\sigma_0)$. Since $g^{(r)}(\lambda_j) = f^{(r)}(\lambda_j)$ for $0 \le r < \nu_j$ and $1 \le j \le k$, by Theorem 6.4, $f(T)E(\sigma_0) = g(E)E(\sigma_0)$. □

Corollary 6.1. *Let f, f_n, $n = 1, 2, \cdots, \in \mathcal{H}(T)$ be such that*

$$\lim_{n \to \infty} f_n(T) = f(T)$$

weakly. Then for every pole λ of T with order ν,

$$f^{(r)}(\lambda) = \lim_{n \to \infty} f_n^{(r)}(\lambda), \quad \forall\, r = 0, 1, \cdots, \nu - 1.$$

Proof Let T_λ be the restriction of T to $R(E(\lambda))$. Then λ is a pole of T_λ with order ν, by Theorem 6.6. Theorem 6.5 shows that λ has index ν with respect to T_λ. Since $(T - \lambda I)^{\nu-1}E(\lambda) \ne 0$, we can find x such that $(T - \lambda I)^{\nu-1}E(\lambda)x \ne 0$. Let $y = E(\lambda)x$, and for $k = 0, 1, \cdots, \nu - 1$ let $y_k = (T - \lambda I)^k y$. Then, using Theorem 6.7 with $\sigma_0 = \{\lambda\}$, we obtain that $f_n(T)y_{\nu-1} = f_n(\lambda)y_{\nu-1}$ and $f(T)y_{\nu-1} = f(\lambda)y_{\nu-1}$. Thus, the assumption that $f_n(T) \to f(T)$ weakly implies that $f_n(\lambda) \to f(\lambda)$. By the same token, we have the equalities $f_n(T)y_{\nu-2} = f_n(\lambda)y_{\nu-2} + f_n'(\lambda)y_{\nu-1}$ and $f(T)y_{\nu-2} = f(\lambda)y_{\nu-2} + f'(\lambda)y_{\nu-1}$, hence $f_n'(\lambda) \to f'(\lambda)$ as $n \to \infty$. By means of mathematical induction, we can deduce that $\lim_{n \to \infty} f_n^{(r)}(\lambda) = f^{(r)}(\lambda)$ for every $r = 0, 1, \cdots, \nu - 1$. □

Theorem 6.8. *Let λ be a pole of T with order ν and let $\sigma_0 = \sigma(T) \setminus \{\lambda\}$. Then $X = X_\lambda \oplus X_{\sigma_0}$, where*

$$X_\lambda = N((T - \lambda I)^\nu) \quad and \quad X_{\sigma_0} = R((T - \lambda I)^\nu).$$

Proof It is obvious that $X = X_\lambda \oplus X_{\sigma_0}$ since $E(\lambda)$ and $E(\sigma_0) = I - E(\lambda)$ are projections. By Theorem 6.6, $T - \lambda I : X_{\sigma_0} \to X_{\sigma_0}$ is onto since $\lambda \in \rho(T_{\sigma_0})$, hence $(T - \lambda I)^\nu X_{\sigma_0} = X_{\sigma_0}$. From Theorem 6.5, $(T - \lambda I)^\nu X_\lambda = 0$. It follows that $X_\lambda \subset N((T - \lambda I)^\nu)$ and

$$(T - \lambda I)^\nu X = (T - \lambda I)^\nu (X_\lambda \oplus X_{\sigma_0}) = X_{\sigma_0}.$$

On the other hand, if $(T - \lambda I)^\nu x = 0$, then

$$0 = (T - \lambda I)^\nu [E(\lambda)x + E(\sigma_0)x] = (T - \lambda I)^\nu E(\sigma_0)x.$$

By Theorem 6.6, $(T - \lambda I)^\nu$ is one-to-one on X_{σ_0}, from which $E(\sigma_0)x = 0$. Thus $E(\lambda)x = x$, so $N((T - \lambda I)^\nu) \subset X_\lambda$. □

6.2 The Spectral Theory of Positive Operators

The spectral theory of nonnegative square matrices as finite dimensional positive operators of $\mathbb{R}^n$ motivates the spectral analysis of more general positive operators defined on Banach lattices. Let $X \neq \{0\}$ be a complex Banach space and let $T \in B(X)$. The set $\{\lambda \in \mathbb{C} : |\lambda| = r(T)\} \cap \sigma(T)$ is called the *peripheral spectrum* of T.

In this section we further assume that X is a Banach lattice when its scalar field is restricted to that of real numbers. Let $T : X \to X$ be a positive operator. By Theorem 5.2, $T \in B(X)$, so all the spectral analysis results in the previous section for bounded linear operators can be applied directly to positive operators.

Theorem 6.1 ensures the convergence of the Laurent series

$$R(\lambda, T) = \sum_{n=0}^{\infty} \frac{T^n}{\lambda^{n+1}}, \quad |\lambda| > r(T). \tag{6.5}$$

Since $T \geq 0$, it follows that $R(\lambda, T) \geq 0$ for all $\lambda > r(T)$. Therefore, Proposition 2.1 is still true for general positive operators. Moreover, $R(\lambda_2, T) \geq R(\lambda_1, T)$ whenever $\lambda_1 > \lambda_2 > r(A)$. Also (6.5) implies that

$$0 \leq TR(\lambda, T) \leq \lambda R(\lambda, T), \quad \forall \lambda > r(T).$$

Our first result in the following shows that like nonnegative matrices, the spectral radius of a positive operator is a spectral point of T, but its proof is based on the resolvent of T.

Proposition 6.2. *Let $T \in B(X)$ be a nonzero positive operator. Then $r(T) \in \sigma(T)$. That is, the spectral radius of T is a spectral point of T.*

Proof By the definition of $r(T)$, Lemma 6.1 ensures that there exists a sequence $\{\lambda_n\}$ of complex numbers such that $|\lambda_n| \downarrow r(T)$ and $\lim_{n\to\infty} \|R(\lambda_n, T)\| = \infty$. The principle of uniform boundedness then implies that there is an $x \in X$ such that $\lim_{n\to\infty} \|R(\lambda_n, T)x\| = \infty$. Now,

$$|R(\lambda_n, T)x| \leq \left(\sum_{n=0}^{\infty} \frac{T^n}{|\lambda_n|^{n+1}} \right) |x| = R(|\lambda_n|, T)|x|$$

and the fact that the norm is monotone imply

$$\lim_{n \to \infty} \|R(|\lambda_n|, T)|x|\| = \infty.$$

Consequently, by Lemma 6.1 again, $r(T) \in \sigma(T)$.   □

From now on we focus on the spectral analysis for composition operators, an important class of positive operators in ergodic theory. Let (X, Σ, μ) be a σ-finite measure space. A transformation $S : X \to X$ is called *measurable* if the inverse image $S^{-1}(A) \in \Sigma$ for every $A \in \Sigma$. A measurable transformation $S : X \to X$ is said to be *nonsingular* if $\mu(A) = 0$ implies $\mu(S^{-1}(A)) = 0$ for all $A \in \Sigma$ and *measure-preserving*, or equivalently, the measure μ is said to be *S-invariant* if

$$\mu(S^{-1}(A)) = \mu(A), \quad \forall\, A \in \Sigma. \tag{6.6}$$

Definition 6.8. Let $S : X \to X$ be a nonsingular transformation. The linear operator U_S defined on a vector subspace V of the vector space of measurable functions on X via

$$U_S f(x) = f(S(x)), \quad \forall\, f \in V \tag{6.7}$$

is called the *composition operator* associated with S.

Because of the non-singularity assumption of S, the corresponding composition operator is well-defined. In various situations where composition operators are used, the domain of the operator is usually taken to be $L^p(X)$ for some $1 \le p \le \infty$ or $C(K)$, where K is a compact space. In the former case, when $p = \infty$, the composition operator will be called the *Koopman operator* which is the dual of the Frobenius-Perron operator to be studied in Chapter 8. In the latter case, which is the topic of Chapter 9, the composition operator $U_S : C(K) \to C(K)$ is well-defined by (6.7) for any continuous transformation $S : K \to K$ viewed as a nonsingular transformation on $(K, \mathcal{B}(K), \mu)$ for any given probability measure μ on K.

The composition operator U_S has the following properties:

(i) U_S is a positive operator;

(ii) U_S is *multiplicative*, that is,

$$U_S(f \cdot g) = (U_S f)(U_S g), \quad \forall\, f, g.$$

Proposition 6.3. *Let* (X, Σ, μ) *be a probability measure space and let* $S : X \to X$ *be a measure-preserving transformation. Then the corresponding composition operator* $U_S : L^p \to L^p$ *is an isometry for any* $1 \le p \le \infty$.

Proof If $p = 1$, then the conclusion is a consequence of the change of variables formula and the definition of measure-preserving. For other finite values of p, apply the proved result to $|f|^p$. The proof for the case of $p = \infty$ is from the formula $\lim_{p \to \infty} \|f\|_p = \|f\|_\infty$. $\qquad\qquad\square$

From Theorem V.4.4 in [Schaefer (1974)], the spectrum of the composition operator on L^p is a cyclic subset of the closed unit disk. Hence the peripheral spectrum of U_S has a group structure. We now impose some "transitivity" condition on S to study the point spectrum of U_S. A set $A \subset X$ is said to be *invariant* under S if it satisfies the equality $S^{-1}(A) = A$, or equivalently, $U_S \chi_A = \chi_A$, where χ_A is the *characteristic function* of A, that is, $\chi_A(x) = 1$ if $x \in A$ and $\chi_A(x) = 0$ if $x \notin A$.

Definition 6.9. A measurable transformation $S : X \to X$ is called *ergodic* if each invariant set A is such that either $\mu(A) = 0$ or $\mu(A^c) = 0$. Namely, S is ergotic if all invariant sets are *trivial*.

We shall investigate the spectral properties of composition operators defined on the L^2 space when S satisfies additional conditions. Any fixed point of the composition operator is called an *invariant function*. First we present a characterization of ergodicity in terms of invariant functions. Let $1 \leq p \leq \infty$ be a fixed number.

Lemma 6.2. *Let (X, Σ, μ) be a finite measure space, let $S : X \to X$ be a nonsingular transformation, and let $U_S : L^p \to L^p$ be the induced composition operator. Then S is ergodic if and only if all the invariant functions of U_S are constant ones.*

Proof Suppose that S is ergodic. If the equation

$$f(S(x)) = f(x), \quad \forall\, x \in X \;\mu-\text{a.e.} \tag{6.8}$$

is satisfied by a nonconstant function $f \in L^p$, then there is some real number a such that the sets $A = f^{-1}((-\infty, a])$ and $B = f^{-1}((a, \infty))$ have positive measures. Since

$$S^{-1}(A) = \{x : S(x) \in A\} = \{x : f(S(x)) \leq a\}$$
$$= \{x : f(x) \leq a\} = A$$

and similarly $S^{-1}(B) = B$, we have a decomposition of X into nontrivial invariant sets A and B, which contradicts the assumption that S is ergodic.

Conversely, assume that S is not ergodic. Then there is a nontrivial invariant set $A \in \Sigma$. Since $S^{-1}(A) = A$,

$$\chi_A(S(x)) = \chi_{S^{-1}(A)}(x) = \chi_A(x), \quad x \in X \;\mu-\text{a.e.},$$

so (6.8) is satisfied by the nonconstant function χ_A. $\qquad\square$

Theorem 6.9. *Let (X, Σ, μ) be a probability measure space, let $S : X \to X$ be a measure-preserving transformation, and let $U_S : L^2(X) \to L^2(X)$ be the induced composition operator. If S is ergodic, then*

(i) *any eigenfunction f associated with $\lambda \in \sigma_p(U_S)$ satisfies $|f(x)| =$ constant, $\forall x \in X$ μ-a.e.;*

(ii) *$\sigma_p(U_S)$ is a subgroup of the unit circle $\partial\mathbb{D}^2$;*

(iii) *$\dim N(\lambda I - U_S) = 1$ for every $\lambda \in \sigma_p(U_S)$;*

(iv) *eigenfunctions associated with different eigenvalues of U_S are mutually orthogonal in $L^2(X)$.*

Proof (i) By Proposition 6.3, $U_S : L^2(X) \to L^2(X)$ is isometric, hence all the eigenvalues of U_S have absolute value 1. Let $\lambda \in \sigma_p(U_S)$, and let $f \in L^2(X)$ be a nonzero function that satisfies

$$U_S f(x) = f(S(x)) = \lambda f(x), \quad \forall x \in X \; \mu - \text{a.e.}$$

Then

$$U_S|f|(x) = |f(S(x))| = |\lambda f(x)| = |f|(x), \quad \forall x \in X \; \mu - \text{a.e.},$$

namely $|f|$ is an invariant function. Since S is ergodic, Lemma 6.2 ensures that $|f|$ is a constant function.

(ii) If $\lambda, \gamma \in \sigma_p(U_S)$, then $f \circ S = \lambda f$ and $g \circ S = \gamma g$ for some nonzero functions $f, g \in L^2(X)$. Then

$$(f\bar{g}) \circ S = (f \circ S)(\bar{g} \circ S) = (\lambda\bar{\gamma})(f\bar{g}),$$

hence $\lambda\gamma^{-1} = \lambda\bar{\gamma}$ is an eigenvalue of U_S since $f\bar{g} \neq 0$ from (i).

(iii) Let $f, g \in L^2(X)$ be eigenfunctions associated with eigenvalue λ. Since $\lambda \neq 0$ and $g(x) \neq 0$ for all $x \in X$ a.e. by (i) and (ii), $(f/g) \circ S = f/g$, namely f/g is an invariant function. Thus $f = cg$ for some constant c.

(iv) Let $\lambda \neq \gamma$ belong to $\sigma_p(U_S)$, and let $U_S f = \lambda f$ and $U_S g = \gamma g$ with nonzero functions $f, g \in L^2(X)$. Since U_S is isometric,

$$(f, g) = (U_S f, U_S g) = (\lambda f, \gamma g) = \lambda\bar{\gamma}(f, g).$$

The fact $\lambda\bar{\gamma} \neq 1$ implies that $(f, g) = 0$. $\qquad\square$

Remark 6.1. Most conclusions of the above theorem can be extended to the composition operator defined on another L^p space with basically the same argument; see, for example, [Ding (1998)].

Definition 6.10. Let (X, Σ, μ) be a probability measure space and let $S : X \to X$ be measure-preserving. If there is an orthonormal basis for $L^2(\mu)$ consisting of eigenfunctions of the composition operator $U_S : L^2(\mu) \to L^2(\mu)$, then S is said to have a *pure-point spectrum*.

Measure-preserving transformations with pure-point spectra have many important properties; see [Cornfeld *et al.* (1982); Walters (1982)]. We give an example of continuous transformation which has a pure-point spectrum.

Example 6.1. Let $S : \partial\mathbb{D}^2 \to \partial\mathbb{D}^2$ be a rotation of the unit circle defined by $S(x) = ax$, where a is not a root of unity. Then S preserves the *Harr measure* of $\partial\mathbb{D}^2$, which is the normalized Lebesgue measure of the circle. Moreover, as will be seen from Example 8.12, S is ergodic. For each integer n let $f_n : \partial\mathbb{D}^2 \to \mathbb{C}$ be defined by $f_n(z) = z^n$. Then

$$U_S f_n(z) = f_n(S(z)) = f_n(az) = a^n z^n = a^n f_n(z),$$

so f_n is an eigenfunction of U_S associated with eigenvalue a^n. Since $\{f_n\}_{n=-\infty}^{\infty}$ is a canonical orthonormal basis for $L^2(\partial\mathbb{D}^2)$, the rotation S has a pure-point spectrum.

At the end of this section we list several general results on the spectrum and the peripheral spectrum of positive operators; their proofs are referred to the monograph [Schaefer (1974)]. First we have

Theorem 6.10. *Let X be a Banach lattice. Then the spectrum of any lattice homomorphism $T \in B(X)$ is cyclic.*

It was shown in Chapter 2 that the peripheral spectrum of a nonnegative matrix is cyclic, but the corresponding problem for arbitrary positive operators on Banach lattices is difficult to solve in general, but many results related to various contexts exist. For example, in [Albeverio and Hoegh-Krohn (1978)], the cyclic structure of the spectrum of irreducible nonnegative matrices has been extended to the case of positive operators of von Neumann algebras. Spectral properties of irreducible positive operators of finite-dimensional C^*-algebras have been described in [Evans and Hoegh-Krohn (1978)], in which a version of the Perron-Frobenius theorem was given. An ergodic decomposition theory for C^*-finitely correlated translation invariant states on a quantum spin chain has been developed in [Fannes *et al.* (1992)], giving the group structure of the peripheral point spectrum of some ergodic positive operator. In the following we just introduce two classes of abstract positive operators with cyclic peripheral spectrum [Schaefer (1974)] to finish this topic.

Definition 6.11. Let X be a Banach lattice. A positive operator $T : X \to X$ is said to satisfy the *growth condition* if the set $\{(\lambda - r(T))R(\lambda, T) : \lambda > r(T)\}$ is bounded in $B(X)$.

Theorem 6.11. *A positive operator T on a Banach lattice X satisfying the growth condition has a cyclic peripheral spectrum.*

Definition 6.12. Let T be a positive operator on a Banach lattice X. An ideal I of X is called a T-*ideal* if $T^n I \subset I$ for all $n \geq 1$. A T-ideal is called *maximal* if it is maximal among the T-ideals $\neq X$ with respect to the inclusion relation. T is said to be *irreducible* if $\{0\}$ is a maximal T-ideal.

Theorem 6.12. *Let $T \neq 0$ be an irreducible positive operator on a Banach lattice X such that $r(T) = 1$. If T has an eigenvalue λ with $|\lambda| = r(T)$ and T' has an eigenvector $\xi \in X'_+$ associated with eigenvalue 1, then*

    (i) *$\dim N(I - T) = 1$;*

    (ii) *$\sigma_p(T) \cap \partial \mathbb{D}^2$ is a subgroup of $\partial \mathbb{D}^2$;*

    (iii) *every peripheral eigenvalue λ of T is a simple one which satisfies the equality $\sigma(T) = \lambda \sigma(T)$.*

    (iv) *1 is the only eigenvalue of T with a positive eigenvector.*

6.3   Ergodic Theorems of Bounded Linear Operators

Many of the ergodic theorems for positive operators are direct consequences of those for bounded linear operators. So in this section we present basic ergodic theory for general bounded linear operators. Let X be a complex Banach space and let $T \in B(X)$. In this section some necessary and sufficient conditions on the bounded linear operator T will be given for various types of convergence of the *Cesáro averages*

$$A_n = A_n(T) \equiv \frac{1}{n} \sum_{k=0}^{n-1} T^k$$

for the iterates of T. These general conditions will then be applied to L^p spaces, which will be used in the subsequent study of some special classes of positive operators.

Let $\mathcal{H}(T)$ be as defined in Section 6.1, let $f \in \mathcal{H}(T)$, and let $\{f_n\}$ be a sequence of functions in $\mathcal{H}(T)$. It is obvious that the convergence of the operator sequence $\{f_n(T)\}$ under any topology of $B(X)$ implies that of the operator sequence $\{f(T)f_n(T)\}$ in the same topology.

The so-called *Tauberian theory* [Dunford and Schwartz (1957)] is concerned with the converse problem: Under what conditions on $f, f_n,$ and T does the convergence of $\{f(T)f_n(T)\}$ imply the convergence of $\{f_n(T)\}$? The main application of this theory in our book is related to the mean ergodic theorem in which $f(\lambda) = 1 - \lambda$ and $f_n(\lambda) = n^{-1}\sum_{k=0}^{n-1}\lambda^k$. Therefore we are only interested in a specialized version of the general Tauberian theory. The following results extend those for matrices.

We first consider the converse problem for the norm convergence of $A_n(T)$ in $B(X)$, which is simpler to study, and then we discuss the same question for the weak convergence and strong convergence of the operator sequence. In the remainder of this section we always denote $f(\lambda) = 1 - \lambda$ and $f_n(\lambda) = n^{-1}\sum_{k=0}^{n-1}\lambda^k$. It is obvious that $f(T) = I - T$ and $f_n(T) = A_n(T)$. A useful identity, which was already used in Section 2.3, is

$$f(\lambda)f_n(\lambda) \equiv \frac{1}{n}(1 - \lambda^n). \tag{6.9}$$

Theorem 6.13. *Suppose that* $\lim_{n\to\infty} n^{-1}T^n = 0$ *under the norm of* $B(X)$. *If* $\lambda = 1$ *is a pole of* T, *then* $\lim_{n\to\infty}\|A_n(T) - E(1)\| = 0,$ *where* $E(1) : X \to X$ *is the projection onto the fixed point space* $N(I - T)$ *of* T *as defined in Section 6.1.*

Proof We first claim that 1 is a pole of T with order 1. Let $g_n = f \cdot f_n$. Then $g_n(T) = n^{-1}(I - T^n)$ from (6.9) and so converge to zero in norm. If the order $\nu(1)$ of 1 is greater than 1, then by Corollary 6.1, $\lim_{n\to\infty} g'_n(1) = 0$. But this is impossible since $g'_n(1) = -1$ for all n. Therefore, $\nu(1) = 1$.

Let $\sigma_0 = \sigma(T) \setminus \{1\}$. From Theorem 6.7,

$$A_n = A_n E(\sigma_0) + f_n(1)E(1) = E(\sigma_0)A_n + E(1). \tag{6.10}$$

Since 1 is isolated in $\sigma(T)$, we can find $h \in \mathcal{H}(T)$ such that h is equal to $(1 - \lambda)^{-1}$ in a neighborhood of σ_0 and is identically zero in a neighborhood of $\lambda = 1$. Then $h(T)(I - T) = E(\sigma_0)$. Since $(I - T)A_n = n^{-1}(I - T^n) \to 0$ in norm as $n \to \infty$,

$$\lim_{n\to\infty} E(\sigma_0)A_n = \lim_{n\to\infty} h(T)(I - T)A_n = 0$$

in norm. It follows from (6.10) that under the norm of $B(X)$,

$$\lim_{n\to\infty} A_n(T) = E(1). \square$$

The central hypothesis in the above theorem is the assumption that the number $\lambda = 1$ is a pole of the resolvent $R(\lambda, T)$. This assumption is not needed for the analogues of the above theorem for the weak or the strong convergence of $A_n(T)$, which we study next.

Theorem 6.14. *Suppose that* $\lim_{n\to\infty} n^{-1}T^n = 0$ *under the weak topology of* $B(X)$. *If the sequence* $\{A_n(T)x\}$ *is weakly pre-compact for each* $x \in X$, *then* $\{A_n(T)\}$ *converges weakly in* $B(X)$. *Moreover,*

$$X = \overline{R(I - T)} \oplus N(I - T). \tag{6.11}$$

Proof Let $x \in N(I - T)$. Then $x = Tx$, so $A_n x = x$. Similarly, if $\xi \in N(I - T')$, then $A_n(T')\xi = \xi$. Since $\{A_n x\}$ is weakly pre-compact for every $x \in X$, the principle of uniform boundedness (Theorem B.14) ensures that the sequence $\{\|A_n\|\}$ is bounded. By (6.9) and the assumption, $A_n(I - T) = n^{-1}(I - T^n) \to 0$ weakly, so $A_n x \to 0$ weakly for $x \in R(I - T)$. Then a limit argument implies that $A_n x \to 0$ weakly for all $x \in \overline{R(I - T)}$. Since $A_n = I$ on $N(I - T)$, we conclude that

$$\overline{R(I - T)} \cap N(I - T) = \{0\}. \tag{6.12}$$

Next we show that $\{A_n x\}$ converges weakly for each $x \in X$. If not, since the sequence $\{A_n x\}$ is weakly pre-compact, we can extract two subsequences $\{A_{n_k} x\}$ and $\{A_{m_k} x\}$, such that

$$\lim_{k\to\infty} A_{n_k} x = y, \quad \lim_{k\to\infty} A_{m_k} x = z, \quad y \neq z,$$

both limits being in the sense of weak convergence. Since $(I - T)A_n \to 0$ weakly, $(I - T)(y - z) = 0$. Thus $y - z \in N(I - T)$. Since $y - z \neq 0$, (6.12) ensures that $y - z \notin \overline{R(I - T)}$. Then Corollary B.5 implies that there exists a $\xi_0 \in X'$ such that $\xi_0(\overline{R(I - T)}) = 0$ and $\xi_0(y - z) \neq 0$. It follows that $(I - T')\xi_0 = (I - T)'\xi_0 = \xi_0(I - T) = 0$, and therefore $\xi_0 \in N(I - T')$. Since $A_n(T') = I$ on $N(I - T')$, we have $\xi_0(A_n x) = (A_n(T')\xi_0)(x) = \xi_0(x)$ for all n. Therefore

$$\xi_0(y) = \lim_{k\to\infty} \xi_0(A_{n_k} x) = \lim_{k\to\infty} \xi_0(A_{m_k} x) = \xi_0(z),$$

contradicting the inequality $\xi_0(y - z) \neq 0$. This proves that the sequence $\{A_n x\}$ converges weakly for each $x \in X$.

Finally we prove the direct sum (6.11). Let

$$Ex = \lim_{n\to\infty} A_n x, \quad \forall x \in X$$

in the sense of weak convergence. Then, by applying Theorem B.14, we see that $E \in B(X)$. Since $(I - T)A_n = n^{-1}(I - T^n)$, we have $TE = E$, so $A_n E = E$ for all n. Thus, $E^2 = E$. Therefore E is a projection onto

$N(I - T)$. On the other hand, $\overline{R(I - T)} \subset N(E)$ since $E(I - T) = 0$. Now let $\xi \in X'$ be such that $\xi(I - T) = 0$. Then $\xi = \xi T = \xi A_n = \xi E$, so $\xi(I - E) = 0$. It follows that $N(E) \subset \overline{R(I - T)}$. Therefore, $N(E) = \overline{R(I - T)}$ and decomposition (6.11) is obtained. □

The corresponding result for the strong convergence of $\{A_n(T)\}$ follows from what we have just proved.

Theorem 6.15. *Suppose that* $\lim_{n\to\infty} n^{-1}T^n = 0$ *under the strong topology of* $B(X)$. *If the vector sequence* $\{A_n(T)x\}$ *is weakly pre-compact for every* $x \in X$, *then the operator sequence* $\{A_n(T)\}$ *converges strongly in* $B(X)$. *Moreover, (6.11) is true.*

Proof Because of Theorem 6.14, we only need to prove the strong convergence of $\{A_n\}$. It is easy to see that $A_n x = x$ for all $x \in N(I - T)$. The proof of Theorem 6.14 shows that the sequence $\{\|A_n\|\}$ is bounded. Since by assumption the sequence $\{A_n x\}$ converges in the norm of X for $x \in R(I - T)$, it follows from the uniform boundedness of $\|A_n\|$ that $\{A_n x\}$ converges for all $x \in \overline{R(I - T)}$. Our result now follows from (6.11). □

Theorem 6.15 is the basis of the following abstract mean ergodic theorem. First we need a lemma which is easy to verify.

Lemma 6.3. *The sequence* $\{A_n(T)\}$ *satisfies the identity*

$$\frac{T^{n-1}}{n} = A_n - \left(1 - \frac{1}{n}\right) A_{n-1}, \quad \forall\, n. \tag{6.13}$$

Theorem 6.16. (The abstract mean ergodic theorem) *Suppose that the sequence* $\{A_n(T)\}$ *is bounded in* $B(X)$. *Then the set of those points* x *in* X *for which the sequence* $\{A_n(T)x\}$ *is convergent in norm is a closed subspace of* X, *which is exactly the same as the set of all the points* x *for which the sequence* $\{A_n(T)x\}$ *is weakly pre-compact and* $\lim_{n\to\infty} n^{-1}T^n x = 0$.

Proof Let X_0 be the set of all $x \in X$ for which the sequence $\{A_n(T)x\}$ is weakly pre-compact and $\lim_{n\to\infty} n^{-1}T^n x = 0$. Since $\{A_n(T)\}$ is bounded, the set of those x for which $\{A_n(T)x\}$ is weakly pre-compact is a closed subspace. Identity (6.13) shows that $\{n^{-1}T^n\}$ is bounded, so the set of x for which $\lim_{n\to\infty} n^{-1}T^n x = 0$ is a closed subspace. Thus X_0 is a closed subspace of X. Since a continuous linear operator maps weakly convergent sequences into weakly convergent sequences, $TX_0 \subset X_0$. It follows from Theorem 6.15 applied to the restriction of T to the Banach space X_0 that

$\{A_n(T)x\}$ is convergent for every $x \in X_0$. Conversely, if, for some $x \in X$, the sequence $\{A_n(T)x\}$ is convergent, then it is weakly pre-compact, and (6.13) shows that $\lim_{n \to \infty} n^{-1}T^n x = 0$.   □

Corollary 6.2. *When the strong limit $E = \lim_{n \to \infty} A_n(T)$ exists, it is the projection of X onto $N(I - T)$ along $\overline{R(I - T)}$.*

Corollary 6.3. *If $\{A_n(T)\}$ is bounded, then it converges strongly if and only if $\lim_{n \to \infty} n^{-1}T^n x = 0$ for x in a fundamental set and $\{A_n(T)x\}$ is weakly pre-compact for x in a fundamental set.*

Proof From the proof of Theorem 6.16, both the set of all those vectors x for which $\lim_{n \to \infty} n^{-1}T^n x = 0$ and the set of x for which $\{A_n x\}$ is weakly pre-compact are closed subspaces. Thus both sets are X and Theorem 6.16 shows that $\{A_n x\}$ converges for every $x \in X$.   □

Corollary 6.4. *If X is separable and reflexive, then the sequence $\{A_n(T)\}$ converges strongly if and only if it is bounded and $\lim_{n \to \infty} n^{-1}T^n x = 0$ for every x in a fundamental set.*

Proof This follows from Corollary 6.3 and Theorem B.20.   □

Our last corollary is related to operators on L^1-spaces.

Corollary 6.5. *Let (X, Σ, μ) be a finite measure space and let $T : L^1(X) \to L^1(X)$ be a linear operator such that T maps $L^\infty(X)$ into itself. Suppose that the Cesáro averages $A_n(T)$ are uniformly bounded as operators on $L^1(X)$ and $L^\infty(X)$, respectively. Then $A_n(T)$ converge strongly if and only if $n^{-1}T^n f \to 0$ in $L^1(X)$ for every f in a fundamental set of $L^1(X)$.*

Proof Given any $A \in \Sigma$, since the sequence $\{A_n \chi_A\}$ is bounded in $L^\infty(X)$, it is weakly pre-compact in $L^1(X)$ by Theorem B.4, The desired conclusion follows from Corollary 6.3.   □

Applications of the general ergodic theorems of this section to positive operators and more special Markov operators will be presented in the next section and the forthcoming chapters.

6.4   Ergodic Theorems of Positive Operators

A fundamental mathematical problem in the statistical study of dynamical systems concerns the existence of certain types of *time means*. This problem may be formulated and studied in the abstract eogodic theory of positive operators. In this section, we shall discuss the behavior of the averages of the iterates of a positive operator on a Banach lattice, but here we mainly restrict ourselves to some specific positive operators defined on L^p spaces since the general ergodic theorems of the previous section can be rephrased for arbitrary positive operators.

We assume that (X, Σ, μ) is a σ-finite measure space and $S : X \to X$ is a measure-preserving transformation. We first present the mean ergodic theorem for L^p functions that was first proved by von Neumann in 1931 for the special case of $p = 2$, which is a consequence of the following *abstract mean ergodic theorem for reflexive spaces*.

Lemma 6.4. *Let X be a separable reflexive Banach space and let $T \in B(X)$ satisfy the condition that $\lim_{n \to \infty} n^{-1} T^n = 0$ under the strong topology of $B(X)$. Then the Cesáro averages $A_n(T)x$ converge in norm to a fixed point of T for all vectors $x \in X$.*

Proof The sequence $\{A_n x\}$ is bounded, so it is weakly pre-compact by Theorem B.20. The conclusion follows from Theorem 6.15. □

Theorem 6.17. (von Neumann's mean ergodic theorem) *Let $1 < p < \infty$ and let $L^p(X)$ be separable. Then for any $f \in L^p(X)$, there is $\tilde{f} \in L^p(X)$ such that $\tilde{f}(S(x)) = \tilde{f}(x)$, $x \in X$ μ − a.e. and*

$$\lim_{n \to \infty} \left\| \frac{1}{n} \sum_{k=0}^{n-1} f \circ S^k - \tilde{f} \right\|_p = 0.$$

Proof $L^p(X)$ is a separable reflexive Banach space and the composition operator U_S satisfies the condition of Lemma 6.4. □

Remark 6.2. The theorem is still valid for $p = 1$ when $\mu(X) < \infty$, but a counterexample with $\mu(X) = \infty$ is: $X = \mathbb{R}$, μ = Lebesgue measure, $S(x) = x + 1$, and $f = \chi_{[0,1)}$. The conclusion may not hold for $p = \infty$ even if $\mu(X) = 1$ [Krengel (1985)].

We study the pointwise convergence of the Cesáro averages

$$A_n f = \frac{1}{n} \sum_{k=0}^{n-1} U_S^k f$$

of the iterates of the composition operator U_S at $x \in X$. For this purpose we need the following *maximum ergodic theorem* for general positive contractions of L^1 spaces.

Lemma 6.5. *Let* $T : L^1(X) \to L^1(X)$ *be a positive operator such that* $\|T\|_1 \leq 1$ *and for a real-valued function* $f \in L^1(X)$ *denote* $X_\infty = \bigcup_{n=1}^\infty X_n$, *where* $X_n = \{x \in X : \max\{A_1(T)f(x), \cdots, A_n(T)f(x)\} \geq 0\}$. *Then*

$$\int_{X_n} f d\mu \geq 0 \quad and \quad \int_{X_\infty} f d\mu \geq 0.$$

Proof $\int_X Tg d\mu \leq \int_X g d\mu$ for all $g \in L^1(X)_+$. Denote $S_n f = \sum_{k=0}^{n-1} T^k f$ and $M_n f = \max\{S_1 f, \cdots, S_n f\}$. Then $X_n = \{x \in X : M_n f(x) \geq 0\}$. Since $(M_n f)^+ \geq S_k f$ for $k = 1, \cdots, n$, we have $f + T(M_n f)^+ \geq f + T S_k f = S_{k+1} f$. It follows that $f \geq S_k f - T(M_n f)^+$ for $k = 1, \cdots, n$. The definition of M_n then ensures the inequality $f \geq M_n f - T(M_n f)^+$. Now

$$\int_{X_n} f d\mu \geq \int_{X_n} (M_n f - T(M_n f)^+) d\mu$$

$$= \int_{X_n} ((M_n f)^+ - T(M_n f)^+) d\mu$$

$$= \int_X (M_n f)^+ d\mu - \int_{X_n} T(M_n f)^+ d\mu$$

$$\geq \int_X (M_n f)^+ d\mu - \int_X T(M_n f)^+ d\mu \geq 0.$$

This proves the first inequality of the lemma, from which the second one follows by letting $n \to \infty$. $\qquad\square$

Lemma 6.5 implies the following *maximum ergodic inequality*.

Lemma 6.6. *Let* $\hat{M}_n f = \max\{A_1 f, \cdots, A_n f\}$ *for any real-valued function* $f \in L^1(X)$ *and positive integer* n. *Then*

$$\mu(\{x \in X : \hat{M}_n f(x) \geq a\}) \leq \frac{1}{a} \|f\|_1, \quad \forall\, a > 0.$$

Proof It is enough to assume that $\mu(\{x \in X : \hat{M}_n f(x) \geq a\}) < \infty$. Let $A \subset \{x \in X : \hat{M}_n f(x) \geq a\}$ be measurable and denote

$$\hat{A} = \{x \in X : \hat{M}_n(f(x) - a\chi_A(x)) \geq 0\}.$$

Then $\int_{\hat{A}} (f - a\chi_A) d\mu \geq 0$ by Lemma 6.5. For any $x \in \hat{A}$, there is $k \leq n$ such that $A_k f(x) \geq a$, which implies $S_k(f - a)(x) \geq 0$ and $S_k(f - a\chi_A)(x) \geq 0$. Therefore $A \subset \hat{A}$, and hence

$$\|f\|_1 \geq \int_{\hat{A}} f d\mu \geq a \int_{\hat{A}} \chi_A d\mu = a\mu(A).$$

Thus, $\mu(A) \leq \|f\|_1/a$, which proves the lemma. $\qquad\square$

Theorem 6.18. (Birkhoff's pointwise ergodic theorem) *For every function* $f \in L^1(\mu)$, *there exists a function* $\tilde{f} \in L^1(\mu)$ *such that* $\tilde{f}(S(x)) = \tilde{f}(x)$, $x \in X$ μ-*a.e.*, $\|\tilde{f}\|_1 \leq \|f\|_1$, *and*

$$\lim_{n \to \infty} \frac{1}{n} \sum_{k=0}^{n-1} f(S^k(x)) = \tilde{f}(x), \quad \forall\, x \in X \ \mu - \text{a.e.} \tag{6.14}$$

Moreover, for each $A \in \Sigma$ *such that* $\mu(A) < \infty$ *and* $S^{-1}(A) = A$,

$$\int_A \tilde{f} d\mu = \int_A f d\mu. \tag{6.15}$$

In particular, if $\mu(X) < \infty$, *then* $\int_X \tilde{f} d\mu = \int_X f d\mu$.

Proof We only prove the theorem when $\mu(X) < \infty$ and refer to [Cornfeld *et al.* (1982)] or [Walters (1982)] for the general case. It is enough to consider a real-valued function $f \in L^1(\mu)$. Since

$$A_{n+1} f = \frac{1}{n+1} S_{n+1} f = \frac{1}{n+1} f + \frac{n}{n+1} \left(\frac{1}{n} S_n f \right) \circ S,$$

the upper limit function $f_u(x) = \limsup_{n \to \infty} A_n f(x)$ and the lower limit function $f_l(x) = \liminf_{n \to \infty} A_n f(x)$ satisfy $f_u(S(x)) = f_u(x)$ and $f_l(S(x)) = f_l(x)$ for all $x \in X$. First we show that they are both finite number valued for $x \in X$ μ-a.e.

Let $\hat{M}_n f = \max\{A_1 f, \cdots, A_n f\}$. Then for any real number $b > 0$, the set $D_b = \{x \in X : f_u(x) > b\}$ is contained in the union of the increasing sequence of the sets $\{x \in X : \hat{M}_n f(x) \geq b\}$. By Lemma 6.6,

$$\mu(\{x \in X : \hat{M}_n f(x) \geq b\}) \leq b^{-1} \|f\|_1.$$

So, $\mu(D_b) \leq b^{-1} \|f\|_1$, from which $f_u(x) < \infty, x \in X$ a.e. by letting $b \to \infty$. Similarly, $f_l(x) > -\infty$, $\forall\, x \in X$ a.e. Moreover, for $a < 1$,

$$\mu(\{x \in X : f_l(x) < a\})$$
$$= \mu(-\limsup_{n \to \infty} A_n(-f)(x) < a) \leq \frac{\|f\|_1}{|a|}.$$

Next, we show the convergence of $A_n f(x)$ a.e. in (6.14). If not, then there are rational numbers $a < b$ such that $\mu(B) > 0$, where $B = \{x \in X : f_l(x) < a < b < f_u(x)\}$. The invariance of f_u and f_l implies $S^{-1}(B) = B$. So if $f_b = (f - b)\chi_B$, then $(f_b \circ S^k)(x) = 0$ for $x \notin B$ and $k \geq 0$, and $B = \{x \in X : S_n f_b(x) > 0 \text{ for some } n \geq 1\}$. Lemma 6.5 ensures that $b\mu(B) \leq \int_B f d\mu$. By letting $f_a = (a - f)\chi_B$, we obtain

$$a\mu(B) \geq \int_B f d\mu \geq b\mu(B).$$

Since $\mu(B) > 0$, $a \geq b$, a contradiction to $a < b$. Hence, $f_u = f_l \equiv \tilde{f}$ a.e.

To show $\|\tilde{f}\|_1 \leq \|f\|_1$, it is enough to assume $f \in L^1(X)_+$. Then, since S preserves μ, Fatou's lemma implies that

$$\int_X \tilde{f} d\mu = \int_X \liminf_{n \to \infty} A_n f d\mu \leq \liminf_{n \to \infty} \int_X A_n f d\mu = \int_X f d\mu.$$

Finally, for the proof of (6.15), we may also assume $f \geq 0$. Given any $\epsilon > 0$, there is $c_\epsilon \geq 0$ such that $\|f - f \wedge c_\epsilon\|_1 < \epsilon$. Since

$$\int_X (A_n f - c_\epsilon)^+ d\mu \leq \int_X A_n(f - f \wedge c_\epsilon) d\mu < \epsilon, \quad \forall\, n \geq 1,$$

the sequence $\{A_n f\}$ is *uniformly integrable* (see Definition 3.13 of [Krengel (1985)]). Hence $\lim_{n \to \infty} \|A_n f - \tilde{f}\|_1 = 0$, so

$$\int_X \tilde{f} d\mu = \lim_{n \to \infty} \int_X A_n f d\mu = \int_X f d\mu. \square$$

There are many abstract and deep results on the pointwise convergence of the averages sequence $\{A_n(T)f\}$ for a general positive operator T on an L^p-space. We list two famous ones in the following and refer their proofs to [Dunford and Schwartz (1957); Krengel (1985)].

Theorem 6.19. (Dunford-Schwartz) *Let $T : L^1 \to L^1$ be a positive operator such that $\|T\|_1 \leq 1$ and $\sup\{\|Tf\|_\infty : f \in L^1 \cap L^\infty, \|f\|_\infty \leq 1\} \leq 1$. Then $\lim_{n \to \infty} A_n(T)f(x)$ exists for $x \in X$ μ-a.e. and all $f \in L^1$.*

Theorem 6.20. (Akcoglu's ergodic theorem) *Let $1 < p < \infty$. Suppose that $T : L^1 \to L^1$ is positive such that $\|T\|_p \leq 1$. Then $\lim_{n \to \infty} A_n(T)f(x)$ exists for $x \in X$ μ-a.e. and all $f \in L^p$.*

From the next chapter on we shall turn our attention to several important classes of concrete positive operators defined on either an $L^1(X)$ space or a $C(K)$ space. More specifically, we shall investigate various structural and ergodic properties of Markov operators, Frobenius-Perron operators, and composition operators. They are widely used in various subjects of mathematical and physical sciences which are related to deterministic dynamical systems and stochastic analysis.

6.5   Notes and Remarks

1. The comprehensive treatise [Dunford and Schwartz (1957)] is always the standard reference for linear operators, including the spectral theory of bounded linear operators presented in the first section of this chapter. An excellent textbook which contains most of the materials presented in this chapter is [Conway (1990)].

2. The spectral theory for general positive operators of vector lattices is quite abstract and many of its results are deep and difficult to understand for beginners, in addition to that part of the theory which is directly inherited from the one for bounded linear operators. We only touched a partial spectral theory for a particular class of positive operators on function spaces in Section 6.2, but since such composition operators are defined through measure-preserving transformations, they are very important in physical sciences such as thermodynamics and statistical physics [Bowen (1975); Beck and Schlögl (1993)]. For many general spectral results of positive operators, [Schaefer (1974); Zaanen (1997)] can be consulted.

3. The ergodic theory of bounded linear operators with norm 1 has many applications in various subjects of mathematical and physical sciences [Cornfeld *et al.* (1982); Beck and Schlögl (1993); Lasota and Mackey (1994)] that need the consideration of average processes for the desired convergence of sequences or other purposes. The last chapter of the the large volume monograph [Dunford and Schwartz (1957)] on applications of linear operators basically provides an essential part of the classic ergodic theory after all preparatory knowledge on the theory of linear operators has been studied, part of which is adapted for the presentation of Section 6.3.

The ergodic theorems proved in Section 6.3 are only the prototype of the general theorems in the Tauberian theory which concerns the conditions that guarantee the convergence of the sequence $\{f_n(T)\}$ from that of the sequence $\{f(T)f_n(T)\}$. For a given analytic function $f \in \mathcal{H}(T)$ with $T \in B(X)$, if the operator $f(T)$ has a bounded inverse $f(T)^{-1} \in B(X)$, the above problem is trivially solved since $f_n(T) = f(T)^{-1}[f(T)f_n(T)]$. In the Tauberian theory, one encounters the situation that $f \in \mathcal{H}(T)$ vanishes somewhere in the spectrum $\sigma(T)$.

For example, a generalization (Theorem VII.7.4 of [Dunford and Schwartz (1957)]) of Theorem 6.15 is

Theorem 6.21. *Let X be a Banach space, let $T \in B(X)$, and let $f, f_n \in \mathcal{H}(T)$, $\forall\, n = 1, 2, \cdots$, be such that $\lim_{n \to \infty} f(T)f_n(T) = 0$ under the strong topology of $B(X)$. Suppose that the vector sequence $\{f_n(T)x\}$ is weakly precompact for every $x \in X$, and that f has finitely many roots in $\sigma(T)$. If each root λ_0 of f has a finite order $\nu(\lambda_0)$, and if $\lim_{n \to \infty} f_n^{(r)}(\lambda_0)$ exists* *for* $0 \leq r < \nu(\lambda_0)$ *and is nonzero for* $r = 0$, *then the operator sequence* $\{f_n(T)\}$ *converges strongly in* $B(X)$. *Moreover,*

$$X = \overline{R(f(T))} \oplus N(f(T)).$$

4. Von Neumann's mean ergodic theorem and Birkhoff's pointwise ergodic theorem were published almost the same time at the beginning of the 1930s, which have many applications, such as the proofs of the Kronecker-Weyl theorem and Borel's normal number theorem (Exercise 6.20) [Choe (2005)]. These ergodic theorems can be extended to groups or semigroups of measure-preserving transformations which depend on the continuous time. There are numerous generalizations of these classic results to positive operators of unit norm defined on Banach lattices or more general weak contractions on Banach spaces. [Schaefer (1974); Krengel (1985)] cover many abstract and concrete ergodic theorems in the more general settings.

Exercises

6.1 Let $T \in B(X)$ be a complex Banach space X. Prove:

$$\sigma(T) = \{\lambda \in \mathbb{C} : N(\lambda I - T) \neq \{0\} \text{ or } R(\lambda I - T) \neq X\}.$$

6.2 Let (X, Σ, μ) be a σ-finite measure space and let $g \in L^\infty$. Define the *multiplication operator* $M_g : L^1 \to L^1$ by

$$M_g f(x) = g(x) f(x), \quad \forall x \in X.$$

   (i) Calculate $\|M_g\|_1$.
   (ii) Study the spectrum $\sigma(M_g)$ of M_g.
   (iii) Find the resolvent $R(\lambda, M_g)$ for $\lambda \in \rho(M_g)$.

6.3 Let $T \in B(X)$ and $f \in \mathcal{H}(T)$. Show that if $f(\lambda) = \lambda^n$, then $f(T) = T^n$ for any integer n.

6.4 Let $T \in B(X)$ and let $G \subset \mathbb{C}$ be an open set containing $\sigma(T)$. Show that if $f, f_n, \forall n \geq 1$ are all analytic in G and $\lim_{n \to \infty} f_n(\lambda) = f(\lambda)$ uniformly on compact subsets of G, then

$$\lim_{n \to \infty} \|f_n(T) - f(T)\| = 0.$$

6.5 Let $T, S \in B(T)$ with $TS = ST$. Show: $f(T)S = Sf(T), \forall f \in \mathcal{H}(T)$.

6.6 Let $T \in B(X)$ and $\sigma(T) = F_1 \cup F_2$, where F_1, F_2 are disjoint nonempty closed sets. Show that there is $E \in B(T)$ such that $E^2 = E$ and

(i) if $TS = ST$ for some $S \in B(X)$, then $ES = SE$;

(ii) if $T_1 = TE$ and $T_2 = T - T_1$, then $T_1 T_2 = T_2 T_1 = 0$;

(iii) $\sigma(T_1) = F_1 \cup \{0\}$ and $\sigma(T_2) = F_2 \cup \{0\}$.

6.7 Under the condition of Exercise 6.6, show that there are topologically complementary subspaces X_1 and X_2 of X such that:

(i) $SX_j \subset X_j$ for $S \in B(X)$ with $ST = TS$, $j = 1, 2$;

(ii) $\sigma(T_j) = F_j$ for $T_j = T|_{X_j}$, $j = 1, 2$.

6.8 Suppose that $T^n = 0$ for some $n > 0$ and $T \in B(X)$. Let $f, g \in \mathcal{H}(T)$. Find a necessary and sufficient condition on f and g for $f(T) = g(T)$.

6.9 Let $T \in B(X)$ and let M be a closed subspace of X such that

$$(\lambda I - T)^{-1} M \subset M, \ \forall \lambda \in \rho(T).$$

Show that $f(T)M \subset M$, $\forall f \in \mathcal{H}(T)$.

6.10 Let X be a Hilbert space and $T \in B(X)$ a *normal operator*, namely $TT^* = T^*T$. Show that $f(T)$ is also normal for any $f \in \mathcal{H}(T)$.

6.11 Let $N \subset B(X)$ be an ideal and $T \in N$. Show that if $f \in \mathcal{H}(T)$ such that $f(0) = 0$, then $f(T) \in N$.

6.12 Let $T \in B(X)$. Show that if $\lambda_0 \in \sigma(T)$ is an isolated point of $\sigma(T)$, then λ_0 is an eigenvalue of T.

6.13 Let $T_A : \mathbb{R}^n \to \mathbb{R}^n$ be given by the $n \times n$ matrix

$$A = [e_2, \cdots, e_n, 0].$$

Find the index ν of 0 with respect to T_A.

6.14 Let (X, Σ, μ) be a σ-finite measure space, let $S : X \to X$ be a nonsingular transformation, and let $A \in \Sigma$. Prove that

(i) if $\mu(A \cap (S^{-1}(A))^c) = 0$ or $\mu(S^{-1}(A) \cap A^c) = 0$, then A is S-invariant;

(ii) if $\mu(X) < \infty$ and $\mu(A) \leq \mu(S^{-1}(A))$, $\forall A \in \Sigma$, then μ is S-invariant; is the assumption $\mu(X) < \infty$ essential for the invariance of μ?

(iii) A is S-invariant if and only if $\chi_A \circ S = \chi_A$;

(iv) if $\mu(A) = 0$ or $\mu(A^c) = 0$, then A is S-invariant.

6.15 Let $S(x) = 2x$, $\forall x \in \mathbb{R}$. Show that S is not measure-preserving with respect to the Lebesgue measure on $\mathbb{R}$. If we let $S(x, y) = (2x, y/2)$, then S is measure-preserving with respect to the Lebesgue measure on $\mathbb{R}^2$.

6.16 Let $S : X \to X$ be a measurable transformation on a measure space (X, Σ, μ). Show: the set function $\nu(A) = \mu(S^{-1}(A))$, $\forall A \in \Sigma$ is a measure.

6.17 Let $S : X \to X$ be a measurable transformation on a measure space (X, Σ, μ). Show that if μ is an S^n-invariant measure for some positive integer n, then the set function ν defined by

$$\nu = \frac{1}{n} \sum_{k=0}^{n-1} \mu \circ S^{-k}$$

is an S-invariant measure.

6.18 Let (X, Σ, μ) be a probability measure space and $S : X \to X$ measure-preserving. Show that the following are equivalent:

(i) S is ergodic.

(ii) For every set $A \in \Sigma$ such that $\mu(A) > 0$, we have

$$\mu \left(\sum_{n=1}^{\infty} S^{-n}(A) \right) = 1.$$

(iii) $\mu(A \cap S^{-n}(B)) > 0$ for an $n > 0$, $\forall A, B$ with $\mu(A) > 0, \mu(B) > 0$.

6.19 Let $T \in B(X)$ be an isometry. Show that

(i) $\sigma(T) = \mathbb{D}^2$ if $0 \in \sigma(T)$;

(ii) $\sigma(T) \subset \partial\mathbb{D}^2$ if $0 \in \rho(T)$.

6.20 Applying Birkhoff's pointwise ergodic theorem to the mapping $S :$ $[0, 1] \to [0, 1]$ defined by $S(x) = 2x(\mathrm{mod}\ 1)$ to show that almost all numbers in $[0, 1)$ are *normal* to the base 2, in other words, for all $x \in [0, 1)$ a.e., the frequency of 1's in the binary expansion of x is $1/2$.

6.21 Discuss Birkhoff's pointwise ergodic theorem applied to a finite space $X = \{a_1, a_2, \cdots, a_k\}$ with the counting measure μ.

6.22 Let X be a reflexive Banach space and $T \in B(X)$ be such that the sequence $\{\|T^n\|\}$ is bounded. Show that for any $x \in X$ the averages $A_n x$ converge in norm to a fixed point of T.

6.23 Let X be a Banach space and let $T \in B(X)$ be such that the sequence $\{\|T^n\|\}$ is bounded. Show that $\lim_{n \to \infty} A_n x$ exists in norm for all $x \in X$ if and only if for any ξ, $\eta \in N(I - T')$ with $\xi \neq \eta$, there is $x \in N(I - T)$ such that $\xi(x) \neq \eta(x)$.

6.24 Let X be a Banach space and let $T \in B(X)$ be such that the number sequence $\{\|T^n\|\}$ is bounded. Show that $\lim_{n \to \infty} A_n = E \in B(X)$ in the operator norm of $B(X)$ if and only if $R(I - T)$ is a closed subspace of X.

6.25 Let X be a Banach space and let $T \in B(X)$ be such that the number sequence $\{\|T^n\|\}$ is bounded. Show that $\lim_{n \to \infty} A_n = E \in B(X)$ in the operator norm of $B(X)$ if and only if either $1 \in \rho(T)$ or 1 is a pole of $R(\lambda, T)$ with order 1.

Chapter 7

Markov Operators

We have studied abstract positive operators between vector lattices. In this chapter we introduce a class of positive operators defined on L^1 spaces, called Markov operators which are concrete stochastic operators and widely used in stochastic analysis. A subclass of Markov operators, referred to as Frobenius-Perron operators associated with nonsingular transformations, are extremely important in modern ergodic theory and its applications to various applied fields in physical, biological, and engineering sciences, and they will be studied in the next chapter.

Markov operators map density functions onto themselves, and so they are closely related to probability theory and arise in Markov processes which are special stochastic processes. In the statistical and stochastic study of deterministic dynamical systems, Markov operators describe the evolution of probability density functions governed by the deterministic dynamics.

In Section 7.1, the definition and some examples of Markov operators are given, and various properties of Markov operators will be studied in Section 7.2. Section 7.3 is devoted to the ergodic theory of Markov operators, and we shall explore the asymptotic behavior of the iteration sequences of Markov operators in Section 7.4. [Foguel (1969); Lasota and Mackey (1994)] are two excellent monographs on Markov operators.

7.1 Definition and Examples of Markov Operators

Let (X, Σ, μ) be a σ-finite measure space. As a convention, when we mention a measure space, we always assume that it is σ-finite without mentioning it explicitly. Also we assume that the measure space (X, Σ, μ) is *complete*, that is, all subsets of measurable sets of X with μ-measure zero are measurable. The Lebesgue measure space $(\Omega, \mathcal{L}, m)$, which is the completion of the Borel measure space $(\Omega, \mathcal{B}, m)$, is complete, where Ω is an open subset of $\mathbb{R}^n$ and m denotes the Lebesgue measure on $\mathbb{R}^n$.

Let $L^p(X, \Sigma, \mu)$ be a *real L^p-space* with $1 \le p \le \infty$, which is a Banach lattice under the L^p-norm $\| \ \|_p$. Denote

$$\mathcal{D} \equiv \mathcal{D}(X, \Sigma, \mu) = \{f \in L^1(X, \Sigma, \mu) : \ f \ge 0, \ \|f\|_1 = 1\}.$$

Sometimes we write $\mathcal{D} = \mathcal{D}(X)$. Any function $f \in \mathcal{D}$ is called a *density*. If for some $f \in \mathcal{D}$, $\mu_f(A) = \int_A f d\mu$, $\forall A \in \Sigma$, then f is called the density of the probability measure μ_f with respect to the measure μ.

Definition 7.1. A linear operator $P : L^1(X) \to L^1(X)$ is called a *Markov operator* if $P\mathcal{D}(X) \subset \mathcal{D}(X)$.

In other words, P is a Markov operator if and only if P is a positive operator from $L^1(X)$ into itself and preserves the integral of nonnegative and integrable functions.

Remark 7.1. Since f and Pf are L^1 functions, $f \ge 0$ and $Pf \ge 0$ are understood to hold almost everywhere (a.e.) with respect to the underlying measure μ. As a consequence, all the equalities and inequalities among measurable functions are in the sense of "a.e." with respect to μ without further explanations.

Definition 7.2. Let P be a Markov operator on $L^1(X)$. Any $f \in \mathcal{D}$ such that $Pf = f$ is called a *stationary density* of P.

We shall study the existence and computation of stationary densities of Markov operators in the future since such problems are very important in ergodic theory, stochastic analysis, partial differential equations, and many other fields. But first we give several examples of Markov operators. More examples will be provided in the exercise set and in the subsequent chapters. The first example below clearly indicates the relation of stochastic matrices and finite dimensional Markov operators.

Example 7.1. Let $X = \{1, 2, \cdots, n\}$ with the *counting measure*, so that $L^1(X) = \mathbb{R}^n$. Then any $n \times n$ stochastic matrix A defines a Markov operator $P : L^1(X) \to L^1(X)$ via

$$(Px)_j = \sum_{i=1}^n a_{ij} x_i, \ \ j = 1, 2, \cdots, n$$

for all vectors $x = (x_1, x_2, \cdots, x_n)^T \in \mathbb{R}^n$, which will play the major role in the structure-preserving finite dimensional approximation of Markov operators in Chapter 10.

Example 7.2. More generally, let (X, Σ, μ) be a measure space and let $w : X \times X \to \mathbb{R}_+$ be measurable such that $\int_X w(x, y)\mu(dx) = 1$ for all $y \in X$. Then the linear operator

$$Pf(x) = \int_X w(x, y)f(y)\mu(dy)$$

is a Markov operator on $L^1(X)$ with a *stochastic kernel* w.

Example 7.3. (Evolution equations) As a concrete case of Example 7.2, we consider an *evolution differential equation*

$$\frac{\partial u}{\partial t}(t, x) = Lu(t, x),$$

where L is a partial differential operator with respect to x. Its solution can sometimes be expressed as a *semigroup of Markov operators* $\{P_t\}_{t \in \mathbb{R}_+}$. A simple concrete case is the one dimensional *heat equation*

$$\frac{\partial u}{\partial t} = \frac{a^2}{2}\frac{\partial^2 u}{\partial x^2}, \quad t > 0, \ x \in \mathbb{R}$$

with the initial value condition

$$u(0, x) = f(x), \quad x \in \mathbb{R},$$

for which the solution $u(t, x)$ can be written as

$$u(t, x) = P_t f(x) = \int_{-\infty}^{\infty} w(t, x, y)f(y)dy, \quad P_0 f(x) = f(x),$$

where the stochastic kernel

$$w(t, x, y) = \frac{1}{\sqrt{2\pi a^2 t}}\exp\left(-\frac{(x - y)^2}{2a^2 t}\right) \geq 0$$

satisfies the Markov property

$$\int_{-\infty}^{\infty} w(t, x, y)dx = 1, \quad \forall \, y \in \mathbb{R}, \ t > 0.$$

Example 7.4. (The Ruelle operator) We consider a problem related to Gibbs measures [Bowen (1975); Baladi (2000)]. Let

$$K = \prod_{n=1}^{\infty}\{1, 2, \cdots, k\}$$

be the *one-sided symbolic space* of sequences $x = (x_1, x_2, \cdots)$ of symbols $1, 2, \cdots, k$, endowed with the compact product topology of the discrete topology of $\{1, 2, \cdots, k\}$, and let $\phi \in C(K)$ satisfy some conditions. Then the existence of a Gibbs measure on K is closely related to the maximal eigenvalue problem of the *Ruelle operator* $L_\phi : C(K) \to C(K)$ defined by

$$L_\phi f(x) = \sum_{y \in S^{-1}(\{x\})} \exp(\phi(y)) f(y), \ \forall f \in C(K),$$

where $S : K \to K$ is the *one-sided shift*, that is,

$$S(x_1, x_2, x_3, \cdots) = (x_2, x_3, x_4, \cdots).$$

Let L'_ϕ be the dual operator of L_ϕ defined on the the space of all Borel measures of K. The Ruelle theorem [Ruelle (1968)] asserts that there is a constant $r > 0$, a positive function $h \in C(K)$, and an S-invariant probability measure ν such that

$$L_\phi h = rh, \ L'_\phi \nu = r\nu, \ \int_K h d\nu = 1.$$

Denote $P = r^{-1} L_\phi$. Then for all $f \in L^1(\nu)$,

$$\int_K P f d\nu = \frac{1}{r} \int_K L_\phi f d\nu = \frac{1}{r} \int_K f d(L'_\phi \nu) = \int_K f d\nu.$$

Hence, $P : L^1(\nu) \to L^1(\nu)$ is a Markov operator and the function h is a stationary density of P.

Example 7.5. (Wavelets construction) The *scaling equation* in the theory of wavelets is the function equation of the form:

$$\psi(x) = \sum_{n \in I} a_n \psi(2x - n), \ x \in \mathbb{R},$$

where ψ is the *scaling function* for the multi-resolution analysis in the wavelets theory, I is a finite or countable index set, and $\sum_{n \in I} a_n = 2$. If we define an operator P by

$$P f(x) = \sum_{n \in I} a_n f(2x - n), \ \forall x \in \mathbb{R},$$

then the solution ψ of the above scaling equation is just a fixed point of P. If $a_n \geq 0$ for all $n \in I$, then from

$$\int_{-\infty}^{\infty} P f(x) dx = \sum_{n \in I} a_n \int_{-\infty}^{\infty} f(2x - n) dx$$

$$= \frac{1}{2} \sum_{n \in I} a_n \int_{-\infty}^{\infty} f(x) dx = \int_{-\infty}^{\infty} f(x) dx,$$

$P : L^1(\mathbb{R}) \to L^1(\mathbb{R})$ is a Markov operator.

Example 7.6. (Conditional expectation) Let (X, Σ, μ) be a measure space and let Σ_0 be a σ-subalgebra of Σ. Suppose that the measure space (X, Σ_0, μ) is also σ-finite. Given any $f \in L^1(X, \Sigma, \mu)$, the set function

$$\mu_f(A) = \int_A f d\mu, \ \forall \, A \in \Sigma_0$$

defines a real (or complex) measure on the measurable space (X, Σ_0), which is absolutely continuous with respect to μ. By the Radon-Nikodym theorem, there exists a unique function in $L^1(X, \Sigma_0, \mu) \subset L^1(X, \Sigma, \mu)$, which is denoted as $E(\cdot \, | \Sigma_0)f$, such that

$$\int_A E(\cdot \, | \Sigma_0) f d\mu = \mu_f(A) = \int_A f d\mu, \ \forall \, A \in \Sigma_0.$$

The resulting operator $E \equiv E(\cdot \, | \Sigma_0) : L^1(\Sigma) \to L^1(\Sigma_0)$ is well-defined and is clearly linear, which is referred to as the *conditional expectation* associated with the σ-subalgebra Σ_0 of Σ.

If $f \geq 0$, then $Ef \geq 0$ and $\|Ef\|_1 = \|f\|_1$. Hence E is a Markov operator. Moreover, $Ef = f$ if and only if $f \in L^1(\Sigma)$ is Σ_0-measurable. Thus E is a positive projection from $L^1(X, \Sigma, \mu)$ onto $L^1(X, \Sigma_0, \mu)$ with norm 1. We note that $L^1(X, \Sigma, \mu) = R(E) \oplus N(E)$. See [Walters (1982)] for more properties of E.

In particular, if $S : X \to X$ is nonsingular such that the measure space $(X, S^{-1}(\Sigma), \mu)$ is also σ-finite, then the corresponding conditional expectation is closely related to the Frobenius-Perron operator associated with S in the next example.

Example 7.7. (Frobenius-Perron operators) For a nonsingular transformation $S : X \to X$, the corresponding *Frobenius-Perron operator* $P_S : L^1(X) \to L^1(X)$ is defined implicitly by

$$\int_A P_S f d\mu = \int_{S^{-1}(A)} f d\mu, \ \forall \, A \in \Sigma, \ f \in L^1(X).$$

The existence of $P_S f \in L^1$ for any $f \in L^1$ is guaranteed by the Radon-Nikodym theorem because of the non-singularity assumption of S. Clearly P_S is a positive operator, and letting $A = X$ in the definition shows that P_S is a Markov operator. Suppose that the measure space $(X, S^{-1}(\Sigma), \mu)$ is also σ-finite. Since the measure $\nu \equiv \mu \circ S^{-1}$ is absolutely continuous with respect to the measure μ, by the Radon-Nikodym theorem, there is a nonnegative Σ-measurable function h such that

$$\nu(A) = \int_A h d\mu, \ \forall \, A \in \Sigma.$$

It can be shown [Ding and Hornor (1994)] that

$$P_S f(x) = h(x) E f(x), \ \forall \, x \in X,$$

where E is the conditional expectation associated with the σ-subalgebra $S^{-1}(\Sigma)$, as defined in the previous example. The next chapter will be devoted to the study of Frobenius-Perron operators in detail.

Let $P' : L^\infty(X) \to L^\infty(X)$ be the dual operator of a Markov operator $P : L^1(X) \to L^1(X)$. The *transition function* $p : X \times \Sigma \to \mathbb{R}_+$ associated with the Markov operator P is defined by

$$p(x, A) = P' \chi_A(x), \ \forall \, x \in X, \ A \in \Sigma.$$

Proposition 7.1. *The transition function p satisfies:*
(i) *$p(\cdot, A)$ is Σ-measurable for each fixed $A \in \Sigma$;*
(ii) *$p(x, \cdot)$ is a probability measure on X for each fixed $x \in X$;*
(iii) *the probability measure $p(x, \cdot)$ is absolutely continuous with respect to the measure μ for each fixed $x \in X$.*

Proof Since $P' \chi_A \in L^\infty$, it is measurable, and (i) is proved. Clearly $0 \le p(x, A) \le 1$ since $P' \chi_A \ge 0$ and $\|P' \chi_A\|_\infty \le \|P'\|_\infty \|\chi_A\|_\infty \le 1$. Further $p(x, X) = 1$ since $P'1 = 1$. Now let $\{A_k\}$ be a mutually disjoint sequence of measurable subsets of X. Then for any $0 \le f \in L^1$, we have

$$\langle f, p\, (x, \cup_{k=1}^\infty A_k) \rangle = \langle f, P' \chi_{\cup_{k=1}^\infty A_k} \rangle = \langle Pf, \chi_{\cup_{k=1}^\infty A_k} \rangle$$

$$= \sum_{k=1}^\infty \langle Pf, \chi_{A_k} \rangle = \sum_{k=1}^\infty \langle f, P' \chi_{A_k} \rangle$$

$$= \left\langle f, \sum_{k=1}^\infty P' \chi_{A_k} \right\rangle = \left\langle f, \sum_{k=1}^\infty p(x, A_k) \right\rangle,$$

where the left equality of the last line is from Fatou's lemma. The same equality extends to arbitrary $f \in L^1$ using $f = f^+ - f^-$, thus giving (ii). To prove (iii), let $A \in \Sigma$ be any measurable set such that $\mu(A) = 0$. Then $\chi_A = 0$, so $P' \chi_A = 0$. Therefore $p(x, A) = 0$. $\qquad \square$

Actually, the above properties (i) to (iii) uniquely determine a Markov operator. Suppose that a real-valued function p defined on $X \times \Sigma$ satisfies (i)-(iii) of Proposition 7.1. Given $0 \le f \in L^1$. Then the measure μ_f defined by $\mu_f(A) = \int_A f d\mu, \ \forall \, A \in \Sigma$ satisfies $\mu_f \ll \mu$. Let

$$\nu(A) = \int_X p(x, A) \mu_f(dx), \ \forall \, A \in \Sigma.$$

Property (iii) above ensures that $\nu \ll \mu$. Now set

$$Pf = \frac{d\nu}{d\mu},$$

the Radon-Nikodym derivative of the measure ν with respect to the measure μ. If $f \in L^1$, then we define $Pf = Pf^+ - Pf^-$, and the resulting linear operator $P : L^1 \to L^1$ is indeed a Markov operator whose corresponding transition function is the original p. Moreover, its dual operator $P' : L^\infty \to L^\infty$ has the expression

$$P'g(x) = \int_X g(y)p(x, dy), \ \forall\, g \in L^\infty.$$

7.2   Properties of Markov Operators

We study general properties of Markov operators in this section. Of course all the results on positive operators are valid for Markov operators, so some of the propositions in the following are simply consequences of the respective ones in the previous two chapters. We start with a simple result.

Proposition 7.2. *The class of all Markov operators* $P : L^1(X) \to L^1(X)$ *is a convex set.*

Proof. Let P and Q be Markov operators and let $a \in [0, 1]$. Then the linear operator $aP + (1 - a)Q$ is positive and

$$\int_X (aP + (1 - a)Q)f d\mu = a \int_X Pf d\mu + (1 - a) \int_X Qf d\mu = \int_X f d\mu$$

for $0 \le f \in L^1(X)$. Hence, $aP + (1 - a)Q$ is a Markov operator.   □

Several elementary inequalities of general Markov operators are listed in the following proposition.

Proposition 7.3. *Let* $f \in L^1(X)$. *Then*

(i) $(Pf)^+ \le Pf^+$;
(ii) $(Pf)^- \le Pf^-$;
(iii) $|Pf| \le P|f|$;
(iv) $\|Pf\|_1 \le \|f\|_1$.

Proof Properties (i)-(iii) are just direct consequences of Proposition 5.7. Property (iv) is from

$$\|Pf\|_1 = \int_X |Pf| d\mu \le \int_X P|f| d\mu = \int_X |f| d\mu = \|f\|_1.$$ □

Remark 7.2. Property (iv) means that P is a *weak contraction*, and so its operator norm $\|P\|_1 = 1$ since $P\mathcal{D} \subset \mathcal{D}$.

Proposition 7.4. *Suppose that $f \in L^1(X)$. Then $Pf = f$ if and only if both $Pf^+ = f^+$ and $Pf^- = f^-$.*

Proof The sufficiency part of the proposition is obvious. For the necessity part, suppose that $Pf = f$. Then

$$f^+ = (Pf)^+ \le Pf^+ \text{ and } f^- = (Pf)^- \le Pf^-$$

by Proposition 7.3. Hence

$$\int_X (Pf^+ - f^+)d\mu + \int_X (Pf^- - f^-)d\mu$$

$$= \int_X (Pf^+ + Pf^-)d\mu - \int_X (f^+ + f^-)d\mu$$

$$= \int_X P|f|d\mu - \int_X |f|d\mu = \|P|f|\|_1 - \||f|\|_1 = 0.$$

Since both the integrands $Pf^+ - f^+$ and $Pf^- - f^-$ are nonnegative, it follows that $Pf^+ = f^+$ and $Pf^- = f^-$. $\quad\square$

Corollary 7.1. *Let $f \in L^1(X)$. If $Pf = f$, then $P|f| = |f|$.*

We are interested in when $\|Pf\|_1 = \|f\|_1$ for a given $f \in L^1$. In order to obtain an equivalent condition for the actual equality, we need the concept of the support of a function.

Definition 7.3. The *support* of a function f is defined as

$$\operatorname{supp} f = \{x : f(x) \ne 0\}.$$

Proposition 7.5. *Let $f \in L^1(X)$. Then the equality $\|Pf\|_1 = \|f\|_1$ is valid if and only if $\operatorname{supp} Pf^+ \cap \operatorname{supp} Pf^- = \emptyset$.*

Proof Since $|a - b| < a + b$ if and only if $a > 0$ and $b > 0$ for given two real numbers $a \ge 0$ and $b \ge 0$, from the inequality

$$|Pf^+(x) - Pf^-(x)| \le Pf^+(x) + Pf^-(x), \ \forall \, x \in X,$$

we see that an equivalent condition for the equality

$$\int_X |Pf^+ - Pf^-|d\mu = \int_X Pf^+ d\mu + \int_X Pf^- d\mu$$

is that there is no set $A \in \Sigma$ with $\mu(A) > 0$ on which $Pf^+ > 0$ and $Pf^- > 0$, namely Pf^+ and Pf^- have disjoint supports. Since $f = f^+ - f^-$

and $|f| = f^+ + f^-$, the left-hand side integral is simply $\|Pf\|_1$, and the right-hand side is $\|Pf^+\|_1 + \|Pf^-\|_1 = \|f^+\|_1 + \|f^-\|_1 = \|f\|_1$, hence the proposition is proved. $\qquad\square$

A Markov operator P gives rise to a family of subsets

$$\Sigma_d = \{A \in \Sigma : (P')^n \chi_A = \chi_{B_n} \text{ for some } B_n \in \Sigma, \ n = 1, 2, \cdots\}.$$

That is, $(P')^n \chi_A$ is a characteristic function for any n for $A \in \Sigma_d$. A Markov operator is called *deterministic* if $\Sigma_d = \Sigma$. For example, the Frobenius-Perron operator in Example 7.7 is a deterministic Markov operator since $P'\chi_A = \chi_{S^{-1}(A)}$, which will be proved in the next chapter.

Proposition 7.6. Σ_d *is a σ-ring. Furthermore, if $A \in \Sigma_d$ then $P'\chi_A = \chi_B$ for some $B \in \Sigma_d$.*

Proof Let $\Sigma_n = \{A \in \Sigma : (P')^n \chi_A = \chi_B \text{ for some } B \in \Sigma\}$, $n = 1, 2, \cdots$. Then $\Sigma_d = \bigcap_{n=1}^{\infty} \Sigma_n$, so it is enough to show that each Σ_n is a σ-ring. Let $A_1, A_2 \in \Sigma_n$. We show $A_1 \cup A_2 \in \Sigma_n$ and $A_1 \setminus A_2 \in \Sigma_n$. Set $(P')^n \chi_{A_1} = \chi_{B_1}$ and $(P')^n \chi_{A_2} = \chi_{B_2}$. Then

$$\chi_{B_1} + \chi_{B_2} = (P')^n(\chi_{A_1} + \chi_{A_2}) \geq (P')^n \chi_{A_1 \cup A_2}$$
$$\geq \max\{(P')^n \chi_{A_1}, (P')^n \chi_{A_2}\} = \chi_{B_1 \cup B_2}.$$

Thus, since $(P')^n(\chi_{A_1} + \chi_{A_2}) \leq 1$, $x \in B_1 \cup B_2$ implies $(P')^n(\chi_{A_1} + \chi_{A_2})(x) = 1$ and $x \in (B_1 \cup B_2)^c$ implies $(P')^n(\chi_{A_1} + \chi_{A_2})(x) = 0$. Hence, $(P')^n(\chi_{A_1} + \chi_{A_2})$ is a characteristic function of $B_1 \cup B_2$, so $A_1 \cup A_2 \in \Sigma_n$.

Next, since $A_1 \setminus A_2 = (A_1 \cup A_2) \setminus A_2$, without loss of generality we may assume $A_1 \supset A_2$, so that $\chi_{A_1 \setminus A_2} = \chi_{A_1} - \chi_{A_2}$. Then

$$(P')^n \chi_{A_1 \setminus A_2} = \chi_{B_1} - \chi_{B_2} = \chi_{B_1 \setminus B_2}$$

since $B_1 \supset B_2$. Thus Σ_n is a ring. It remains to show that Σ_n is countably additive. Let $\{A_k\}$ be a sequence of monotone increasing sets in Σ_n that converges to A. Then $(P')^n \chi_A = \lim_{k \to \infty} (P')^n \chi_{A_k}$ exists and is a characteristic function.

To prove the second conclusion, we let $A \in \Sigma_d$. Then $P'\chi_A = \chi_B$ for some $B \in \Sigma$, so $(P')^n \chi_B = (P')^{n+1} \chi_A$ are characteristic functions for all $n = 1, 2, \cdots$. Therefore, $B \in \Sigma_d$. $\qquad\square$

Remark 7.3. Since X does not need to belong to Σ_d, the σ-ring Σ_d is not necessarily a σ-algebra in general.

For a given Markov operator P, a set $A \in \Sigma$ such that $P' \chi_A = \chi_A$ is called an *invariant set* of P. Denote by Σ_i the set of all the invariant sets of P. Then the same reasoning as in the proof to Proposition 7.6 gives

Proposition 7.7. Σ_i *is a σ-ring and $P = I$ on $L^1(X, \Sigma_i, \mu)$.*

7.3   Ergodic Theorems of Markov Operators

We continue to assume that $P : L^1 \to L^1$ is a Markov operator. At the first glance, it seems that finding a stationary density f^* of P can be done via repeated actions of P to an initially chosen density $f \in \mathcal{D}$. Indeed, given any $f \in L^1$, if the iterated function sequence $\{P^n f\}$ converges, then its limit must be a fixed point of P since P is continuous. The sequence $\{P^n f\}$, however, is not always convergent in general, even in the weak topology. What we need is an averaging process for the iterates $f, Pf, P^2 f, \cdots$. In other words, for $f \in \mathcal{D}$, we want to explore the asymptotic behavior of the sequence of the Cesáro averages

$$A_n f = A_n(P)f \equiv \frac{1}{n} \sum_{k=0}^{n-1} P^k f.$$

Clearly, if $\lim_{n \to \infty} A_n f = f^*$, then $Pf^* = f^*$. Since $\|P\|_1 = 1$,

$$\|A_n\|_1 \leq \frac{1}{n} \sum_{k=0}^{n-1} \|P^k\|_1 \leq \frac{1}{n} \sum_{k=0}^{n-1} \|P\|_1^k = 1.$$

As a matter of fact, $\|A_n\|_1 \equiv 1$ for all n since $A_n f \in \mathcal{D}$ for any $f \in \mathcal{D}$. Thus, Theorem 6.16 is applicable, which gives the following *abstract mean ergodic theorem for Markov operators.*

Theorem 7.1. *If for a given function $f \in L^1$, the sequence $\{A_n f\}$ is weakly pre-compact in L^1, then it converges strongly in L^1 to some function $f^* \in L^1$. In other words,*

$$\lim_{n \to \infty} \|A_n f - f^*\|_1 = 0.$$

The limit function f^* is a fixed point of P. Furthermore, if $f \in \mathcal{D}$, then $f^* \in \mathcal{D}$, so that f^* is a stationary density of P.

Several corollaries follow immediately by the criteria for the weak pre-compactness in Section B.4 of Appendix B.

Corollary 7.2. *If for some density $f \in L^1$ there is a function $g \in L^1$ and a positive integer n_0 such that*

$$P^n f \leq g, \quad \forall\, n \geq n_0,$$

then there exists a stationary density f^* for P.

Corollary 7.3. *If for some density $f \in L^1$ there are two real numbers $b > 0$ and $p > 1$ and an integer $n_0 > 0$ such that*

$$\|P^n f\|_p \leq b, \quad \forall\, n \geq n_0,$$

then P has a stationary density f^*.

Theorem 7.1 provides a powerful sufficient condition for the existence of a stationary density of a Markov operator. That is, for a density f, the weak pre-compactness of $\{A_n f\}$ implies its norm convergence to a stationary density. There is another way to prove the existence of stationary densities of Markov operators, based on the concept of quasi-compactness. This approach gives a sufficient condition for the strong pre-compactness of the sequence $\{A_n f\}$, which implies the weak one of $\{A_n f\}$.

Definition 7.4. Let X be a Banach space. A bounded linear operator $T : X \to X$ is said to be *quasi-compact* if there is a positive integer k and a compact operator $S : X \to X$ such that

$$\|T^k - S\| < 1.$$

Theorem 7.2. *Suppose that V is a subspace of L^1, which is also invariant under P, and suppose that $(V, \|\ \|_V)$ is a Banach space. If $P : (V, \|\ \|_V) \to (V, \|\ \|_V)$ is quasi-compact and the number sequence $\{\|P^n\|_V\}$ is uniformly bounded, then for any density $f \in V \cap \mathcal{D}$, there exists a stationary density $f^* \in V \cap \mathcal{D}$ of P such that*

$$\lim_{n \to \infty} \|A_n f - f^*\|_V = 0.$$

Proof By Definition 7.4, there is an integer $k > 0$ and a compact operator $S : (V, \|\ \|_V) \to (V, \|\ \|_V)$ such that $\|P^k - S\|_V < 1$. Denote $Q = P^k - S$. Then the assumption $\|Q\|_V < 1$ implies that the inverse $(I - Q)^{-1} : V \to V$ is bounded with the expression

$$(I - Q)^{-1} = \sum_{n=0}^{\infty} Q^n,$$

which is absolutely convergent. It follows that

$$A_n = (I - Q)^{-1} S A_n + (I - Q)^{-1}(A_n - P^k A_n). \tag{7.1}$$

Since the sequence $\{\|P^n\|_V\}$ is bounded, the identity

$$A_n - P^k A_n = \frac{I + P + \cdots + P^{k-1}}{n}(I - P^n)$$

leads to

$$\lim_{n\to\infty} \|A_n - P^k A_n\|_V$$

$$= \lim_{n\to\infty} \frac{1 + \|P\|_V + \cdots + \|P^{k-1}\|_V}{n}(1 + \|P^n\|_V) = 0.$$

Thus for any density $f \in V \cap \mathcal{D}$, we have

$$\lim_{n\to\infty} (I - Q)^{-1}(A_n - P^k A_n)f = 0.$$

On the other hand, since the operator $(I - Q)^{-1}S$ is compact and $\|A_n f\|_V$ are uniformly bounded, $\{(I-Q)^{-1}SA_n f\}$ has a limit point f^* in V, and by (7.1) f^* is also a limit point of $\{A_n f\}$. It follows that $\{A_n f\}$ is pre-compact and so weakly pre-compact in $(V, \|\ \|_V)$. Therefore Theorem 6.16 implies the conclusion. $\qquad\square$

The following corollary is a consequence of the Ionescu-Tulcea and Marinescu theorem [Boyarsky and Góra (1997)].

Corollary 7.4. *Under the additional condition that* $\|f\|_1 \leq M\|f\|_V$, $\forall f \in V$ *for some constant M, if there exist two positive numbers $\alpha < 1$ and C and a positive integer k such that*

$$\|P^k f\|_V \leq \alpha\|f\|_V + C\|f\|_1, \ \forall f \in V,$$

then P has a stationary density $f^* \in V$.

Proof By the Ionescu-Tulcea and Marinescu theorem, the operator $P : (V, \|\ \|_V) \to (V, \|\ \|_V)$ is quasi-compact. For any function $f \in V$, write a positive integer $n = rk + s$ with $0 \leq s < k$, then

$$\|P^n f\|_V = \|P^{rk+s} f\|_V = \|(P^k)^r P^s f\|_V$$
$$\leq \alpha\|(P^k)^{r-1} P^s f\|_V + C\|(P^k)^{r-1} P^s f\|_1$$
$$\leq \alpha^r \|P^s f\|_V + C \sum_{j=0}^{r-1} \alpha^j \|P^s f\|_1,$$

and hence $\|P^n f\|_V$ can be bounded by

$$\|P^s\|_V \|f\|_V + \frac{C}{1 - \alpha}\|f\|_1,$$

which is less than or equal to

$$\left(\max_{0 \le s < k} \|P^s\|_V + \frac{CM}{1-\alpha}\right) \|f\|_V.$$

So $\{\|P^n\|_V\}$ is bounded and Theorem 7.2 is applicable. □

We have proved that *the pre-compactness of $\{A_n(P)f\}$ for some $f \in \mathcal{D}$ implies the existence of a stationary density of P* for any Markov operator P. The next theorem gives a partial answer to an opposite question whether the existence of a stationary density for a Markov operator P gives some information about the asymptotic properties of the sequence $\{A_n(P)f\}$.

Theorem 7.3. *If a Markov operator P has a unique stationary density f^* such that $f^*(x) > 0$ on X a.e., then*

$$\lim_{n \to \infty} A_n f = f^*, \ \forall f \in \mathcal{D}.$$

Proof We first assume that $f \le cf^*$ for a positive constant c. Then, since A_n are positive operators for all n, we have

$$A_n f \le c A_n f^* = cf^*, \ \forall n.$$

Hence the function sequence $\{A_n f\}$ is weakly pre-compact and, by Theorem 7.1, is convergent to a stationary density which must be f^* since f^* is the unique stationary density of P. Thus, the theorem is proved when the quotient f/f^* is a bounded function.

In the general case, define a function f_c by letting

$$f_c(x) = \min\{f(x), cf^*(x)\}$$

for $x \in X$ with a constant $c > 0$. Since $f^*(x) > 0$ on X,

$$\lim_{c \to \infty} f_c(x) = f(x), \ \forall x \in X.$$

Thus, by the fact that $f_c \le f$ and the Lebesgue dominated convergence theorem, $\|f_c - f\|_1 \to 0$ and $\|f_c\|_1 \to \|f\|_1 = 1$ as $c \to \infty$. We now write

$$f = \frac{f_c}{\|f_c\|_1} + r_c, \tag{7.2}$$

where the remainder

$$r_c = \left(1 - \frac{1}{\|f_c\|_1}\right) f_c + f - f_c$$

converges strongly to the zero function as $c \to \infty$. Given any $\epsilon > 0$, there is $c > 0$ such that $\|r_c\|_1 < \epsilon$, and so

$$\|A_n r_c\|_1 \le \|r_c\|_1 < \frac{\epsilon}{2}. \tag{7.3}$$

On the other hand, since $\|f_c\|_1^{-1} f_c$ is a density which is bounded above by $c\|f_c\|_1^{-1} f^*$, the first part of the proof guarantees that there is a positive integer n_0 such that

$$\left\| A_n \left(\frac{f_c}{\|f_c\|_1} \right) - f^* \right\|_1 < \frac{\epsilon}{2}, \ \ \forall\, n \geq n_0. \tag{7.4}$$

Combining inequalities (7.3) and (7.4) with the decomposition (7.2), we immediately have the inequality

$$\|A_n f - f^*\|_1 < \epsilon, \ \ \forall\, n \geq n_0. \square$$

7.4 Asymptotic Periodicity of Markov Operators

We have studied some fundamental convergence theorems of the Cesáro averages sequence $\{A_n f\}$ for a Markov operator P, but nothing was said about the convergence problem of the sequence $\{P^n f\}$ of the iterates of f. Here we study the asymptotic properties of this iterated sequence. We shall state that, under some conditions, the sequence $\{P^n f\}$ is *asymptotically periodic*. In this section we always assume that (X, Σ, μ) is a finite measure space and $P : L^1 \to L^1$ is a Markov operator.

Definition 7.5. P is said to be *constrictive* if there exist two positive numbers δ and $\epsilon < 1$ such that for every density $f \in \mathcal{D}$, a positive integer $n_0(f)$ can be found such that

$$\int_A P^n f d\mu \leq \epsilon, \ \ \forall\, n \geq n_0(f), \ \ A \in \Sigma \text{ with } \mu(A) \leq \delta. \tag{7.5}$$

Since the left-hand side integral of the above inequality is always less than or equal to 1, condition (7.5) for constrictiveness means that eventually this integral cannot be close to 1 for sufficiently small sets A. Thus, for a constrictive Markov operator, it cannot happen that $P^n f$ will be eventually concentrated on a set of very small or vanishing measure.

The following two propositions provide sufficient conditions for the constrictiveness of Markov operators.

Proposition 7.8. *If there are $p > 1$ and $b > 0$ such that for every $f \in \mathcal{D}$, we have $P^n f \in L^p$ for sufficiently large n and*

$$\limsup_{n \to \infty} \|P^n f\|_p \leq b, \tag{7.6}$$

then P is constrictive.

Proof Condition (7.6) implies that there is $n_0(f)$ such that

$$\|P^n f\|_p \leq b + 1, \quad \forall\, n \geq n_0(f).$$

Thus, the family $\{P^n f\}$, for all $n \geq n_0(f)$ and $f \in \mathcal{D}$, is weakly pre-compact. Given any number $\epsilon \in (0,1)$, by Theorem B.21, there exists a number $\delta > 0$ such that

$$\int_A P^n f d\mu < \epsilon, \quad \forall\, n \geq n_0(f), \ A \in \Sigma \text{ with } \mu(A) < \delta. \quad\square$$

Proposition 7.9. *If there exist $h \in L^1$ and $\alpha < 1$ such that*

$$\limsup_{n \to \infty} \|(P^n f - h)^+\|_1 \leq \alpha, \quad \forall\, f \in \mathcal{D}, \tag{7.7}$$

then P is constrictive.

Proof Let $\epsilon = 4^{-1}(1 - \alpha) < 1$. There is $\delta > 0$ such that

$$\int_A h d\mu < \epsilon, \quad \forall\, A \in \Sigma \text{ with } \mu(A) < \delta.$$

From condition (7.7), for any $f \in \mathcal{D}$, there is $n_0(f)$ such that

$$\|(P^n f - h)^+\|_1 \leq \alpha + \epsilon, \quad \forall\, n \geq n_0(f).$$

Consequently, for all $A \subset X$ with $\mu(A) < \delta$ and $n \geq n_0(f)$,

$$\int_A P^n f d\mu \leq \int_A h d\mu + \alpha + \epsilon < \frac{1 + \alpha}{2} < 1. \quad\square$$

The geometric meaning of Proposition 7.9 is that, for those regions where $P^n f > h$, if the area of the difference between $P^n f$ and h is bounded above by $\alpha < 1$, then P is constrictive.

Remark 7.4. It is not necessary to check conditions (7.5)-(7.7) for all $f \in \mathcal{D}$ in the above results. Rather, it is sufficient to verify them for all the densities in a dense subset $\mathcal{D}_0$ of $\mathcal{D}$.

The following main result of this section on the asymptotic periodicity of constrictive Markov operators was proved in [Komornik and Lasota (1987)].

Theorem 7.4. (Komornik-Lasota) *Let $P : L^1(X) \to L^1(X)$ be a constrictive Markov operator. Then there is a positive integer r, two sequences of nonnegative functions $g_k \in \mathcal{D}$ and $h_k \in L^\infty, k = 1, \cdots, r$, and a bounded linear operator $Q : L^1(X) \to L^1(X)$ such that for each $f \in L^1(X)$, Pf may be written in the form of*

$$Pf(x) = \sum_{k=1}^{r} \left(\int_X f h_k d\mu \right) g_k(x) + Qf(x), \quad \forall\, x \in X. \tag{7.8}$$

Moreover, $g_1, \cdots, g_r$ and Q enjoy the following properties:

(i) supp $g_k \cap$ supp $g_j = \emptyset$ *for all $k \neq j$;*

(ii) *there is a permutation $\pi : \{1, \cdots, r\} \to \{1, \cdots, r\}$ such that $Pg_k = g_{\pi(k)}$ for each k. That is, P permutes all g_k;*

(iii) $\lim_{n \to \infty} \|P^n Qf\|_1 = 0$ *for every $f \in L^1(X)$.*

The decomposition (7.8) in Theorem 7.4 implies that for any positive integer n, $P^{n+1}f$ can be represented by

$$P^{n+1}f(x) = \sum_{k=1}^{r} \left(\int_X fh_k d\mu \right) g_{\pi^n(k)}(x) + Q_n f(x), \qquad (7.9)$$

where $Q_n = P^n Q$ and $\lim_{n \to \infty} \|Q_n f\|_1 = 0$. The terms of the summation in (7.9) are just permuted with actions of P, so

$$\sum_{k=1}^{r} \left(\int_X fh_k d\mu \right) g_{\pi^n(k)}(x) \qquad (7.10)$$

must be periodic with period $\leq r!$. Since π^n is a permutation of $\{1, \cdots, r\}$, the summation (7.10) can be rewritten as

$$\sum_{k=1}^{r} \left(\int_X fh_{\pi^{-n}(k)} d\mu \right) g_k(x),$$

where π^{-n} is the inverse permutation of π^n. Thus, the representation (7.9) has an equivalent form

$$P^{n+1}f(x) = \sum_{k=1}^{r} \left(\int_X fh_{\pi^{-n}(k)} d\mu \right) g_k(x) + Q_n f(x). \qquad (7.11)$$

In the above representation of P^{n+1}, since the basis functions $g_1, \cdots, g_r$ in the summation are fixed with disjoint supports, it is clear to see that each application of P re-distributes the coefficient of each g_k with $\int_X fh_{\pi^{-n}(k)} d\mu$.

The powers sequence $\{P^n\}$ for P which satisfies formula (7.8) will be called *asymptotically periodic*. Theorem 7.4 just says that if P is a constrictive Markov operator, then the sequence $\{P^n\}$ is asymptotically periodic.

A direct consequence of the above theorem is the following existence result of stationary densities.

Proposition 7.10. *A constrictive Markov operator $P : L^1(X) \to L^1(X)$ has a stationary density.*

Proof Define a density f^* by

$$f^*(x) = \frac{1}{r} \sum_{k=1}^{r} g_k(x),$$

where r and g_k are as given in Theorem 7.4. Because P permutes the densities $g_1, \cdots, g_r$, we have

$$Pf^*(x) = \frac{1}{r} \sum_{k=1}^{r} g_{\pi(k)}(x) = \frac{1}{r} \sum_{k=1}^{r} g_k(x) = f^*(x). \quad \square$$

When the Markov operator has a constant stationary density, we have the following corollary of Theorem 7.4.

Proposition 7.11. *Let* $P : L^1(X) \to L^1(X)$ *be a constrictive Markov operator such that* $P1 = 1$. *Then for any* n,

$$P^{n+1}f = \sum_{k=1}^{r} \left(\int_X f h_{\pi^{-n}(k)} d\mu \right) 1_k + Q_n f, \quad \forall f \in L^1(X), \quad (7.12)$$

where

$$1_k = \frac{\chi_{A_k}}{\mu(A_k)}, \quad k = 1, \cdots, r,$$

and the measurable sets $A_1, \cdots, A_r$ *form a finite partition of* X *with an additional property that* $\mu(A_{\pi^n(k)}) = \mu(A_k)$ *for each* k.

Proof From the expression (7.11) for $P^{n+1}f$,

$$1 = P^{n+1}1(x) = \sum_{k=1}^{r} \left(\int_X h_{\pi^{-n}(k)} d\mu \right) g_k(x) + Q_n 1(x). \quad (7.13)$$

Let l be the period of the summation portion of (7.13). Then $\pi^{-nl}(k) = k$ for all k, and therefore,

$$1 = P^{(n+1)l}1(x) = \sum_{k=1}^{r} \left(\int_X h_k d\mu \right) g_k(x) + Q_{nl}1(x).$$

Letting $n \to \infty$, we obtain

$$1 = \sum_{k=1}^{r} \left(\int_X h_k d\mu \right) g_k(x). \quad (7.14)$$

If we define $A_k = \text{supp } g_k$ for each k, then they are mutually disjoint. From (7.14), $\bigcup_{k=1}^{r} A_k = X$ and each g_k must be a constant function. In fact, it is easy to see that

$$g_k = \frac{\chi_{A_k}}{\int_X h_k d\mu}.$$

Since the functions g_k are densities, we have

$$\int_{A_k} g_k d\mu = 1 = \frac{\mu(A_k)}{\int_X h_k d\mu}.$$

Therefore, $\mu(A_k) = \int_X h_k d\mu$ and

$$g_k = \frac{\chi_{A_k}}{\mu(A_k)}.$$

This proves (7.12). Applying P^n to equation (7.14) gives

$$1 \equiv P^n 1(x) = \sum_{k=1}^{r} \left(\int_X h_k d\mu \right) P^n g_k(x)$$

$$= \sum_{k=1}^{r} \left(\int_X h_k d\mu \right) g_{\pi^n(k)}(x),$$

and the same argument as above implies that

$$g_{\pi^n(k)}(x) \equiv \frac{1}{\int_X h_k d\mu}, \quad \forall\, x \in A_{\pi^n(k)}.$$

Thus, g_k and $g_{\pi^n(k)}$ must have the same constant on their respective supports. So $\mu(A_{\pi^n(k)}) = \mu(A_k)$ for all n and k. $\qquad\square$

A stronger notion than the asymptotic periodicity of the function sequence $\{P^n f\}$ in the above is the asymptotic stability of the operator sequence $\{P^n\}$ in the following definition.

Definition 7.6. The sequence $\{P^n\}$ is said to be *asymptotically stable* if there exists $f^* \in \mathcal{D}$ such that

$$\lim_{n \to \infty} \|P^n f - f^*\|_1 = 0, \quad \forall\, f \in \mathcal{D}. \qquad (7.15)$$

Remark 7.5. If $\{P^n\}$ is asymptotically stable, then f^* in (7.15) is a unique stationary density of P.

The next theorem is a direct consequence of Theorem 7.4.

Theorem 7.5. *Let $P : L^1 \to L^1$ be constrictive. If there is a set A of positive measure with the property that for every $f \in \mathcal{D}$, there is a positive integer $n_0(f)$ such that*

$$P^n f(x) > 0, \quad \forall\, x \in A \text{ a.e.}, \ n \geq n_0(f), \qquad (7.16)$$

then $\{P^n\}$ is asymptotically stable.

Proof Since P is constrictive, the representation (7.8) is valid. We first deduce that $r = 1$ in (7.8) .

Assume $r > 1$. The asymptotic periodicity of $\{P^n f\}$ and the assumption (7.16) imply that there is an index $k \in \{1, \cdots, r\}$ such that A is not contained in supp g_k. Let l be the period of the permutation π. Then we have $P^{nl} g_k(x) = g_k(x)$ for all $n \geq 1$, which is a contradiction since $P^{nl} g_k(x)$ cannot be positive on A from the above. Thus, $r = 1$ and (7.11) becomes

$$P^{n+1} f(x) = \left(\int_X f h_1 d\mu \right) g_1(x) + Q_n f(x),$$

which implies that

$$\lim_{n \to \infty} P^n f = \left(\int_X f h_1 d\mu \right) g_1, \ \forall f \in L^1.$$

So, $\int_X f h_1 d\mu = 1$ if $f \in \mathcal{D}$. Hence, $\lim_{n \to \infty} P^n f = g_1$ for all $f \in \mathcal{D}$. Letting $f^* = g_1$ completes the proof. $\square$

If we introduce the concept of lower-bound functions, the constrictiveness assumption in Theorem 7.5 can be avoided.

Definition 7.7. A function $h \in L^1$ is called a *lower-bound function* for a Markov operator $P : L^1 \to L^1$ if

$$\lim_{n \to \infty} \|(P^n f - h)^-\|_1 = 0, \ \forall f \in \mathcal{D}. \tag{7.17}$$

If in addition we have $h \geq 0$ and $\|h\|_1 > 0$, then the lower-bound function h is said to be *nontrivial*.

Theorem 7.6. $\{P^n\}$ *is asymptotically stable if and only if there is a nontrivial lower-bound function for P.*

Proof If P is asymptotically stable, then

$$\lim_{n \to \infty} \|(P^n f - f^*)^-\|_1 \leq \lim_{n \to \infty} \|P^n f - f^*\|_1 = 0,$$

so $h = f^*$ is a nontrivial lower-bound function for P.

Conversely, suppose that (7.17) is satisfied and h is nontrivial. Pick an $f \in \mathcal{D}$ and choose a positive integer $n_0(f)$ such that

$$\|(P^n f - h)^-\|_1 < \frac{\|h\|_1}{4}, \ \forall n \geq n_0(f).$$

From $|a - b| = a - b + 2(a - b)^-$ we have

$$\|(P^n f - h)^+\|_1 \leq \|P^n f - h\|_1$$
$$\leq \|P^n f\|_1 - \|h\|_1 + 2\|(P^n f - h)^-\|_1$$
$$= 1 - \|h\|_1 + 2\|(P^n f - h)^-\|_1 \leq 1 - \|h\|_1 + \frac{\|h\|_1}{2}$$

and hence
$$\|(P^n f - h)^+\|_1 \le 1 - \frac{\|h\|_1}{2}, \quad \forall\, n \ge n_0(f),$$
which together with Proposition 7.9 yields that P is constrictive. Thus Theorem 7.4 holds. We show that $r = 1$ in the representation (7.8).

If $r \ge 2$, then there are at least two basis functions g_1 and g_2 in (7.8). Since the integer $r!$ is a multiple of the period of the permutation π, we see that $P^{nr!} g_k = g_k$ for $k = 1, 2$. Then (7.17) implies that
$$g_k = P^{nr!} g_k \ge h - (P^{nr!} g_k - h)^-, \quad k = 1, 2.$$
Since $h \ge 0$ and $h \ne 0$, the above inequality implies that the product function $g_1 g_2$ is positive on a nontrivial set, which is a contradiction to the property that $\operatorname{supp} g_1 \cap \operatorname{supp} g_2 = \emptyset$ in Theorem 7.4. Hence $r = 1$, and the asymptotic stability follows from (7.8) with $f^* = g_1$. $\qquad\square$

Proposition 7.12. *If the sequence $\{P^n\}$ is asymptotically stable and $Pf = f$ for some $f \in L^1$ with $\|f\|_1 = 1$, then $f = f^*$ or $f = -f^*$, where f^* is the unique stationary density of P.*

Proof By Proposition 7.4, both functions f^+ and f^- are fixed points of P. Assume that $f^+ \ne 0$. Then $\|f^+\|_1^{-1} f^+$ is a stationary density of P, therefore $\|f^+\|_1^{-1} f^+ = f^*$. Hence,
$$f^+ = \|f^+\|_1 f^*,$$
which is also true if $f^+ = 0$. Similarly, $f^- = \|f^-\|_1 f^*$. Thus,
$$f = f^+ - f^- = (\|f^+\|_1 - \|f^-\|_1) f^*.$$
The assumption $\|f\|_1 = 1 = \|f^*\|_1$ then implies that
$$\left| \|f^+\|_1 - \|f^-\|_1 \right| = 1,$$
and this completes the proof. $\qquad\square$

Finally, we introduce ergodicity, mixing, and exactness for Markov operators with a constant stationary density.

Definition 7.8. Let (X, Σ, μ) be a probability measure space and let $P : L^1 \to L^1$ be a Markov operator such that $P1 = 1$.

(i) P is said to be *ergodic* if the sequence $\{P^n f\}$ is Cesáro convergent to 1 for all $f \in \mathcal{D}$;

(ii) P is said to be *mixing* if the sequence $\{P^n f\}$ is weakly convergent to 1 for all $f \in \mathcal{D}$;

(iii) P is said to be *exact* if the sequence $\{P^n f\}$ is strongly convergent to 1 for all $f \in \mathcal{D}$.

As a further application of Theorem 7.4, we can determine the properties of ergodicity, mixing, and exactness for constrictive Markov operators which can be expressed in the form of the spectral decomposition (7.8). Recall that a permutation $\pi : \{1, \cdots, r\} \to \{1, \cdots, r\}$ is said to be *cyclical* if π has no proper invariant subset.

Theorem 7.7. *Let* (X, Σ, μ) *be a probability measure space and let* $P : L^1 \to L^1$ *be constrictive such that* $P1 = 1$. *Then* P *is ergodic if and only if the permutation* π *in (7.8) is cyclical.*

Proof Suppose that π is cyclical. Then, for the sequence of the Cesáro averages $A_n = n^{-1} \sum_{k=0}^{n-1} P^k$, with the help of the representation (7.12),

$$A_n f(x) = \sum_{j=1}^{r} \left(\frac{1}{n} \sum_{k=0}^{n-1} \int_X f \, h_{\pi^{-k}(j)} d\mu \right) 1_{A_j}(x) + \hat{Q}_n f(x),$$

where the remainder $\hat{Q}_n f$ is given by

$$\hat{Q}_n f = \frac{1}{n} \sum_{k=0}^{n-1} Q_k f, \quad Q_0 f = Pf - \sum_{j=1}^{r} \left(\int_X f \, h_j d\mu \right) 1_{A_j}.$$

Since for each j, the number sequence $\{\int_X f h_{\pi^{-k}}(j) d\mu\}$ is periodic as we have shown above, the sequence of the coefficients

$$\frac{1}{n} \sum_{k=0}^{n-1} \int_X f \, h_{\pi^{-k}(j)} d\mu \to c_j(f)$$

as $n \to \infty$, where the limit $c_j(f)$ must be independent of j because the summations are permutations of the same numbers for different j. This common limit is denoted by $c(f)$. Hence,

$$\lim_{n \to \infty} A_n f = \sum_{j=1}^{r} c(f) 1_{A_j}.$$

Since π is cyclical, Proposition 7.11 implies that $\mu(A_j) = r^{-1}$ and then $1_{A_j} = r\chi_{A_j}$ for all j. It follows that

$$\lim_{n \to \infty} A_n f = rc(f),$$

which implies that $\{P^n f\}$ is Cesáro convergent to 1 for all $f \in \mathcal{D}$, that is, P is ergodic by Definition 7.8.

Conversely, let P be ergodic. If π is not cyclical, then π has a non-empty proper invariant subset $J \subset \{1, \cdots, r\}$. Define

$$f(x) = \sum_{j \in J} 1_{A_j}(x) \neq 0.$$

Then

$$\lim_{n \to \infty} A_n f = \sum_{j \in J} c_j(f) 1_{A_j} \neq 0.$$

Since $\mu(\bigcup_{j \in J} A_j) < \mu(X) = 1$, the limit of $A_n f$ is not a constant function of x, therefore P cannot be ergodic, a contradiction. $\qquad\square$

Theorem 7.8. *Let (X, Σ, μ) be a probability measure space and let $P : L^1(X) \to L^1(X)$ be a constrictive Markov operator such that $P1 = 1$. Then P is exact if and only if $r = 1$ in (7.8).*

Proof We first prove the sufficiency part. Since $r = 1$,

$$P^{n+1} f(x) = \left(\int_X f \, h d\mu \right) 1 + Q_n f(x)$$

from (7.12). So

$$\lim_{n \to \infty} P^{n+1} f = \int_X f \, h d\mu.$$

In particular, $\{P^n f\}$ is strongly convergent to 1 for any $f \in \mathcal{D}$. Therefore, P is exact by Definition 7.8.

The necessity part can be proved with a weaker assumption that P is mixing. Suppose that $r > 1$ in (7.8). Let

$$f(x) = \frac{1}{\mu(A_1)} \chi_{A_1}(x).$$

Then $f \in \mathcal{D}$ and, from (7.9),

$$P^n f(x) = \frac{1}{\mu(A_1)} \chi_{A_{\pi^n(1)}}(x).$$

Since P is mixing, $\lim_{n \to \infty} P^n f = 1$ weakly. On the other hand,

$$\langle P^n f, \chi_{A_1} \rangle = \begin{cases} \frac{1}{\mu(A_1)}, & \text{if } \pi^n(1) = 1, \\ 0, & \text{if } \pi^n(1) \neq 1. \end{cases}$$

Hence $\{P^n f\}$ will converge weakly to 1 only if there is a natural number n_0 such that $\pi^n(1) = 1$ for all $n \geq n_0$. Since P is ergodic, π is a cyclical permutation by Theorem 7.7. Thus, $\pi^n(1)$ cannot always be 1 for sufficiently large n. This is a contradiction. $\qquad\square$

Remark 7.6. Theorem 7.8 means that for constrictive Markov operators, the concepts of mixing and exactness are equivalent.

7.5   Notes and Remarks

1. The concept and applications of Markov operators appear in almost all subjects in which probability and stochastic analysis are involved. For example, Markov processes, whether they are continuous time or discrete time ones, and whether their phase spaces are countable or continuous, define Markov operators in a natural way. A concise and important book on Markov operators is [Foguel (1969)], and more recent ones devoted to the study of Markov operators and related topics include [Lasota and Mackey (1994); Baladi (2000); Ding and Zhou (2009)].

2. The Ruelle operator as defined in Example 7.4, which will be reduced to the usual Frobenius-Perron operator when the potential function ϕ there is chosen to be $\phi(x) = -\ln|S'(x)|$ in the case of $K = [0,1]$, as well as the related notions of Gibbs measures and Sinai-Ruelle-Bowen measures are all important mathematical tools in the modern investigation of chaotic dynamical systems arising in physical sciences, such as thermodynamics of chaotic transformations [Beck and Schlögl (1993)], and have intimate connections to other useful concepts and quantities such as topological pressure and topological entropy. They are very useful in the study of equilibrium states of such physically significant systems [Bowen (1975); Ruelle (1978); Baladi (2000); Terhesiu and Froyland (2008)].

3. The mean ergodic theorem, Theorem 7.1 in Section 7.3, for general Markov operators is a main mathematical tool for establishing the convergence of the averages sequence of a density to a stationary density of a given Markov operator. There are many corollaries to this fundamental result in the study of Markov operators and more general bounded linear operators with unit norm; see, for example, Chapter 5 of [Lasota and Mackey (1994)] and the monograph [Krengel (1985)].

4. The definition and properties of quasi-compact operators were given in, for example, [Dunford and Schwartz (1957); Krengel (1985)]. A sufficient condition for the quasi-compactness is given by a classic result called the Ionescu-Tulcea and Marinescu theorem [Boyarsky and Góra (1997)].

5. Periodicity of Markov operators and more general positive operators which are weak contractions have been studied in such papers as [Lasota *et al.* (1984); Schaefer (1980); Keller (1982)]. Most results of Section 7.4 were contained in the monograph [Lasota and Mackey (1994)].

Exercises

7.1 The boundary value problem

$$-u'' + u = f(x), \ 0 \le x \le 1, \ u'(0) = u'(1) = 0$$

has a unique solution $u = u(x)$ defined on $[0,1]$ for every $f \in L^1(0,1)$. Show that the correspondence from the right-hand side function f to the solution u defines a Markov operator on $L^1(0,1)$.

7.2 Let $L^1(X)$ be a complex L^1-space and let $P : L^1(X) \to L^1(X)$ be a Markov operator. Show that $\|P\|_1 = 1$.

7.3 Let $P : L^1(X) \to L^1(X)$ be a Markov operator. Let $B \in \Sigma$ with $\mu(B) > 0$ be such that $P' \chi_{B^c} = 0$ on B. Show that the inclusion supp $f \subset B$ implies the inclusion supp $Pf \subset B$. As a consequence, $P : L^1(B) \to L^1(B)$ is a Markov operator.

7.4 Describe a general form of the matrix (a_{ij}) in Example 7.1 which corresponds to a deterministic Markov operator.

7.5 If P_1 and P_2 are both deterministic Markov operators, is their product $P_1 P_2$ also deterministic Markov operator? How about their convex combination $aP_1 + (1-a)P_2$, $0 < a < 1$?

7.6 Show that $P : L^1(0,1) \to L^1(0,1)$ defined by

$$Pf(x) = \frac{1}{2} f(x) + \frac{1}{4} f\left(\frac{x}{2}\right) + \frac{1}{4} f\left(\frac{x}{2} + \frac{1}{2}\right)$$

is not a deterministic Markov operator.

7.7 Let $P : L^1 \to L^1$ be a Markov operator. Prove that for any two nonnegative functions $f, g \in L^1$ the condition supp $f \subset$ supp g implies that supp $Pf \subset$ supp Pg.

7.8 Show that the converse of Corollary 7.1 is not necessarily true by providing a counterexample.

7.9 Let (X, Σ, μ) be a finite measure space and let $P : L^1 \to L^1$ be a Markov operator. Show that if there is a compact subset $\mathcal{F}$ of L^1 such that $\lim_{n\to\infty} \text{dist}(P^n f, \mathcal{F}) = 0$ for all $f \in \mathcal{D}$, where $\text{dist}(g, \mathcal{F}) = \sup_{f \in \mathcal{F}} \|g - f\|_1$, then P is constrictive.

7.10 Show that the converse of Exercise 7.9 is also true. That is, if $P : L^1 \to L^1$ is a constrictive Markov operator, then there exists a compact set $\mathcal{F} \subset L^1$ such that $\lim_{n\to\infty} \text{dist}(P^n f, \mathcal{F}) = 0$ for all $f \in \mathcal{D}$.

7.11 Let $\mathcal{D}_0$ be a dense subset of $\mathcal{D}$ and let $\mathcal{F}$ be a subset of $L^1(X)$. Show that if $\lim_{n\to\infty} \text{dist}(P^n f, \mathcal{F}) = 0$ for all densities $f \in \mathcal{D}_0$, then the same is true for all densities $f \in \mathcal{D}$.

7.12 Let $P : L^1 \to L^1$ be a Markov operator. Show that if P^k is constrictive for some $k > 1$, then P is constrictive.

7.13 Let (X, Σ, μ) be a probability measure space and $P : L^1 \to L^1$ a Markov operator such that $P1 = 1$. Assume $k > 1$. Show that if P^k is ergodic, mixing, or exact, then so is P, respectively.

7.14 Let X be a Banach space. Show that $T \in B(X)$ is quasi-compact if and only if there exists a sequence $\{T_n\}$ of compact operators such that $\lim_{n\to\infty} \|T^n - T_n\| = 0$.

7.15 Show that any power T^k of a quasi-compact operator $T \in B(X)$ is also quasi-compact.

7.16 Let X be a Banach space and let a bounded linear operator $T \in B(X)$ be quasi-compact. Show that $\lim_{n\to\infty} A_n(T) = E$ exists for some $E \in B(X)$ under the operator norm of $B(X)$.

7.17 Show that a Markov operator $P : L^1(X) \to L^1(X)$ is continuous not only in the strong topology but also in the weak topology of $L^1(X)$.

7.18 Let $P : L^1(\mu) \to L^1(\mu)$ be a Markov operator, let $g \in \mathcal{D}$, and let $\epsilon \in (0, 1)$. Define an operator $P_\epsilon : L^1(\mu) \to L^1(\mu)$ by

$$P_\epsilon f(x) = (1 - \epsilon)Pf(x) + \epsilon \int_X f d\mu \cdot g(x), \quad \forall f \in L^1(\mu).$$

Show that

(i) P_ϵ is a Markov operator;

(ii) P_ϵ has a unique stationary density

$$f_\epsilon = \epsilon \sum_{n=0}^{\infty} (1 - \epsilon)^n P^n g;$$

(iii) $\{P_\epsilon^n\}$ is asymptotically stable;

(iv) if $\lim_{\epsilon \to 0} f_\epsilon = f^*$ in the weak topology or in the norm of $L^1(\mu)$, then f^* is a stationary density of P.

7.19 Let $S : \mathbb{R}^n \to \mathbb{R}^n$ be a nonsingular transformation and let a kernel function w be defined by $w(x, y) = g(x - S(y))$, where $g \in \mathcal{D}(\mathbb{R}^n) \cap L^\infty(\mathbb{R}^n)$. Define an operator $P : L^1(\mathbb{R}^n) \to L^1(\mathbb{R}^n)$ by

$$Pf(x) = \int_{\mathbb{R}^n} w(x, y)f(y)dy, \quad \forall f \in L^1(\mathbb{R}^n).$$

Show that

(i) P is a Markov operator;

(ii) $Pf(x) = \int_{\mathbb{R}^n} P_S f(x - y)g(y)dy$, where P_S is the Frobenius-Perron operator associated with S, as defined in Example 7.7.

7.20 Let a Markov operator $P : L^1(\mathbb{R}) \to L^1(\mathbb{R})$ be defined by

$$Pf(x) = \int_{-\infty}^{\infty} w(x,y)f(y)dy, \quad \forall f \in L^1(\mathbb{R}),$$

where the stochastic kernel

$$w(x,y) = \frac{1}{\sqrt{2\pi}} \exp\left(-\frac{(x-y)^2}{2}\right).$$

(i) Find the expression for $P^n f$.

(ii) Show that $\lim_{n \to \infty} P^n f = 0$ for all $f \in \mathcal{D}$.

(iii) Explain why P has no stationary density.

7.21 Let $S : \mathbb{R}_+ \to \mathbb{R}_+$ be a positive continuous function and $g \in \mathcal{D}(\mathbb{R}_+)$. Define $P : L^1(\mathbb{R}_+) \to L^1(\mathbb{R}_+)$ by

$$Pf(x) = \int_0^{\infty} w(x,y)f(y)dy, \quad \forall f \in L^1(\mathbb{R}_+),$$

where the stochastic kernel

$$w(x,y) = g\left(\frac{x}{S(y)}\right)\frac{1}{S(y)}.$$

Show that P is a Markov operator.

7.22 Let $E(\cdot \,|S^{-1}\Sigma)$ be the conditional expectation with respect to the σ-subalgebra $S^{-1}\Sigma$. Show that the Frobenius-Perron operator P_S defined in Example 7.7 satisfies the equality

$$(P_S f) \circ S = E(f|S^{-1}\Sigma).$$

7.23 Show that if f is a stationary density of the Frobenius-Perron operator P_S associated with a nonsingular transformation $S : X \to X$ and if $S^{-1}(A) = A$ is satisfied by a measurable subset A of X, then the function $f \cdot \chi_A$ is a fixed point of P_S.

7.24 Let (X, Σ, μ) be a σ-finite measure space, let $S : X \to X$ be a nonsingular transformation, and let $P_S : L^1 \to L^1$ be the corresponding Frobenius-Perron operator. Show that the associated transition function $p(x, A) = \chi_{S^{-1}(A)}(x)$ and the dual operator

$$P_S' g(x) = \int_X g(y)P(x,dy) = g(S(x))$$

is the same as the composition operator U_S with domain L^∞.

Chapter 8

Frobenius-Perron Operators

In this chapter we shall concentrate on Frobenius-Perron operators which constitute a particularly important subclass of Markov operators. The Frobenius-Perron operator associated with a nonsingular transformation governs the evolution of the density functions under the action of the given transformation, and its stationary densities give rise to absolutely continuous invariant probability measures which describe the statistical properties of the deterministic dynamical system.

In Section 8.1 we introduce basic concepts of ergodic theory that motivate the definition of Frobenius-Perron operators. In Section 8.2 we study general properties of Frobenius-Perron operators. Several existence results of stationary densities of Frobenius-Perron operators for different classes of transformations will be presented in Section 8.3, and Section 8.4 is devoted to the ergodic properties of such invariant measures.

8.1 Measure-preserving Transformations

We assume that (X, Σ, μ) is a σ-finite measure space and $S : X \to X$ is a measurable transformation such that the measure space $(X, S^{-1}\Sigma, \mu)$ is σ-finite. Recall that if (6.6) is satisfied, then μ is called S-invariant or S preserves μ. If (6.6) is valid for all measurable sets in a subclass π of Σ which is closed under the intersection operation of its members and which *generates* Σ in the sense that Σ is the smallest σ-algebra containing π, then μ is S-invariant [Billingsley (1968)]. The subclass π with the above property is called a *π-system*. Another equivalent condition for the S-invariance of μ is given by the following result, the proof of which is based on the *change of variables formula* to be presented by Lemma 8.1 later on and which is also a consequence of the relationship between Frobenius-Perron operators and Koopman operators in the forthcoming section.

Proposition 8.1. *Let* (X, Σ, μ) *be a finite measure space. Then* $S : X \to X$ *is* μ-preserving if and only if

$$\int_X g(x)\mu(dx) = \int_X g(S(x))\mu(dx), \ \forall \, g \in L^\infty.$$

We give examples of measure-preserving transformations.

Example 8.1. The identity transformation I of any measurable space (X, Σ) is measure-preserving with respect to any measure μ on X.

Example 8.2. The *tent map* $T : [0, 1] \to [0, 1]$ defined by

$$T(x) = \begin{cases} 2x, & 0 \le x \le \frac{1}{2}, \\ 2(1 - x), & \frac{1}{2} < x \le 1 \end{cases}$$

is measure-preserving with respect to the Lebesgue measure m.

Example 8.3. Let $k \ge 2$ be an integer. Then the *k-adic map* $S : [0, 1] \to [0, 1]$ defined by $S(x) = kx \pmod 1$ is m-preserving.

Proof It is easy to see that the family of all closed intervals $[a, b] \subset [0, 1]$ is a π-system and generates the Borel σ-algebra of $[0, 1]$. The inverse image $S^{-1}([a, b])$ is a disjoint union of k subintervals of length $(b - a)/k$, so $m(S^{-1}([a, b])) = m([a, b])$. Thus, m is S-invariant. □

Example 8.4. The *translation map* $S(x) = x + a \pmod 1$ preserves the Lebesbue measure of the interval $[0, 1]$. This can be shown by using the same argument as in the above example.

Example 8.5. The quadratic mapping $S(x) = 4x(1 - x)$, $x \in [0, 1]$, referred to as the *logistic model*, preserves the absolutely continuous probability measure μ on $[0, 1]$, given by

$$\mu(A) = \int_A \frac{1}{\pi\sqrt{x(1 - x)}} dx, \ \forall \, m - \text{measurable } A \subset [0, 1].$$

Proof We use Proposition 8.1 since verification of (6.6) is tedious even for $A = [a, b]$. A direct computation shows that

$$\int_0^1 g(x)\mu(dx) = \int_0^1 g(x)\frac{1}{\pi\sqrt{x(1 - x)}} dx = \frac{2}{\pi} \int_0^{\frac{\pi}{2}} g(\sin^2 \theta)d\theta$$

and

$$\int_0^1 g(S(x))\mu(dx) = \int_0^1 g(4x(1-x))\frac{1}{\pi\sqrt{x(1-x)}}dx$$

$$= \frac{2}{\pi}\int_0^{\frac{\pi}{2}} g(\sin^2(2\theta))d\theta = \frac{1}{\pi}\int_0^\pi g(\sin^2 t)dt = \frac{2}{\pi}\int_0^{\frac{\pi}{2}} g(\sin^2 t)dt,$$

therefore the result follows. □

Example 8.6. Given $a \in \partial\mathbb{D}^2$, then the *rotation map* $R(z) = az$ preserves the *Harr measure* on the unit circle $\partial\mathbb{D}^2$.

Proof This is from the facts that the class of closed sub-arcs of $\partial\mathbb{D}^2$ is a π-system and generates the Borel σ-algebra of $\partial\mathbb{D}^2$ and that the rotation preserves the length of the arcs. □

Remark 8.1. By the same token, any *translation map* $S(x) = ax$ on a compact group G preserves the Harr measure on G, where a is a fixed element of G. More generally, any continuous isomorphism of G preserves the Harr measure on G.

Example 8.7. (Hénon's map) Let $X = [0,1] \times [0,1]$ and let $\beta > 0$. The mapping $S : X \to X$ defined by

$$S(x,y) = (4x(1-x) + y, \beta x) \pmod{1}$$

is called a *Hénon's map*. It can be shown that Hénon's map preserves the Lebesgue measure of the unit square (see Remark 8.2 and Exercise 8.5).

In this chapter we are mainly interested in nonsingular transformations for the purpose of defining Frobenius-Perron operators in the next section. By its definition in Section 6.2, a measurable transformation $S : X \to X$ is nonsingular if $\mu(S^{-1}(A)) = 0$ for all $A \in \Sigma$ such that $\mu(A) = 0$. Note that every measure-preserving transformation is nonsingular with respect to the invariant measure μ. For a nonsingular transformation $S : (X, \Sigma, \mu) \to (X, \Sigma, \mu)$, the problem that we shall be concerned with is to find an invariant finite measure that is absolutely continuous with respect to μ. By the Radon-Nikodym theorem (Theorem B.4), a finite measure ν is absolutely continuous with respect to μ if and only if there exists $f \in L^1(\mu)_+$ such that $\nu(A) = \int_A f d\mu$ for all $A \in \Sigma$.

In order to motivate the definition of the Frobenius-Perron operator associated with S, we assign to each measure ν on X a new measure $\nu \circ S^{-1}$ defined by $\nu \circ S^{-1}(A) = \nu(S^{-1}(A))$. This defines an operator from measures onto measures. It is clear that every fixed point of this operator is an invariant measure of S. Since we are only interested in absolutely continuous invariant finite measures which are most important in applications, we restrict this operator to the set of all finite measures on X which are absolutely continuous with respect to μ. From the Radon-Nikodym theorem, this set is equivalent to L^1_+. This motivates the definition of Frobenius-Perron operators in the next section.

8.2   Frobenius-Perron Operators

Throughout the section we assume that $S : X \to X$ is a nonsingular transformation. The corresponding Frobenius-Perron operator to be defined in the following gives the evolution of densities governed by the underlying deterministic discrete dynamical system.

Any $f \in L^1(X)_+$ defines a set function

$$\mu_f(A) = \int_{S^{-1}(A)} f d\mu, \ \ \forall \, A \in \Sigma.$$

Since S is nonsingular, $\mu(A) = 0$ implies $\mu_f(A) = 0$. Thus the Radon-Nikodym theorem ensures that there exists a unique function $\hat{f} \in L^1(X)_+$, which will be denoted as $P_S f$, such that

$$\mu_f(A) = \int_A \hat{f} d\mu, \ \ \forall \, A \in \Sigma.$$

The resulting mapping from $L^1(X)_+$ into itself is obviously additive, so Theorem 5.5 guarantees its extension as a positive operator on $L^1(X)$.

Definition 8.1. The operator $P_S : L^1(X) \to L^1(X)$ defined by

$$\int_A P_S f d\mu = \int_{S^{-1}(A)} f d\mu, \ \ \forall \, A \in \Sigma, \ f \in L^1 \tag{8.1}$$

is called the *Frobenius-Perron operator* associate with S.

From its definition, it is straightforward to see that the Frobenius-Perron operator P_S has the following properties:

(i) P_S is a positive operator;

(ii) P_S is *integral-preserving*, that is,

$$\int_X P_S f d\mu = \int_X f d\mu, \ \ \forall \, f \in L^1;$$

(iii) $P_{S_1 \circ S_2} = P_{S_1} P_{S_2}$ for nonsingular transformations S_1 and S_2, so $P_{S^n} = (P_S)^n$ for any positive integer n.

Properties (i) and (ii) show that Frobenius-Perron operators are Markov operators. The importance of introducing the concept of Frobenius-Perron operators is reflected in the following theorem.

Theorem 8.1. *Let $f \in L^1(\mu)_+$. The finite measure μ_f defined by*

$$\mu_f(A) = \int_A f d\mu, \quad \forall A \in \Sigma$$

is invariant under S if and only if f is a fixed point of P_S.

Proof μ_f is S-invariant, that is,

$$\mu_f(A) = \mu_f(S^{-1}(A)), \quad \forall A \in \Sigma$$

if and only if, by the definition of μ_f,

$$\int_A f d\mu = \int_{S^{-1}(A)} f d\mu, \quad \forall A \in \Sigma,$$

which is true if and only if, from the definition (8.1) of P_S,

$$\int_A f d\mu = \int_A P_S f d\mu, \quad \forall A \in \Sigma.$$

The above integral equality is valid if and only if $P_S f = f$ from real analysis [Natanson (1961); Rudin (1986)]. $\qquad \square$

Remark 8.2. Note from the theorem that the original measure μ is S-invariant if and only if $P_S 1 = 1$.

Definition 8.1 of Frobenius-Perron operators is implicit, but sometimes its explicit expression can be obtained. For example, if $X = [a, b]$ and μ is the Lebesgue measure m, then by (8.1),

$$\int_a^x P_S f dm = \int_{S^{-1}([a,x])} f dm, \quad \forall x \in [a, b].$$

Taking derivative with respect to x gives

$$P_S f(x) = \frac{d}{dx} \int_{S^{-1}([a,x])} f dm, \quad \forall x \in [a, b] \text{ a.e.} \tag{8.2}$$

If in addition S is differentiable and monotonic, then we have

$$P_S f(x) = f(S^{-1}(x)) \left| \frac{d}{dx} S^{-1}(x) \right|. \tag{8.3}$$

Now let X be an N-dimensional rectangle $[a_1, b_1] \times \cdots \times [a_N, b_N]$. Then, differentiating the following equality

$$\int_{\prod_{k=1}^N [a_k, x_k]} P_S f dm = \int_{S^{-1}(\prod_{k=1}^N [a_k, x_k])} f dm$$

with respect to $x_1, \cdots, x_N$ successively gives

$$P_S f(x_1, \cdots, x_N) = \frac{\partial^N}{\partial x_1 \cdots \partial x_N} \int_{S^{-1}(\prod_{k=1}^N [a_k, x_k])} f \, dm. \qquad (8.4)$$

Before studying more properties of Frobenius-Perron operators, we give expressions of Frobenius-Perron operators associated with some well-known nonsingular transformations.

Example 8.8. (The logistic model) Let $S : [0,1] \to [0,1]$ be the logistic model $S(x) = 4x(1-x)$ from Example 8.5. Then

$$S^{-1}([0,x]) = \left[0, \frac{1}{2}\left(1 - \sqrt{1-x}\right)\right] \bigcup \left[\frac{1}{2}\left(1 + \sqrt{1-x}\right), 1\right],$$

thus the related Frobenius-Perron operator P_S is given by

$$P_S f(x) = \frac{1}{4\sqrt{1-x}}\left\{ f\left(\frac{1 - \sqrt{1-x}}{2}\right) + f\left(\frac{1 + \sqrt{1-x}}{2}\right)\right\}.$$

It was shown in [Ulam and von Neumann (1947)] that P_S has a unique stationary density which has the expression

$$f^*(x) = \frac{1}{\pi\sqrt{x(1-x)}},$$

hence the absolutely continuous probability measure μ_{f^*} defined by

$$\mu_{f^*}(A) = \int_A \frac{dx}{\pi\sqrt{x(1-x)}}, \quad \forall\, A \in \mathcal{B}$$

is invariant under the quadratic mapping $S(x) = 4x(1-x)$.

Example 8.9. (The baker transformation). Let $X = [0,1] \times [0,1]$. The *baker transformation* $S : X \to X$ is defined by

$$S(x,y) = \begin{cases} \left(2x, \frac{1}{2}y\right), & 0 \le x < \frac{1}{2}, \ 0 \le y \le 1, \\ \left(2x-1, \frac{1}{2}(y+1)\right), & \frac{1}{2} \le x \le 1, \ 0 \le y \le 1. \end{cases}$$

Since $S^{-1}([0,x] \times [0,y]) = [0, x/2] \times [0, 2y]$ for $0 \le y < 1/2$,

$$P_S f(x,y) = \frac{\partial^2}{\partial x \partial y} \int_0^{\frac{1}{2}x} ds \int_0^{2y} f(s,t)\,dt = f\left(\frac{1}{2}x, 2y\right)$$

for $0 \le y < 1/2$ by (8.4). If $1/2 \le y \le 1$, then we have

$$S^{-1}([0,x] \times [0,y])$$

$$= \left(\left[0, \frac{1}{2}x\right] \times [0,1]\right) \cup \left(\left[\frac{1}{2}, \frac{1}{2}(1+x)\right] \times [0, 2y-1]\right),$$

and it follows from (8.4) that

$$P_S f(x,y)$$

$$= \frac{\partial^2}{\partial x \partial y} \left\{ \int_0^{\frac{1}{2}x} ds \int_0^1 f(s,t)dt + \int_0^{\frac{1}{2}(1+x)} ds \int_0^{2y-1} f(s,t)dt \right\}$$

$$= f\left(\frac{1}{2}(1+x), 2y-1 \right), \quad \frac{1}{2} \le y \le 1.$$

In summary, we obtain the explicit expression

$$P_S f(x,y) = \begin{cases} f\left(\frac{1}{2}x, 2y \right), & 0 \le y < \frac{1}{2}, \\ f\left(\frac{1}{2}(1+x), 2y-1 \right), & \frac{1}{2} \le y \le 1. \end{cases}$$

Since $P_S 1 = 1$, the Lebesgue measure m is invariant under S.

Example 8.10. (The Anosov diffeomorphism). The mapping

$$S(x,y) = (x+y, x+2y) \pmod 1$$

is called the *Anosov diffeomorphism*. It is invertible and

$$S^{-1}(x,y) = (2x-y, y-x) \pmod 1.$$

Hence S preserves the Lebesgue measure since

$$P_S f(x,y) = f(2x - y \pmod 1), y - x \pmod 1).$$

Example 8.11. (Fractal geometry) Many well-known fractals, such as the Sierpiński triangle, are generated by the iterated functions systems (IFS). An IFS is a set of r contractions $\{S_1, \cdots, S_r\}$ with corresponding probabilities $\{p_1, \cdots, p_r\}$ on a metric space (X, d). It gives a *random dynamical system*: $x_{n+1} = S_{\alpha_n}(x_n)$ is determined with probability p_{α_n} for $n \ge 0$. Asymptotic statistical properties of such iterates can be determined by an invariant probability measures μ^* of the *Markov operator on measures* $M \equiv \sum_{k=1}^r p_k M_k$, where $M_k \mu = \mu \circ S_k^{-1}$ is the *Frobenius-Perron operator on measures*, which will be an important tool for Section 9.2. Applying Banach's fixed point theorem to the Hutchinson metric [Barnsley (1988)] proves the existence and the uniqueness of μ^*. Moreover,

$$\operatorname{supp} \mu^* = \bigcup_{k=1}^r S_k(\operatorname{supp} \mu^*).$$

In other words, the support is exactly the *invariant set* of the IFS. The invariant set of an IFS is often a fractal. For example, a Sierpiński pedal triangle is the invariant set of the IFS consisting of three simple affine contraction mappings determined by the three angles of the initial triangle. In many cases the invariant measure of M is absolutely continuous, so its density is a fixed point of the Markov operator $P = \sum_{k=1}^r p_k P_{S_k}$, where P_{S_k} is the Frobenius-Perron operator associated with S_k.

The following result relates the support of $P_S f$ to that of f.

Proposition 8.2. *Let $A \in \Sigma$ and let $f \in L^1$ be nonnegative. Then $f(x) = 0$ for all $x \in S^{-1}(A)$ if and only if $P_S f(x) = 0$ for all $x \in A$, and in particular, $S^{-1}(\text{supp } P_S f) \supset \text{supp } f$.*

Proof From the definition (8.1) of P_S,

$$\int_X \chi_A P_S f d\mu = \int_X \chi_{S^{-1}(A)} f d\mu.$$

Suppose that $f \in L^1$ is nonnegative. Then $P_S f = 0$ on A implies that $f = 0$ on $S^{-1}(A)$, and vice versa. Since

$$\int_{\text{supp } f} f d\mu = \int_{\text{supp } P_S f} P_S f d\mu = \int_{S^{-1}(\text{supp } P_S f)} f d\mu,$$

we have $S^{-1}(\text{supp } P_S f) \supset \text{supp } f$. $\square$

The simple formula (8.3) above provides an explicit expression of the Frobenius-Perron operator for one dimensional invertible mappings. For a general invertible transformation S, based on the following change of variables formula, we can also obtain an explicit expression for the corresponding Frobenius-Perron operator.

Lemma 8.1. *Let (X, Σ, μ) be a measure space. If f is a nonnegative measurable function or $f \in L^1(\mu)$, then for every $A \in \Sigma$,*

$$\int_{S^{-1}(A)} f \circ S \, d\mu = \int_A f d(\mu \circ S^{-1}) = \int_A f h d\mu,$$

where h is the Radon-Nikodym derivative of $\mu \circ S^{-1}$ with respect to μ, in other words,

$$\mu \circ S^{-1}(A) = \int_A h d\mu, \quad \forall A \in \Sigma.$$

Proof Let $f = \chi_B$ with $B \in \Sigma$. Then $f \circ S = \chi_{S^{-1}(B)}$, and

$$\int_{S^{-1}(A)} \chi_B \circ S \, d\mu = \int_X \chi_{S^{-1}(A)} \chi_{S^{-1}(B)} d\mu$$
$$= \mu(S^{-1}(A) \cap S^{-1}(B)) = \mu(S^{-1}(A \cap B)).$$

Note that the second integral of the lemma may be rewritten as

$$\int_A \chi_B d(\mu \circ S^{-1}) = \int_X \chi_A \chi_B d(\mu \circ S^{-1}) = \mu(S^{-1}(A \cap B))$$

and the last integral of the lemma becomes

$$\int_A \chi_B h d\mu = \int_{A \cap B} h d\mu = \mu(S^{-1}(A \cap B)).$$

Hence, the lemma is true for all simple functions, and the limit process gives rise to the required result. □

Proposition 8.3. *Let $S : X \to X$ be invertible such that S^{-1} is also nonsingular. Then for every $f \in L^1$,*

$$P_S f(x) = f(S^{-1}(x))h(x), \qquad (8.5)$$

where h is the Radon-Nikodym derivative of the measure $\mu \circ S^{-1}$ with respect to the measure μ.

Proof Let $A \in \Sigma$. Then, from the definition of P_S,

$$\int_A P_S f(x)\mu(dx) = \int_{S^{-1}(A)} f(x)\mu(dx).$$

Letting $x = S^{-1}(y)$ in the right-hand side integral gives

$$\int_{S^{-1}(A)} f(x)\mu(dx) = \int_A f(S^{-1}(y))h(y)\mu(dy).$$

Thus,

$$\int_A P_S f(x)\mu(dx) = \int_A f(S^{-1}(x))h(x)\mu(dx),$$

which implies (8.5) since $A \in \Sigma$ is arbitrary. □

The composition operator with $L^\infty(X)$ as its domain is closely related to the Frobenius-Perron operator and it was named after Koopman who first used it in [Koopman (1931)].

Definition 8.2. The operator $U_S : L^\infty(X) \to L^\infty(X)$ defined by

$$U_S g(x) = g(S(x)), \ \forall g \in L^\infty(X)$$

is called the *Koopman operator* with respect to S.

Because of the non-singularity of S, the Koopman operator is well-defined since $f_1(x) = f_2(x)$ μ-a.e. implies that $f_1(S(x)) = f_2(S(x))$ μ-a.e. Some basic properties of U_S are listed below:

(i) U_S is a positive operator;
(ii) $\|U_S f\|_\infty \leq \|f\|_\infty$ for all $f \in L^\infty(X)$;
(iii) $U_{S_1 \circ S_2} = U_{S_2} U_{S_1}$. In particular, $U_{S^n} = (U_S)^n$.

Proposition 8.4. $U_S = P'_S$. That is,

$$\int_X (P_S f)g d\mu = \int_X f U_S g d\mu, \ \forall f \in L^1(X), g \in L^\infty(X).$$

Proof Let $f \in L^1(X)$. For any $A \in \Sigma$, we have

$$\int_X (P_S f) \chi_A d\mu = \int_A P_S f d\mu = \int_{S^{-1}(A)} f d\mu$$

$$= \int_X f \chi_A \circ S \, d\mu = \int_X f U_S \chi_A \, d\mu,$$

which means that

$$\int_X (P_S f) g d\mu = \int_X f U_S g d\mu$$

is true for all simple functions $g \in L^\infty$. Now, let $g \in L^\infty$. Then $g = \lim_{n\to\infty} g_n$ for a sequence of simple functions $g_n \in L^\infty$. So,

$$\int_X (P_S f) g d\mu = \lim_{n\to\infty} \int_X (P_S f) g_n d\mu$$

$$= \lim_{n\to\infty} \int_X f U_S g_n d\mu = \int_X f U_S g d\mu. \quad \square$$

At the end of this section we state two decomposition theorems for Frobenius-Perron operators and Koopman operators, which show that such operators are *partially isometric* on their respective domains. Their proofs are referred to [Ding and Hornor (1994); Ding (1997)].

Theorem 8.2. *Let* $P_S : L^1(\Sigma) \to L^1(\Sigma)$ *be the Frobenius-Perron operator associated with* $S : X \to X$. *Then*

$$L^1(\Sigma) = N(P_S) \oplus L^1(S^{-1}\Sigma),$$

and P_S *is isometric on* $L^1(S^{-1}\Sigma)$.

Corollary 8.1. $R(P_S)$ *is a closed subspace of* $L^1(\Sigma)$.

Corollary 8.2. P_S *is isometric on* $L^1(\Sigma)$ *if and only if* $S^{-1}\Sigma = \Sigma$.

Theorem 8.3. *Let* $U_S : L^\infty(X) \to L^\infty(X)$ *be the Koopman operator with respect to* $S : X \to X$. *Then*

$$L^\infty(X) = N(U_S) \oplus L^\infty(\text{supp } h),$$

and U_S *is isometric on* $L^\infty(\text{supp } h)$.

Corollary 8.3. $R(U_S)$ *is a closed subspace of* $L^\infty(X)$.

Corollary 8.4. U_S *is isometric on* $L^\infty(\mu)$ *if and only if* $\mu \cong \mu \circ S^{-1}$.

Thus, the following relation between P_S and U_S is obvious.

Theorem 8.4. *The Frobenius-Perron operator $P_S : L^1 \to L^1$ is one-to-one if and only if the Koopman operator $U_S : L^\infty \to L^\infty$ is onto, and P_S is onto if and only if U_S is one-to-one.*

Proof Since $U_S = P_S'$, and since $R(P_S)$ and $R(U_S)$ are closed, the theorem follows from Theorem B.23. □

8.3 Existence of Absolutely Continuous Invariant Measures

We investigate the problem on the existence of an absolutely continuous invariant finite measure associated with a given nonsingular transformation, which is equivalent to the existence of a stationary density of the corresponding Frobenius-Perron operator. Since the Frobenius-Perron operator is also a Markov operator, Theorem 7.1 provides an abstract approach to this problem. So here we apply that theorem to present some existence results for several classes of one dimensional nonsingular transformations, including the classic Lasota-Yorke theorem. For a more complete coverage and discussion of various approaches to the existence problem of stationary densities of Frobenius-Perron operators, see, for example, the monograph [Lasota and Mackey (1994)] and the textbook [Boyarsky and Góra (1997)].

From the general result on the existence of stationary densities of Markov operators based on the Cesáro averages for the iterates of the operator in Section 7.3, for a given nonsingular transformation S, if it can be shown that the Cesáro averages sequence $\{n^{-1} \sum_{k=0}^{n-1} P_S^k f\}$ is weakly pre-compact in L^1 for a given density f, then the corresponding Frobenius-Perron operator P_S has a stationary density which is the strong limit of the Cesáro sequence. In the following, we demonstrate three existence results based on the above argument.

In 1973, a fundamental existence result for a class of *piecewise C^2 and stretching mappings* of the interval $[0,1]$ was proved in [Lasota and Yorke (1973)], which also answered a question posed by Ulam, in his book entitled "A Collection of Mathematical Problems" [Ulam (1960)], on the existence of absolutely continuous invariant measures for some "simple" mappings such as piecewise linear ones. For the proof of the now classic Lasota-Yorke theorem, we need the concept of variation.

By definition, the *variation* of a function f defined on a finite interval

$[a, b]$ is the number (including ∞)

$$\bigvee_a^b f = \sup\left\{\sum_{k=1}^n |f(x_k) - f(x_{k-1})| : a = x_0 < \cdots < x_n = b\right\}.$$

If $\bigvee_a^b f < \infty$, we say that f is of *bounded variation* on $[a, b]$. The *variation* of a function $f \in L^1(a, b)$ is defined to be

$$\bigvee_{[a,b]} f = \inf\left\{\bigvee_a^b g : \; g(x) = f(x), \; x \in [a, b] \text{ a.e.}\right\}.$$

It can be shown [Giusti (1984)] that the variation of $f \in L^1(0, 1)$ can also be expressed as (see Definition 8.3)

$$\bigvee_{[a,b]} f = \sup\left\{\int_a^b f(x)g'(x)dx : g \in C_0^1(a, b), \|g\|_\infty \leq 1\right\}.$$

The space $BV(a, b)$ of all functions in $L^1(a, b)$ with bounded variation is a Banach space under the *BV-norm*

$$\|f\|_{BV} = \|f\|_1 + \bigvee_{[a,b]} f.$$

The following result shows that the closed unit ball of $(BV(a, b), \| \; \|_{BV})$ is a compact subset of $(L^1(a, b), \| \; \|_1)$.

Lemma 8.2. (Helly's lemma) *Let $\{f_n\}$ be a sequence of functions on $[a, b]$ such that $\|f_n\|_{BV} \leq C$ for a constant C and all n. Then there is a subsequence $\{f_{n_k}\} \subset \{f_n\}$ such that $\lim_{k\to\infty} \|f_{n_k} - f\|_1 = 0$ for some $f \in BV(a, b)$ with $\|f\|_{BV} \leq C$.*

The proof of the next lemma can be seen from [Lasota and Mackey (1994)] or the textbook [Natanson (1961)].

Lemma 8.3. *Let f, g be two functions defined on $[a, b]$. If f is of bounded variation and g has a continuous derivative on $[a, b]$, then the function fg is of bounded variation on $[a, b]$, and*

$$\bigvee_a^b fg \leq \sup_{x \in [a,b]} |g(x)| \bigvee_a^b f + \int_a^b |f(x)g'(x)|dx.$$

Lemma 8.4. (Yorke's inequality) *Let f be a function defined on $[0, 1]$ and be of bounded variation on $[a, b] \subset [0, 1]$. Then the function $f\chi_{[a,b]}$ is of bounded variation on $[0, 1]$, and*

$$\bigvee_0^1 f\chi_{[a,b]} \leq 2 \bigvee_a^b f + \frac{2}{b-a}\int_a^b |f(x)|dx.$$

Proof Pick $c \in [a, b]$ so that

$$|f(c)| \le \frac{2}{b-a} \int_a^b |f(x)| dx.$$

Then, we have

$$\bigvee_0^1 f\chi_{[a,b]} \le \bigvee_a^b f + |f(a)| + |f(b)|$$

$$\le \bigvee_a^b f + |f(a) - f(c)| + |f(c) - f(b)| + 2|f(c)|$$

$$\le 2\bigvee_a^b f + \frac{2}{b-a} \int_a^b |f(x)| dx. \quad \Box$$

Theorem 8.5. (Lasota-Yorke) *Suppose that a mapping* $S : [0,1] \to [0,1]$ *satisfies the following conditions:*

(i) *there is a partition* $0 = a_0 < a_1 < \cdots < a_r = 1$ *of* $[0,1]$ *such that for* $k = 1, \cdots, r$, *the restriction of* S *to* (a_{k-1}, a_k) *can be extended to* $[a_{k-1}, a_k]$ *as a* C^2*-function;*

(ii) *there is a constant* $b > 1$ *such that*

$$\inf\{|S'(x)| : x \in [0,1] \setminus \{a_0, \cdots, a_r\}\} \ge b;$$

(iii) *there is a constant* c *such that*

$$\sup\left\{\frac{S''(x)}{[S'(x)]^2} : x \in [0,1] \setminus \{a_0, \cdots, a_r\}\right\} \le c.$$

Then P_S *has a stationary density. Furthermore, for any* $f \in \mathcal{D}$, *the sequence of the Cesáro averages*

$$\frac{1}{n} \sum_{i=0}^{n-1} P_S^i f$$

strongly converges to a stationary density of P_S.

Proof Let $S_k = S|_{(a_{k-1}, a_k)}$ and $g_k = S_k^{-1}$ for $k = 1, \cdots, r$. For each open interval $I_k \equiv S((a_{k-1}, a_k))$, denote

$$J_k(x) = \begin{cases} (a_{k-1}, g_k(x)), & x \in I_k, \ g_k'(x) > 0, \\ (g_k(x), a_k), & x \in I_k, \ g_k'(x) < 0, \\ \emptyset, & x \notin I_k. \end{cases}$$

Then, for any $x \in [0,1]$,

$$S^{-1}((0, x)) = \bigcup_{k=1}^r J_k(x),$$

so from the definition of the Frobenius-Perron operator,

$$P_S f(x) = \frac{d}{dx} \int_{S^{-1}([0,x])} f(t)dt = \sum_{k=1}^{r} \frac{d}{dx} \int_{J_k(x)} f(t)dt,$$

where

$$\frac{d}{dx} \int_{J_k(x)} f(t)dt = \begin{cases} g_k'(x)f(g_k(x)), & x \in I_k, \ g_k'(x) > 0, \\ -g_k'(x)f(g_k(x)), & x \in I_k, \ g_k'(x) < 0, \\ 0, & x \notin I_k. \end{cases}$$

Thus,

$$P_S f(x) = \sum_{k=1}^{r} |g_k'(x)| f(g_k(x)) \chi_{I_k}(x), \ \forall \, x \in [0,1].$$

Let $f \in \mathcal{D} \cap BV(0,1)$. Then from Lemmas 8.3 and 8.4,

$$\bigvee_0^1 P_S f \leq \sum_{k=1}^{r} \bigvee_0^1 [|g_k'|(f \circ g_k)\chi_{I_k}]$$

$$\leq 2 \sum_{k=1}^{r} \bigvee_{I_k} [|g_k'|(f \circ g_k)] + \sum_{k=1}^{r} \frac{2}{m(I_k)} \int_{I_k} |g_k'|(f \circ g_k)dm$$

$$\leq 2 \sum_{k=1}^{r} \left(\sup_{x \in I_k} |g_k'(x)| \bigvee_{I_k} (f \circ g_k) + \int_{I_k} |g_k''|(f \circ g_k)dm \right)$$

$$+ \sum_{k=1}^{r} \frac{2}{m(I_k)} \int_{I_k} |g_k'|(f \circ g_k)dm$$

$$\leq \frac{2}{b} \sum_{k=1}^{r} \bigvee_{I_k} (f \circ g_k) + 2 \sum_{k=1}^{r} \left(c + \frac{1}{m(I_k)} \right) \int_{I_k} |g_k'|(f \circ g_k)dm$$

$$= \frac{2}{b} \sum_{k=1}^{r} \bigvee_{a_{k-1}}^{a_k} f + 2 \sum_{k=1}^{r} \left(c + \frac{1}{m(I_k)} \right) \int_{a_{k-1}}^{a_k} f(y)dy,$$

or

$$\bigvee_0^1 P_S f \leq \frac{2}{b} \bigvee_0^1 f + C \int_0^1 f(y)dy = \frac{2}{b} \bigvee_0^1 f + C,$$

where $C = \max_k 2\left(c + m(I_k)^{-1}\right)$ is a constant independent of f.

First, assume $b > 2$. Let $\alpha = 2b^{-1}$. Then $0 < \alpha < 1$. Using induction, for all positive integers n we have

$$\bigvee_0^1 P_S^n f \leq \alpha^n \bigvee_0^1 f + C \sum_{i=0}^{n-1} \alpha^i < \bigvee_0^1 f + \frac{C}{1-\alpha},$$

thus, for every $f \in \mathcal{D} \cap BV(0,1)$,

$$\bigvee_0^1 A_n f = \bigvee_0^1 \frac{1}{n} \sum_{i=0}^{n-1} P_S^i f < \bigvee_0^1 f + \frac{C}{1-\alpha}, \quad \forall\, n.$$

By Lemma 8.2, $\{A_n f\}$ is pre-compact, hence by Theorem 7.1,

$$\lim_{n \to \infty} A_n f = f^*$$

in $L^1(0,1)$, where f^* is a stationary density of P_S.

Now assume only $b > 1$. Choose a positive integer k such that $b^k > 2$. Denote the mapping $S^k : [0,1] \to [0,1]$ by $\hat{S}$. Then, from the above argument, for any $f \in \mathcal{D} \cap BV(0,1)$, there is a constant M such that

$$\bigvee_0^1 P_{\hat{S}}^n f < M, \quad \forall\, n.$$

Write $n = ik + j$ where $0 \le j < k$. Then

$$\bigvee_0^1 P_S^n f = \bigvee_0^1 P_S^j (P_S^k)^i f = \bigvee_0^1 P_S^j P_{\hat{S}}^i f$$

$$\le \left(\frac{2}{b}\right)^j \bigvee_0^1 P_{\hat{S}}^i f + C \sum_{t=0}^{j-1} \left(\frac{2}{b}\right)^t$$

$$< M \sup_{0 \le j \le k-1} \left(\frac{2}{b}\right)^j + C \sum_{t=0}^{k} \left(\frac{2}{b}\right)^t, \quad \forall\, n,$$

so Lemma 8.2 implies that P_S has a stationary density. $\qquad \square$

Remark 8.3. In fact all the stationary densities of P_S in the above theorem are uniformly bounded in variation; see [Lasota and Mackey (1994)]. In the paper [Li and Yorke (1978)], the structure of the fixed point space of P_S under the condition of Theorem 8.5 was further explored, and in particular it was shown that if $r = 2$ in the above theorem, then P_S has exactly one stationary density. If in addition S is piecewise onto, then from the proof of Theorem 8.5, it is easy to see that

$$\bigvee_0^1 P_S f \le \frac{1}{b} \bigvee_0^1 f + c \|P_S f\|_1.$$

Remark 8.4. It can further be shown that $P_S : (BV(0,1), \|\ \|_{BV}) \to (BV(0,1), \|\ \|_{BV})$ is constrictive, so by Theorem 7.4, $\{P_S^n\}$ is asymptotically periodic. Thus, there are functions $g_j \in \mathcal{D}$ and $h_j \in L^\infty(0,1)$, $j = 1, \cdots, k$, such that the linear operator

$$Qf = P_S f - \sum_{j=1}^{k} \left(\int_0^1 f(t) h_j(t) dt \right) g_j$$

satisfies the properties that $\lim_{n \to \infty} \|P_S^n Q f\|_1 = 0$ for each $f \in L^1(0,1)$, supp g_j are disjoint, and $P_S g_j = g_{\pi(j)}$ for all j, where $\pi : \{1, \cdots, k\} \to \{1, \cdots, k\}$ is a permutation. See [Lasota and Mackey (1994); Boyarsky and Góra (1997)] for more details.

Remark 8.5. The assumption of the Lasota-Yorke theorem can be relaxed to that the mapping S is piecewise C^1 and stretching, and the function $|S'|^{-1}$ is of bounded variation. See [Wong (1978); Boyarsky and Góra (1997)], in which indeed it has been shown that there exist constants $0 < \beta < 1$, C_1, and C_2 such that for any $f \in BV(0,1)$ and all $n \geq 1$,

$$\|P_S^n f\|_{BV} \leq C_1 \beta^n \|f\|_{BV} + C_2 \|f\|_1.$$

The condition $\inf |S'| > 1$ in the Lasota-Yorke theorem is important for the existence of stationary densities for piecewise monotonic mappings, but it is not necessary, as the mapping $S(x) = 4x(1 - x)$ shows. On the other hand, even if this condition is violated at only one point, the corresponding Frobenius-Perron operator may not have a stationary density. One counterexample [Lasota and Mackey (1994)] is the following mapping

$$S(x) = \begin{cases} \frac{x}{1-x}, & x \in \left[0, \frac{1}{2}\right], \\ 2x - 1, & x \in \left(\frac{1}{2}, 1\right], \end{cases}$$

for which $S'(0) = 1$ while $S'(x) > 1$, $\forall x \in (0,1]$.

We now proceed to prove another existence result, the one for *piecewise convex mappings with a strong repellor*, which is also due to Lasota and Yorke [Lasota and Yorke (1982)].

Theorem 8.6. *Suppose that $S : [0,1] \to [0,1]$ satisfies*
    (i) *there is a partition $0 = a_0 < a_1 < \cdots < a_r = 1$ of $[0,1]$ such that $S|_{[a_{k-1}, a_k)}$ is a C^2-function for each $k = 1, \cdots, r$;*
    (ii) *$S'(x) > 0$ and $S''(x) \geq 0$ for all $x \in [0,1)$, where $S'(a_k)$ and $S''(a_k)$ are right derivatives for each k;*
    (iii) *$S(a_k) = 0$ for $k = 1, \cdots, r$;*
    (iv) *$S'(0) > 1$.*
Then P_S has a stationary density f^*. Moreover, f^* is decreasing.

Proof Let S_k be the restriction of S to $[a_{k-1}, a_k)$ and let

$$g_k(x) = \begin{cases} S_k^{-1}(x), & x \in [0, S(a_k^-)), \\ a_k, & x \in [S(a_k^-), 1] \end{cases}$$

for $k = 1, \cdots, r$, where $S(a_k^-) = \lim_{x \to a_k^-} S(x)$. Then

$$S^{-1}([0, x]) = \bigcup_{k=1}^{r} [a_{k-1}, g_k(x)],$$

which together with (8.2) implies

$$P_S f(x) = \sum_{k=1}^{r} g_k'(x) f(g_k(x)).$$

Note that S_k is increasing, so is g_k. And g_k' is decreasing since

$$g_k''(x) = -\frac{S''(x)}{S'(x)^2} \le 0.$$

Thus $P_S f$ is nonnegative and decreasing if so is f, and

$$P_S f(x) = \sum_{k=1}^{r} g_k'(x) f(g_k(x)) \le \sum_{k=1}^{r} g_k'(0) f(g_k(0))$$

$$= g_1'(0) f(0) + \sum_{k=2}^{r} g_k'(0) f(a_{k-1}).$$

Now let $f \in \mathcal{D}$ be a decreasing function. Then

$$1 \ge \int_0^x f(t)dt \ge \int_0^x f(x)dt = xf(x),$$

which implies that

$$f(x) \le \frac{1}{x}.$$

Hence, for $2 \le k \le r$,

$$g_k'(0) f(a_{k-1}) \le \frac{g_k'(0)}{a_{k-1}}.$$

Since $g_1'(0) = S'(0)^{-1}$, by letting $\alpha = S'(0)^{-1} < 1$, we have

$$P_S f(x) \le \alpha f(0) + \sum_{k=2}^{r} \frac{g_k'(0)}{a_{k-1}} = \alpha f(0) + C,$$

where the constant $C = \sum_{k=2}^{r} g_k'(0)/a_{k-1}$. It follows that

$$P_S^n f(x) \le \alpha^n f(0) + \frac{C}{1-\alpha} < f(0) + \frac{C}{1-\alpha}, \quad \forall\, n.$$

Since $\|P_S\|_1 = 1$ and the set

$$\left\{ h \in L^1(0,1) : 0 \le h(x) \le f(0) + \frac{C}{1-\alpha}, \ x \in [0,1] \right\}$$

is weakly compact in $L^1(0,1)$, by Theorem 7.1,

$$\lim_{n \to \infty} \frac{1}{n} \sum_{i=0}^{n-1} P_S^i f = f^*$$

under the L^1-norm, where f^* is a stationary density of P_S. It is obvious that f^* is decreasing since f is decreasing. $\qquad\qquad\square$

Remark 8.6. The point $x = 0$ is called the *strong repellor* of S since the trajectory $\{x_0, S(x_0), S^2(x_0), \cdots\}$, starting from any point $x_0 \in [0, a_1)$, will leave the interval $[0, a_1)$ eventually.

Remark 8.7. It can be shown [Lasota and Yorke (1982)] that f^* is the unique stationary density of P_S and that the sequence $\{P_S^n\}$ is asymptotically stable (Definition 7.6).

Theorem 8.5 can be extended to multi-dimensional transformations with the help of the modern notion ([Giusti (1984); Ziemer (1989)]) of variation for functions of multi-variable. Let Ω be a bounded region of $\mathbb{R}^N$ with a regular (such as piecewise smooth or piecewise Lipschitz) boundary.

Definition 8.3. Let $f \in L^1(\Omega)$. The number (may be ∞)

$$\bigvee(f; \Omega) = \sup\left\{ \int_\Omega f \ \mathrm{div} \ g \ dm : g \in C_0^1(\Omega; \mathbb{R}^N), \|g\|_{0,\infty} \le 1 \right\}$$

is called the *variation* of f over Ω. Here $\mathrm{div} \ g = \sum_{i=1}^N \partial g_i / \partial x_i$ is the *divergence* of g and $\|g\|_{0,\infty} = \sup\{\|g(x)\|_2 : x \in \Omega\}$.

The space $BV(\Omega)$ of all $f \in L^1(\Omega)$ of bounded variation is a Banach space under the BV-norm

$$\|f\|_{BV} = \|f\|_1 + \bigvee(f; \Omega).$$

The multi-variable version of Helly's lemma is the following

Lemma 8.5. *The closed unit ball of the Banach space* $(BV(\Omega), \| \ \|_{BV})$ *is a compact subset of the Banach space* $(L^1(\Omega), \| \ \|_1)$.

We introduce a class of transformations on a bounded region Ω of $\mathbb{R}^N$ with a piecewise smooth boundary $\partial\Omega$.

Definition 8.4. Let $S : \Omega \to \Omega$ and let $\{\Omega_1, \cdots, \Omega_r\}$ be a partition of Ω. For each $k = 1, \cdots, r$ denote $S_k = S|_{\Omega_k}$. We say that S is a *piecewise C^2 and b-expanding* transformation if each S_k is one-to-one on its domain Ω_k, can be extended as a C^2-transformation on $\overline{\Omega_k}$, namely C^2 in a neighborhood of $\overline{\Omega_k}$, and satisfies the inequalities

$$\sup_{x \in S(\Omega_k)} \|(S_k^{-1})'(x)\|_2 \le \frac{1}{b}, \quad \forall\, k = 1, 2, \cdots, r.$$

Using Definitions 8.3 and 8.4 and Lemma 8.5, [Góra and Boyarsky (1989)] proved the following existence theorem.

Theorem 8.7. (Góra-Boyarsky) *Suppose that $S : \Omega \to \Omega$ is a piecewise C^2 and b-expanding transformation. Then for b large enough, $P_S : L^1(\Omega) \to L^1(\Omega)$ has a stationary density f^*.*

8.4 Ergodic Properties of Invariant Measures

The discrete dynamical systems in the previous section are closely related to the concept of *transitivity* of the iteration sequences, which is related to the notion of ergodicity for transformations as given by Definition 6.9. In this section we give the characteristics of ergodicity in terms of Frobenius-Perron operators $P_S : L^1 \to L^1$ and Koopman operators $U_S : L^\infty \to L^\infty$ associated with nonsingular transformations $S : X \to X$. First we have the following characterization theorem for ergodicity, whose proof is exactly the same as that for Lemma 6.2.

Theorem 8.8. *S is ergodic if and only if all the fixed points of the corresponding Koopman operator U_S are constant functions.*

Theorem 8.8 says that S is ergodic if and only if constant functions are its only invariant functions. Thus, in Birkhoff's pointwise ergodic theorem, if S is ergodic with respect to the invariant measure μ, then the limit function $\tilde{f}$ is a constant one, and if in addition $\mu(X) < \infty$, then

$$\tilde{f} = \frac{1}{\mu(X)} \int_X f d\mu.$$

In particular, if $\mu(X) = 1$ and S is ergodic, then

$$\lim_{n \to \infty} \frac{1}{n} \sum_{k=0}^{n-1} \chi_A(S^k(x)) = \mu(A), \quad x \in X \; \mu - \text{a.e.}, \tag{8.6}$$

which means that the iterates of $x \in X$ a.e. enter a given measurable set $A \subset X$ with the asymptotic frequency $\mu(A)$.

The next theorem employs the tool of Frobenius-Perron operators to check the ergodicity of nonsingular transformations.

Theorem 8.9. *If S is ergodic, then there is at most one stationary density of P_S. Conversely, if f^* is a unique stationary density of P and $f^*(x) > 0$ on X a.e., then S is ergodic.*

Proof Assume S is ergodic. Suppose that f_1 and f_2 are different stationary densities of P_S. Set $g = f_1 - f_2$, so $P_S g = g$. Thus

$$P_S g^+ = g^+ \quad \text{and} \quad P_S g^- = g^- \tag{8.7}$$

from Proposition 7.4. Since $f_1 \neq f_2$ and they are both densities, $g^+ \neq 0$ and $g^- \neq 0$. Denote

$$A = \text{supp } g^+ \quad \text{and} \quad B = \text{supp } g^-.$$

Then A and B are disjoint sets and both have positive measures. From (8.7) and Proposition 8.2, we have

$$A \subset S^{-1}(A) \quad \text{and} \quad B \subset S^{-1}(B).$$

By induction, for all n,

$$A \subset S^{-1}(A) \subset S^{-2}(A) \subset \cdots \subset S^{-n}(A)$$

and

$$B \subset S^{-1}(B) \subset S^{-2}(B) \subset \cdots \subset S^{-n}(A).$$

Since $A \cap B = \emptyset$, $S^{-n}(A) \cap S^{-n}(B) = \emptyset$ for all n. Let

$$\hat{A} = \bigcup_{n=0}^{\infty} S^{-n}(A) \quad \text{and} \quad \hat{B} = \bigcup_{n=0}^{\infty} S^{-n}(B),$$

which are also disjoint. Then they are S-invariant since

$$S^{-1}(\hat{A}) = \bigcup_{n=1}^{\infty} S^{-n}(A) = \bigcup_{n=0}^{\infty} S^{-n}(A) = \hat{A}$$

and

$$S^{-1}(\hat{B}) = \bigcup_{n=1}^{\infty} S^{-n}(B) = \bigcup_{n=0}^{\infty} S^{-n}(B) = \hat{B}.$$

On the other hand, neither $\hat{A}$ nor $\hat{B}$ is of zero measure since A and B are not. Thus, $\hat{A}$ and $\hat{B}$ are nontrivial invariant sets, which contradicts the ergodicity assumption of S.

To prove the second part of the theorem, assume that $f^*(x) > 0$ for all $x \in X$ and f^* is the unique stationary density of P_S but that S is not ergodic. Then there exist nontrivial sets A and $B = A^c$ such that $S^{-1}(A) = A$ and $S^{-1}(B) = B$. Write $f^* = \chi_A f^* + \chi_B f^*$, so that

$$\chi_A f^* + \chi_B f^* = P_S(\chi_A f^*) + P_S(\chi_B f^*). \tag{8.8}$$

Since $\chi_A f^*$ is zero on $B = S^{-1}(B)$ and $\chi_B f^*$ is zero on $A = S^{-1}(A)$, Proposition 8.2 implies that $P_S(\chi_A f^*)$ is zero on B and $P_S(\chi_B f^*)$ is zero on A. Thus, equality (8.8) implies that

$$\chi_A f^* = P_S(\chi_A f^*) \quad \text{and} \quad \chi_B f^* = P_S(\chi_B f^*).$$

Since $f^*(x)$ is positive for $x \in A$ and $x \in B$,

$$f_A = \frac{\chi_A f^*}{\|\chi_A f^*\|_1} \quad \text{and} \quad f_B = \frac{\chi_B f^*}{\|\chi_B f^*\|_1}$$

are distinct stationary densities of P_S, a contradiction. $\qquad\square$

Example 8.12. The rotation map of Example 8.5 is not ergodic provided that $a = \exp(i\theta)$ and θ/π is rational. For example, if $\theta = \pi/3$, then the union of the arcs with the end points $\exp(i(k-1)\pi/6)$ and $\exp(ik\pi/6)$ for $k = 1, \cdots, 6$ is invariant under the rotation. When θ/π is irrational, the rotation map becomes ergodic, which can be proved by a later result.

Theorem 8.10. *Let (X, Σ, μ) be a finite measure space and $S : X \to X$ be measure-preserving. Then S is ergodic if and only if*

$$\lim_{n \to \infty} \frac{1}{n} \sum_{k=0}^{n-1} \mu(A \cap S^{-k}(B)) = \mu(A)\mu(B), \quad \forall \, A, B \in \Sigma. \tag{8.9}$$

Proof For the sufficiency part, suppose that S is not ergodic with respect to μ. Then there are A, $B \in \Sigma$ that are nontrivial disjoint invariant sets. Since $S^{-n}(A) \cap B = \emptyset$ for each n, the left-hand side of (8.9) is zero, while the right-hand side $\mu(A)\mu(B) > 0$. The necessity of (8.9) for ergodicity is an easy consequence of Theorem 8.12(i) to be presented later by letting $f = \chi_A$ and $g = \chi_B$. $\qquad\square$

The characterization (8.9) of ergodicity motivates the following definition of mixing, which is stronger than ergodicity.

Definition 8.5. A measure-preserving transformation S on a probability measure space (X, Σ, μ) is said to be *mixing* if

$$\lim_{n \to \infty} \mu(A \cap S^{-n}(B)) = \mu(A)\mu(B), \quad \forall\, A, B \in \Sigma. \tag{8.10}$$

Mixing can be interpreted as meaning that the fraction of points starting in B that ended up in A after n iterations as n approaches infinity is just given by the product of the measures of A and B and is independent of where the subsets A and B are in X.

Example 8.13. Consider the dyadic map $S(x) = 2x \pmod 1$ which preserves the Lebesgue measure m. Let $B = [0, a]$. It is obviously seen that $S^{-n}(B)$ consists of 2^n subintervals

$$\left[\frac{i}{2^n}, \frac{i+a}{2^n} \right], \quad i = 0, 1, \cdots, 2^n - 1.$$

It follows that $m(A \cap S^{-n}(B)) \to m(A)m(B)$ as $n \to \infty$. More generally, the k-adic map of Example 8.3 is mixing.

For the dyadic map, since $S^n(x) = 2^n x \pmod 1$, $S^n([0, a]) = [0, 1]$ eventually for any $a \in (0, 1)$. Thus $\lim_{n \to \infty} m(S^n(A)) = 1$ for any measurable set $A \subset [0, 1]$. This property is made precisely by the following definition due to Rohlin [Rohlin (1964)].

Definition 8.6. Let (X, Σ, μ) be a probability measure space and let $S : X \to X$ be μ-preserving such that $S\Sigma \subset \Sigma$. The transformation S is said to be *exact* if for all $A \in \Sigma$ such that $\mu(A) > 0$,

$$\lim_{n \to \infty} \mu(S^n(A)) = 1. \tag{8.11}$$

It can be shown that *exactness implies mixing*. Thus, the three concepts of exactness, mixing, and ergodicity give a hierarchy of different irregular behaviors of measure-preserving transformations. The next theorem characterizes exactness in terms of a specially defined σ-subalgebra.

Theorem 8.11. *Let (X, Σ, μ) be a probability measure space and let $S : X \to X$ be a measure-preserving transformation such that $S\Sigma \subset \Sigma$. Then S is exact if and only if the σ-subalgebra*

$$\Sigma^S \equiv \bigcap_{n=0}^{\infty} S^{-n} \Sigma$$

of Σ consists only of the sets of μ-measure 0 or 1.

Proof Suppose that there is $A \in \Sigma^S$ with $0 < \mu(A) < 1$. Then $A = S^{-n}A_n$, where $A_n \in \Sigma$ for each n. We have $\mu(A) = \mu(A_n)$ since S preserves μ. Now $S^n(A) = S^n(S^{-n}A_n) \subset A_n$ implies that $\mu(S^n(A)) \leq \mu(A) < 1$ for all n. Hence, S cannot be exact.

Conversely, let Σ^S consist of sets of μ-measure 0 or 1. Suppose that there is $A \in \Sigma$ such that $\mu(A) > 0$ and (8.11) is not true. Then, without loss of generality, assume $\mu(S^n(A)) \leq \alpha$ for some $\alpha < 1$ and all n. Since the set sequence $\{S^{-n}(S^n(A))\}$ is monotonically increasing, the set $B = \bigcup_{n=0}^{\infty} S^{-n}(S^n(A)) \in \Sigma^S$. Now $B \supset A$ implies that $\mu(B) \geq \mu(A) > 0$, hence $\mu(B) = 1$ by the assumption. On the other hand,

$$\mu(B) = \lim_{n \to \infty} \mu(S^{-n}(S^n(A))) = \lim_{n \to \infty} \mu(S^n(A)) \leq \alpha < 1,$$

which contradicts the fact $\mu(B) = 1$. □

Remark 8.8. Invertible transformations cannot be exact. In fact, if S is invertible and preserves μ, then

$$\mu(S(A)) = \mu(S^{-1}(S(A))) = \mu(A).$$

By induction, $\mu(S^n(A)) = \mu(A)$ for all n, which violates (8.11).

Remark 8.9. Exactness means that images of any nontrivial measurable set $A \in \Sigma$ will spread and completely fill X eventually. Mixing means that images of any $B \in \Sigma$ under the iteration of S become independent of any fixed set $A \in \Sigma$ asymptotically. Ergodicity means that the set B becomes independent of the set A on the average.

We can employ the notions of Frobenius-Perron operators and Koopman operators to reformulate the features of ergodicity, mixing, and exactness for measure-preserving transformations in the following results. Much of the proof to the next theorem is adapted from the presentation in the monograph [Lasota and Mackey (1994)].

Theorem 8.12. *Let (X, Σ, μ) be a probability measure space and $S : X \to X$ be a measure-preserving transformation. Then*

(i) *S is ergodic if and only if for all $f \in L^1$ and $g \in L^{\infty}$,*

$$\lim_{n \to \infty} \frac{1}{n} \sum_{k=0}^{n-1} \int_X (P_S^k f) g \, d\mu = \int_X f d\mu \int_X g d\mu; \tag{8.12}$$

(ii) *S is mixing if and only if for all $f \in L^1$ and $g \in L^{\infty}$,*

$$\lim_{n \to \infty} \int_X (P_S^n f) g \, d\mu = \int_X f d\mu \int_X g d\mu; \tag{8.13}$$

(iii) *S is exact if and only if for all $f \in L^1$,*

$$\lim_{n \to \infty} \left\| P_S^n f - \int_X f d\mu \right\|_1 = 0. \tag{8.14}$$

Proof (i) Since S is a measure-preserving transformation with respect to the probability measure μ, we have $P_S 1 = 1$. If S is ergodic, then by Theorem 8.9, the constant function $f^* \equiv 1$ is the unique stationary density of P_S. Hence from Theorem 7.3, we have

$$\lim_{n \to \infty} \frac{1}{n} \sum_{k=0}^{n-1} P_S^k f = 1, \ \forall f \in \mathcal{D}.$$

Since strong convergence implies weak convergence, multiplying any function $g \in L^\infty$ to the both sides of the above limit equality and taking integration give the following limit equality

$$\lim_{n \to \infty} \frac{1}{n} \sum_{k=0}^{n-1} \int_X (P_S^k f) g \, d\mu = \int_X g d\mu, \ \forall f \in \mathcal{D}, \ g \in L^\infty,$$

and thus the formula (8.12) follows immediately from the above.

Conversely, suppose that limit (8.12) holds. Let f be a stationary density of P_S. Then from (8.12), the equality $\int_X f g d\mu = \int_X g d\mu$ is satisfied by all $g \in L^\infty(X)$. it follows that $f(x) \equiv 1$ a.e., which means that $f^* = 1$ is the unique stationary density of P_S. By means of Theorem 8.9, we see that S is an ergodic transformation.

(ii) Assume that S is mixing. Then the condition (8.10) in Definition 8.5 can be written in the integral form as

$$\lim_{n \to \infty} \int_X \chi_A U_S^n \chi_B \, d\mu = \int_X \chi_A d\mu \int_X \chi_B d\mu. \tag{8.15}$$

Since the Koopman operator U_S is the dual operator of the Frobenius-Perron operator P_S, the above limit equality (8.15) in terms of the Koopman operator can be rewritten as

$$\lim_{n \to \infty} \int_X (P_S^n \chi_A) \chi_B \, d\mu = \int_X \chi_A d\mu \int_X \chi_B d\mu$$

in terms of the Frobenius-Perron operator. In other words, the formula (8.13) is satisfied by any two characteristic functions $f = \chi_A$ and $g = \chi_B$, and therefore for any two simple functions $f = \sum_k a_k \chi_{A_k}$ and $g = \sum_j b_j \chi_{B_j}$ because of the linearity property of integration. On the other hand, every function $g \in L^\infty(X)$ is the uniform limit of simple functions

$g_k \in L^\infty(X)$ and each function $f \in L^1(X)$ is the limit of simple functions $f_k \in L^1(X)$ under the L^1 norm. Now

$$\left| \int_X (P_S^n f) g \, d\mu - \int_X f \, d\mu \int_X g \, d\mu \right|$$

$$\leq \left| \int_X (P_S^n f) g \, d\mu - \int_X (P_S^n f_k) g \, d\mu \right|$$

$$+ \left| \int_X (P_S^n f_k) g \, d\mu - \int_X (P_S^n f_k) g_k \, d\mu \right|$$

$$+ \left| \int_X (P_S^n f_k) g_k \, d\mu - \int_X f_k \, d\mu \int_X g_k \, d\mu \right|$$

$$+ \left| \int_X f_k \, d\mu \int_X g_k \, d\mu - \int_X f_k \, d\mu \int_X g \, d\mu \right|$$

$$+ \left| \int_X f_k \, d\mu \int_X g \, d\mu - \int_X f \, d\mu \int_X g \, d\mu \right|$$

$$\leq \|f - f_k\|_1 \|g\|_\infty + \|f_k\|_1 \|g - g_k\|_\infty + \|f_k\|_1 \|g - g_k\|_\infty$$

$$+ \|f - f_k\|_1 \|g\|_\infty + \left| \int_X (P_S^n f_k) g_k \, d\mu - \int_X f_k \, d\mu \int_X g_k \, d\mu \right|$$

$$= 2(\|f - f_k\|_1 \|g\|_\infty + \|f_k\|_1 \|g - g_k\|_\infty)$$

$$+ \left| \int_X (P_S^n f_k) g_k \, d\mu - \int_X f_k \, d\mu \int_X g_k \, d\mu \right|. \qquad (8.16)$$

Given any $\epsilon > 0$, there is k such that

$$\|f - f_k\|_1 \|g\|_\infty + \|f_k\|_1 \|g - g_k\|_\infty < \frac{\epsilon}{4}.$$

With this fixed k, since (8.13) holds for simple functions, there is a positive integer n_0 such that for all $n \geq n_0$,

$$\left| \int_X (P_S^n f_k) g_k \, d\mu - \int_X f_k \, d\mu \int_X g_k \, d\mu \right| < \frac{\epsilon}{2}.$$

Therefore, from (8.16), for all $n \geq n_0$ we have

$$\left| \int_X (P_S^n f) g \, d\mu - \int_X f \, d\mu \int_X g \, d\mu \right| < \epsilon.$$

This completes the proof that mixing implies (8.13). Clearly (8.13) implies mixing by letting $f = \chi_A$ and $g = \chi_B$.

(iii) We only prove that the strong convergence of $\{P_S^n f\}$ to $\int_X f \, d\mu$ implies exactness. Let A be such that $\mu(A) > 0$ and let

$$f_A = \frac{1}{\mu(A)} \chi_A.$$

Since f_A is a density, $\lim_{n\to\infty} \|P_S^n f_A - 1\|_1 = 0$. It follows that

$$\mu(S^n(A)) = \int_{S^n(A)} d\mu$$

$$= \int_{S^n(A)} P_S^n f_A d\mu - \int_{S^n(A)} (P_S^n f_A - 1) d\mu$$

$$\geq \int_{S^n(A)} P_S^n f_A d\mu - \|P_S^n f_A - 1\|_1$$

$$= \int_{S^{-n}(S^n(A))} f_A d\mu - \|P_S^n f_A - 1\|_1$$

$$= 1 - \|P_S^n f_A - 1\|_1 \to 1 \text{ as } n \to \infty.$$

The last equality above is from the fact that $S^{-n}(S^n(A))$ contains A. The proof of the converse that exactness implies (8.14) is referred to [Lin (1971)]. This completes the proof. □

Corollary 8.5. *Let (X, Σ, μ) be a probability measure space and $S : X \to X$ a measure-preserving transformation. Then*

(i) *S is ergodic if and only if*

$$\lim_{n\to\infty} \frac{1}{n} \sum_{k=0}^{n-1} \int_X f U_S^k g \, d\mu = \int_X f d\mu \int_X g d\mu, \ \forall \, f \in L^1, \, g \in L^\infty;$$

(ii) *S is mixing if and only if*

$$\lim_{n\to\infty} \int_X f U_S^n g \, d\mu = \int_X f d\mu \int_X g d\mu, \ \forall \, f \in L^1, \, g \in L^\infty.$$

When $S : (X, \Sigma, \mu) \to (X, \Sigma, \mu)$ is not measure-preserving, we have the following proposition for the exactness of S with respect to an absolutely continuous invariant measure. If the Frobenius-Perron operator corresponding to S is such that $\{P_S^n\}$ is asymptotically stable, then the transformation $S : X \to X$ is said to be *statistically stable*.

Proposition 8.5. *Let $S : X \to X$ be such that $S\Sigma \subset \Sigma$ and P_S has a unique stationary density f^*. If S is statistically stable, then S is exact with respect to the invariant probability measure*

$$\mu_{f^*}(A) = \int_A f^* d\mu, \ \forall \, A \in \Sigma.$$

Proof Suppose that $\mu_{f^*}(A) > 0$. Let

$$f_A(x) = \frac{1}{\mu_{f^*}(A)} f^*(x) \chi_A(x), \ \forall \, x \in X.$$

Then $f_A \in \mathcal{D}$, and $\lim_{n\to\infty} \|P_S^n f_A - f^*\|_1 = 0$ from the assumption. Since $P_S^n f_A$ is supported on $S^n(A)$ by Proposition 8.2,

$$\int_{S^n(A)} P_S^n f_A d\mu = \int_X P_S^n f_A d\mu = 1.$$

Now

$$\mu_{f^*}(S^n(A)) = \int_{S^n(A)} f^* d\mu \geq \int_{S^n(A)} P_S^n f_A d\mu - \|P_S^n f_A - f^*\|_1$$
$$= 1 - \|P_S^n f_A - f^*\|_1 \to 0 \text{ as } n \to \infty,$$

hence $S : (X, \Sigma, \mu_{f^*}) \to (X, \Sigma, \mu_{f^*})$ is exact. $\qquad \square$

Remark 8.10. Theorem 8.12 basically says that

(i) S is ergodic if and only if the sequence $\{P_S^n f\}$ is Cesáro convergent to 1 for all $f \in \mathcal{D}$;

(ii) S is mixing if and only if the sequence $\{P_S^n f\}$ is weakly convergent to 1 for all $f \in \mathcal{D}$;

(iii) S is exact if and only if the sequence $\{P_S^n f\}$ is strongly convergent to 1 for all $f \in \mathcal{D}$.

Remark 8.11. Remark 8.10 shows that ergodicity, mixing, and exactness for a measure-preserving transformation S on a probability measure space are exactly the same as those in Definition 7.8 for P_S as a Markov operator with stationary density 1.

8.5 Notes and Remarks

1. Stanislaw Ulam seems to be among the first to have given the name of Frobenius-Perron operators associated with a given nonlinear mapping in his famous book [Ulam (1960)] in which he posed many mathematical problems for nonlinear science. Frobenius-Perron operators are often called Perron-Frobenius operators and sometimes referred to as Ruelle-Perron-Frobenius (or some other order of the three names) operators in the literature. Some authors even call them transfer operators, especially in the literature of physical sciences.

2. The existence problem of a stationary density for the Frobenius-Perron operators associated with a class of nonsingular transformations is one of the main concerns in modern ergodic theory. Besides [Lasota and

Mackey (1994); Boyarsky and Góra (1997)], this problem is studied in the authors' book [Ding and Zhou (2009)].

3. Historically, the first existence result for multi-dimensional transformations was obtained by Krzyzewski and Szlenk in 1969 via demonstrating the existence of a unique absolutely continuous invariant probability measure for a C^1 expanding mapping S (Theorem 6.8.1 in [Lasota and Mackey (1994)]). In [Jabłoński (1983)], the existence of absolutely continuous invariant measures for a special class of piecewise expanding transformations on $[0, 1]^N$ with rectangular partitions was proved by using the classic Tonnelli definition (see Definition 2.4.4 in [Ding and Zhou (2009)]) of variation, before Góra and Boyarsky [Góra and Boyarsky (1989)] used Definition 8.3 of variation to establish their existence result. See Section 5.4 of [Ding and Zhou (2009)] for the idea of proving Theorem 8.7.

Exercises

8.1 Show that the mapping $S_c : [0, 1] \rightarrow [0, 1]$ defined by

$$S_c(x) = \begin{cases} \frac{x}{c}, & 0 \le x \le c, \\ \frac{1-x}{1-c}, & c \le x \le 1 \end{cases}$$

preserves the Lebesgue measure on $[0, 1]$.

8.2 Let $S : [0, 1] \rightarrow [0, 1]$ be the *Gauss map* defined by

$$S(x) = \begin{cases} 0, & \text{if } x = 0, \\ \left\{\frac{1}{x}\right\}, & \text{if } x \ne 0, \end{cases}$$

where $\{t\}$ is the fractional part of t. Show the S-invariance of the probability measure μ defined by

$$\mu(A) = \frac{1}{\ln 2} \int_A \frac{1}{1+x} dx, \quad \forall A \in \mathcal{B}.$$

8.3 Show property (iii) of the Frobenius-Perron operator associated with a nonsingular transformation in Section 8.2.

8.4 Find the expression of the Frobenius-Perron operator corresponding to each of the following mappings:

(i) $S : [0, 1] \rightarrow [0, 1]$, $S(x) = 4x^2(1 - x^2)$;

(ii) $S : [0, 1] \rightarrow [0, 1]$, $S(x) = \sin \pi x$;

(iii) $S : [0, 1] \rightarrow [0, 1]$, $S(x) = a \tan(bx + c)$;

(iv) $S : [0, 1] \rightarrow [0, 1]$, $S(x) = axe^{-bx}$, $a, b > 0$.

8.5 Find the expression of the Frobenius-Perron operator associated with the Hénon map in Example 8.7.

8.6 Show that the function $f(x) = 2x/(1+x)^2$ is a stationary density for the mapping $S : [0,1] \to [0,1]$ given by

$$S(x) = \begin{cases} \frac{2x}{1-x}, & 0 \le x \le \frac{1}{3}, \\ \frac{1-x}{2x}, & \frac{1}{3} \le x \le 1. \end{cases}$$

8.7 Show that the function $f(x) = px^{p-1}$ with $p > 1$ is a stationary density for the mapping $S : [0,1] \to [0,1]$ given by

$$S(x) = \begin{cases} 2^{\frac{1}{p}}x, & 0 \le x \le \left(\frac{1}{2}\right)^{\frac{1}{p}}, \\ 2^{\frac{1}{p}}(1 - x^p)^{\frac{1}{p}}, & \left(\frac{1}{2}\right)^{\frac{1}{p}} \le x \le 1. \end{cases}$$

8.8 Show that the function $f(x) = 12(x - 1/2)^2$ is a stationary density for the mapping $S : [0,1] \to [0,1]$ given by

$$S(x) = \left(\frac{1}{8} - 2 \left| x - \frac{1}{2} \right|^3 \right)^{\frac{1}{3}} + \frac{1}{2}.$$

8.9 Show that the function $f(x) = (1-x)/2$ is a stationary density for the *cusp map* $S : [-1,1] \to [-1,1]$ given by

$$S(x) = 1 - 2|x|^{\frac{1}{2}}.$$

8.10 Let (X, Σ) be a measurable space with two equivalent σ-finite measures μ and ν, let $S : X \to X$ be a nonsingular transformation, and let $P_{S,\mu} : L^1(\mu) \to L^1(\mu)$ and $P_{S,\nu} : L^1(\nu) \to L^1(\nu)$ be the corresponding Frobenius-Perron operator with respect to μ and ν, respectively. Show that for any $f \in L^1(\nu)$,

$$P_{S,\nu}f = \frac{P_{S,\mu}(h \cdot f)}{h},$$

where $h \in L^1(\mu)$ is the Radon-Nikodym derivative $d\nu/d\mu$.

8.11 Let a mapping $S : [0,1] \to [0,1]$ be nonsingular and let $h : [0,1] \to [0,1]$ be a diffeomorphism. Denote a mapping $\hat{S} : [0,1] \to [0,1]$ defined by

$$\hat{S} = h \circ S \circ h^{-1}$$

and a function $\hat{f}$ defined by

$$\hat{f} = (f \circ h^{-1}) \cdot |(h^{-1})'|.$$

Show that $P_S f = f$ if and only if $P_{\hat{S}}\hat{f} = \hat{f}$.

8.12 Let $S : [0,1] \to [0,1]$ be a transformation satisfying condition (i) of Theorem 8.5, and let P_S be the corresponding Frobenius-Perron operator.

If there exists a positive function $\phi \in C^1[0,1]$ such that, for some real numbers $b > 1$ and c,

$$\frac{|S'(x)|\phi(S(x))}{\phi(x)} \geq b, \quad 0 < x < 1$$

and

$$\left| \frac{1}{\phi(x)} \frac{d}{dx}\left(\frac{1}{p(x)} \right) \right| \leq c, \quad 0 < x < 1,$$

then P_S has a stationary density.

Hint: Define

$$h(x) = \frac{1}{\|\phi\|_1} \int_0^x \phi(t)dt, \quad \forall \, x \in [0,1]$$

and use Theorem 8.5 and Exercise 8.11 with $\hat{S} \equiv h \circ S \circ h^{-1}$.

8.13 Let f and g be two distinct stationary densities of the Frobenius-Perron operator P_S associated with a nonsingular transformation $S :$ $[0,1] \rightarrow [0,1]$ that satisfies the condition of Theorem 8.5. Show that there exist two stationary densities f^* and g^* of P_S such that

$$\text{supp } f^* \cap \text{supp } g^* = \emptyset.$$

8.14 Let $S : [0,1] \rightarrow [0,1]$ be the translation map defined in Example 8.4. Show that S is ergodic with respect to the Lebesgue measure m of $[0,1]$ if and only if a is an irrational number.

8.15 Show that the rotation map in Example 8.6 is ergodic when $a = e^{i\theta}$ such that θ/π is an irrational number.

8.16 Show that the k-adic map $S(x) = kx \pmod{1}$ is mixing on $[0,1]$ for any integer $k > 1$.

8.17 Let $S : X \rightarrow X$ be a measurable transformation on a measurable space (X, Σ). Suppose that μ and ν are two S-invariant probability measures such that S is ergodic with respect to both μ and ν. Show that there exist sets $A, \ B \in \Sigma$ such that $A \cap B = \emptyset$ and $\mu(A) = \nu(B) = 1$.

8.18 Suppose that μ is a unique S-invariant probability measure for a measurable transformation $S : X \rightarrow X$. Show that S is ergodic with respect to the measure μ.

8.19 Let a function $f : [a,b] \rightarrow \mathbb{R}$ be of bounded variation. Let $x, \ y \in [a,b]$ and $x < y$. Prove the following inequality

$$|f(x)| + |f(y)| \leq \bigvee_0^1 f + \frac{2}{y-x} \int_x^y |f(t)|dt.$$

8.20 Show that the equality (8.6) implies the necessity of the condition (8.9) for the ergodicity of a measure-preserving transformation on a probability measure space.

8.21 Prove the conclusions of Remark 8.10 in detail.

Chapter 9

Composition Operators

From Definition 5.14, an abstract composition operator from an M-space X with order unit e into itself is any positive operator $T : X \to X$ such that $Te = e$. From Theorem 5.16, X is lattice isometric to some $C(K)$ for a compact Hausdorff space K. In this chapter we shall study a special class of abstract composition operators from $C(K)$ into itself, which is defined via the composition of continuous functions in $C(K)$ with a given continuous transformation from K into itself.

Composition operators play a fundamental role in the subject of *topological dynamical systems* which studies the asymptotic properties of continuous transformations of topological spaces, as Koopman operators do for measure-preserving transformations of measure spaces that have been studied in the previous chapter.

In Section 9.1 we review some fundamental results for $C(K)$, where K is a compact metric space. We investigate the existence problem of invariant measures for continuous transformations of compact spaces in Section 9.2, and in particular we prove the classic Krylov-Bogolioubov theorem. Section 9.3 is devoted to the elementary ergodic theory of continuous transformations, and Section 9.4 will be focused on the relation of their topological properties and ergodic properties.

9.1 Positive Functionals on $C(K)$

From Theorem 5.16, any M-space can be represented by some $C(K)$, where K is a compact Hausdorff space. In this introductory section we review the basic topological properties of $C(K)$. Throughout the chapter we assume that (K, d) is a compact metric space and $C(K)$ is the Banach lattice of all real-valued continuous functions defined on K with the *max-*

norm $\|f\|_\infty = \max_{x \in K} |f(x)|$, $\forall f \in C(K)$. When $K = [a, b]$, we simply write $C[a, b]$ instead of $C([a, b])$.

Let $\mathcal{B} \equiv \mathcal{B}(K)$ be the σ-algebra of the Borel subsets of K generated by the family of all the open subsets of K. We denote by $\mathcal{M} \equiv \mathcal{M}(K)$ the collection of all probability measures on the measurable space $(K, \mathcal{B})$. Any $\mu \in \mathcal{M}$ is called a *Borel probability measure* on K. The following *Riesz's representation theorem* [Dunford and Schwartz (1957)] basically gives an isomorphic relation between $\mathcal{M}$ and the collection of all positive linear functionals on $C(K)$ with norm 1.

Theorem 9.1. (Riesz's representation theorem) *Let ξ be a positive functional on $C(K)$ such that $\xi(1) = 1$. Then there exists a unique $\mu \in \mathcal{M}$ such that $\xi(f) = \int_K f d\mu$ for all $f \in C(K)$.*

The integral $\int_K f d\mu$ is sometimes written as $\langle f, \mu \rangle$ since the probability measure μ can be viewed as a positive functional on $C(K)$. The uniqueness of the probability measure μ in Theorem 9.1 gives

Corollary 9.1. *If μ and ν are two Borel probability measures on K such that $\int_K f d\mu = \int_K f d\nu$ for all $f \in C(K)$, then $\mu = \nu$.*

From Theorem 9.1, the set $\mathcal{M}$ of all the probability measures on K can be identified with a convex subset of the closed unit ball in the dual space $C(K)'$. The induced topology of $\mathcal{M}$ is referred to as the w'-*topology* of $\mathcal{M}$ inherited from $C(K)'$, which is the smallest topology of $\mathcal{M}$ that makes each of the functionals

$$\langle f, \mu \rangle = \int_K f d\mu, \ \forall \mu \in \mathcal{M}$$

on $\mathcal{M}$ continuous for all $f \in C(K)$.

Thus a sequence $\{\mu_n\}$ of measures in $\mathcal{M}$ converges to a measure $\mu \in \mathcal{M}$ under the w'-topology if and only if

$$\lim_{n \to \infty} \int_K f d\mu_n = \int_K f d\mu, \ \forall f \in C(K).$$

Furthermore, $\lim_{n \to \infty} \mu_n = \mu$ under the w'-topology if and only if $\limsup_{n \to \infty} \mu_n(F) \leq \mu(F)$ for each closed subset F of K, which is equivalent to $\liminf_{n \to \infty} \mu_n(G) \geq \mu(G)$ for each open subset G of K. Moreover, the w'-topology is induced by the metric

$$d(\mu, \nu) = \sum_{n=1}^\infty \frac{|\int_K f_n d\mu - \int_K f_n d\nu|}{2^n \|f_n\|_\infty},$$

where $\{f_n\}_{n=1}^{\infty}$ is a dense subset of $C(K)$ [Walters (1982)]. Therefore, $\mathcal{M}$ can be viewed as a metric space.

By Theorem 9.1, $C(K)'$ is isomorphic to the space $\mathcal{M}_{\text{fin}} \equiv \mathcal{M}_{\text{fin}}(K)$ of all real measures on the Borel measurable space $(K, \mathcal{B}(K))$. Thus the dual operator T' of an abstract composition operator $T : C(K) \to C(K)$ is a positive operator that maps $\mathcal{M}_{\text{fin}}$ into itself. Any $\mu \in \mathcal{M}$ such that $T'\mu = \mu$ is called an *invariant probability measure* of T.

Let $t : K \times \mathcal{B} \to \mathbb{R}_+$ be defined by

$$t(x, A) = T'\delta_x(A), \ \forall \, x \in K, \ A \in \mathcal{B},$$

where δ_x is the Dirac measure centered at x. The function t is called the *transition function* associated with T.

Proposition 9.1. *Let t be the transition function of T. Then*
  (i) *$t(\cdot, A)$ is $\mathcal{B}$-measurable for each fixed $A \in \mathcal{B}$;*
  (ii) *$t(x, \cdot) \in \mathcal{M}$ for each fixed $x \in K$;*
  (iii) *the mapping $t \to t(x, \cdot)$ is continuous from the metric space (K, d) into the topological space $(\mathcal{M}, w')$.*

Proof (i) and (ii) are easy to check. (iii) is from

$$Tf(x) = \langle Tf, \delta_x \rangle = \langle f, T'\delta_x \rangle = \langle f, t(x, \cdot) \rangle$$

and the continuity of the function Tf. $\square$

9.2 Invariant Measures of Continuous Transformations

We now investigate the existence problem of invariant probability measures for continuous transformations on compact matric spaces. Not every measurable transformation on a measure space has an invariant measure, so naturally we ask whether any continuous transformation on a compact space can preserve a measure on the space. An affirmative answer to this question will be provided in the following.

Let K be a compact metric space and let $S : K \to K$ be a continuous transformation. Then the corresponding composition operator $U_S : C(K) \to C(K)$ is a positive operator defined on the M-space $C(K)$ and satisfies the multiplicative property:

$$U_S(f \cdot g) = (U_S f)(U_S g), \ \forall \, f, g \in C(K).$$

Furthermore, U_S is isometric if $S : K \to K$ is continuous and onto, and thus U_S is a multiplicative linear isometry from $C(K)$ onto $C(K)$ if S is a homeomorphism from K onto K.

We first give two important examples [Walters (1982)] of continuous transformations and the corresponding composition operators before considering the existence problem of invariant probability measures.

Example 9.1. Equip the set $\{1, 2, \cdots, k\}$ with the discrete topology, and let $K = \prod_{n=-\infty}^{\infty} \{1, 2, \cdots, k\}$ be the two-sided symbolic space with the corresponding product topology. The topology of this compact Hausdorff space is induced by the metric on K given by

$$d(x, y) = \sum_{n=-\infty}^{\infty} \frac{|x_n - y_n|}{2^{|n|}},$$

where *bi-sequences* $x = \{x_n\}_{n=-\infty}^{\infty}$, $y = \{y_n\}_{n=-\infty}^{\infty} \in K$. Then the *two-sided shift* $S : K \to K$, defined by $S(x) = y$ with $y_n = x_{n+1}$ for all integers n, is a homeomorphism of K, and the corresponding composition operator $U_S : C(K) \to C(K)$ is defined by

$$U_S f(\cdots, x_{-1}, x_0, x_1, \cdots) = f(\cdots, x_0, x_1, x_2, \cdots).$$

If we define $K = \prod_{n=0}^{\infty} \{1, 2, \cdots, k\}$, which is called the *one-sided symbolic space*, with the metric

$$d(x, y) = \sum_{n=0}^{\infty} \frac{|x_n - y_n|}{2^n}$$

for all $x = \{x_n\}_{n=0}^{\infty}$, $y = \{y_n\}_{n=0}^{\infty} \in K$, then the *one-sided shift* $S(\{x_0, x_1, \cdots\}) = \{x_1, x_2, \cdots\}$ is a continuous transformation on K. The inverse image under S of any point consists of k points. Now the composition operator has the expression

$$U_S f(x_0, x_1, \cdots) = f(x_1, x_2, \cdots).$$

Example 9.2. Suppose that $S : K \to K$ is the two-sided shift as above. Let $A = (a_{ij})$ be a $k \times k$ $(0, 1)$-matrix. Denote

$$K_A = \{\{x_n\}_{n=-\infty}^{\infty} \in K : a_{x_n x_{n+1}} = 1, \ \forall \ n\}.$$

In other words, the subset K_A of K in the previous example consists of all the bi-sequences $\{x_n\}_{n=-\infty}^{\infty}$ whose neighboring pairs are "allowed" by the matrix A. Then K_A is a closed subset of K and $S(K_A) = K_A$, so $S : K_A \to K_A$ is a homeomorphism of K_A, which is called the *two-sided topological Markov chain* determined by A. In particular, if $a_{ij} \equiv 1$, then $K_A = K$, and if $A = I$, the identity matrix, then K_A consists of the k points $\{1\}_{n=-\infty}^{\infty}, \{2\}_{n=-\infty}^{\infty}, \cdots, \{k\}_{n=-\infty}^{\infty}$. One can define *one-sided topological Markov chains* similarly. Topological Markov chains are important models in some diffeomorphisms with significant applications in statistical physics and some other subjects [Bowen (1975); Baladi (2000)].

By Alaoglu's theorem, the closed unit ball of $C(K)'$ is compact in the w'-topology [Dunford and Schwartz (1957)]. Since $\mathcal{M}$ is a closed subset of this ball, we have the following important result which plays a key role in the proof of the existence theorem of invariant measures in this section.

Theorem 9.2. *If K is a compact metric space, then $\mathcal{M}(K)$ is a compact metric space under the w'-topology.*

We shall show that for any continuous transformation S from a compact metric space K into itself, there exists an S-invariant probability measure. For this purpose we define, as was done in Example 8.11, an operator $P_S : \mathcal{M} \to \mathcal{M}$ by $P_S\mu = \mu \circ S^{-1}$, namely,

$$P_S\mu(A) = \mu(S^{-1}(A)), \quad \forall\, A \in \mathcal{B},\ \mu \in \mathcal{M}.$$

The correspondence from μ to $P_S\mu$ is called the *Frobenius-Perron operator on measures*, which is more general than the Frobenius-Perron operator defined on all real measures that are absolutely continuous with respect to μ as defined in Section 8.2. It should be noted that the Frobenius-Perron operator on measure as defined here is not a linear operator since its domain $\mathcal{M}$ is not a vector space. The operator P_S is well-defined since the assumption that S is continuous implies the measurability of S with respect to $\mathcal{B}$. Furthermore, it is easy to show that $P_S : \mathcal{M} \to \mathcal{M}$ is continuous with respect to the w'-topology.

Similar to the dual relation between the Frobenius-Perron operator and the Koopman operator expressed in Proposition 8.4, the Frobenius-Perron operator on measures and the composition operator are closely related as the following proposition displays.

Proposition 9.2. *For any $f \in C(K)$ and $\mu \in \mathcal{M}$,*

$$\int_K f d(P_S\mu) = \int_K U_S f d\mu. \tag{9.1}$$

Proof By the definition of P_S we have

$$\int_K \chi_A d(P_S\mu) = \int_K U_S \chi_A d\mu, \quad \forall\, A \in \mathcal{B},$$

which means that the equality (9.1) is valid for all simple functions. By Lebesgue's monotone convergence theorem (Theorem B.2), (9.1) holds for any nonnegative measurable function f via choosing a monotonically increasing sequence of simple functions converging to f. Therefore (9.1) is true for all functions $f \in C(K)$. $\qquad\square$

Corollary 9.2. *A measure $\mu \in \mathcal{M}$ is S-invariant if and only if $\int_K f d\mu = \int_K U_S f d\mu$ for all $f \in C(K)$.*

Now we are ready to prove the following important existence theorem due to Krylov and Bogolioubov.

Theorem 9.3. (The Krylov-Bogolioubov theorem) *If $S : K \to K$ is a continuous transformation of a compact metric space K, then there is an S-invariant probability measure $\mu \in \mathcal{M}$.*

Proof Let $P_S : \mathcal{M} \to \mathcal{M}$ be the corresponding Frobenius-Perron operator on measures. Pick a probability measure $\mu_0 \in \mathcal{M}$ and define a sequence $\{\mu_n\}$ of probability measures in $\mathcal{M}$ by

$$\mu_n = \frac{1}{n} \sum_{i=0}^{n-1} P_S^i \mu_0, \quad \forall\, n.$$

From the compactness property of $\mathcal{M}$ under the w'-topology, there exists a subsequence $\{\mu_{n_k}\}$ of $\{\mu_n\}$ such that $\lim_{k\to\infty} \mu_{n_k} = \mu^* \in \mathcal{M}$. Therefore, for any function $f \in C(K)$,

$$0 \le \left| \int_K f d\mu^* - \int_K U_S f d\mu^* \right| = \lim_{k\to\infty} \left| \int_K f d\mu_{n_k} - \int_K U_S f d\mu_{n_k} \right|$$

$$= \lim_{k\to\infty} \left| \frac{1}{n_k} \int_K \sum_{i=0}^{n_k-1} (U_S^i f - U_S^{i+1} f) d\mu_0 \right|$$

$$= \lim_{k\to\infty} \frac{1}{n_k} \left| \int_K (f - U_S^{n_k} f) d\mu_0 \right| \le \lim_{k\to\infty} \frac{2\|f\|_\infty}{n_k} = 0.$$

By Corollary 9.2, μ^* is S-invariant. $\qquad\square$

9.3   Ergodic Theorems of Composition Operators

Let K be a compact metric space, let $S : K \to K$ be a continuous transformation, and let $\mu \in \mathcal{M}(K)$ be an invariant probability measure of S. We say that μ is *ergodic* or *mixing* if the measure-preserving transformation S of the measure space $(K, \mathcal{B}(K), \mu)$ is ergodic or mixing, respectively.

We need the following theorem [Walters (1982)] that expresses the concepts of ergodicity, mixing, and exactness using L^2-functions, which is not only useful for the proof of Theorem 9.5, but also an alternative way for checking whether a given measure-preserving transformation on a finite measure space possesses mixing properties. Recall that the inner product of a Hilbert space $L^2(X)$ is defined by $(f,g) = \int_X fg d\mu$, $\forall f,g \in L^2(X)$.

Theorem 9.4. *Let (X, Σ, μ) be a probability measure space and let $S : X \to X$ be a measure-preserving transformation. Then*
(i) *S is ergodic if and only if for all f, $g \in L^2(\mu)$,*

$$\lim_{n \to \infty} \frac{1}{n} \sum_{k=0}^{n-1} \int_X (U_S^k f) g \, d\mu = \int_X f d\mu \int_X g d\mu;$$

(ii) *S is mixing if and only if for all f, $g \in L^2(\mu)$,*

$$\lim_{n \to \infty} \int_X (U_S^n f) g \, d\mu = \int_X f d\mu \int_X g d\mu; \tag{9.2}$$

(iii) *with the additional assumption that $S(A) \in \Sigma$ for all $A \in \Sigma$, S is exact if for every $f \in L^2(\mu)$,*

$$\lim_{n \to \infty} \|U_S^n f - (f, 1)\|_2 = 0. \tag{9.3}$$

Proof (i) and (ii) can be proved with the same idea, so we just prove (ii) and (iii), which leads to (i) with a slight modification.

(ii) The sufficiency part follows from Definition 8.5 by putting $f = \chi_A$ and $g = \chi_B$ in (9.2) with A, $B \in \Sigma$.

Conversely, suppose that S is mixing. Definition 8.5 means that (9.2) is satisfied by characteristic functions $f = \chi_A$, $g = \chi_B$ with A, $B \in \Sigma$. Using linear combinations of such functions, we see that the formula (9.2) is true for all simple functions f and g.

Now let $f, g \in L^2(\mu)$ be arbitrary. For any functions s and $\hat{s}$ in $L^2(\mu)$ and all positive integers n,

$$\left| \int_X (U_S^n f) g \, d\mu - \int_X f d\mu \int_X g d\mu \right|$$
$$\leq \left| \int_X (U_S^n f) g \, d\mu - \int_X (U_S^n s) g \, d\mu \right|$$
$$+ \left| \int_X (U_S^n s) g \, d\mu - \int_X (U_S^n s) \hat{s} \, d\mu \right|$$
$$+ \left| \int_X (U_S^n s) \hat{s} \, d\mu - \int_X s d\mu \int_X \hat{s} d\mu \right|$$
$$+ \left| \int_X s d\mu \int_X \hat{s} d\mu - \int_X f d\mu \int_X \hat{s} d\mu \right|$$
$$+ \left| \int_X f d\mu \int_X \hat{s} d\mu - \int_X f d\mu \int_X g d\mu \right|.$$

Given any $\epsilon > 0$, choose simple functions s and $\hat{s}$ such that $\|f - s\|_2 < \epsilon$ and $\|g - \hat{s}\|_2 < \epsilon$, and choose a positive integer $n_0(\epsilon)$ such that

$$\left| \int_X (U_S^n s)\hat{s}\, d\mu - \int_X s\, d\mu \int_X \hat{s}\, d\mu \right| < \epsilon$$

for all $n \geq n_0(\epsilon)$. Then for any such n,

$$\left| \int_X (U_S^n f)g\, d\mu - \int_X f\, d\mu \int_X g\, d\mu \right|$$

$$\leq \left| \int_X [U_S^n(f - s)]g\, d\mu \right| + \left| \int_X (U_S^n s)(g - \hat{s})d\mu \right| + \epsilon$$

$$+ \left| \int_X (s - f)d\mu \int_X \hat{s}\, d\mu \right| + \left| \int_X f\, d\mu \int_X (\hat{s} - g)d\mu \right|$$

$$\leq \|U_S^n(f - s)\|_2 \|g\|_2 + \|U_S^n s\|_2 \|g - \hat{s}\|_2 + \epsilon$$

$$+ \|s - f\|_2 \|\hat{s}\|_2 + \|f\|_2 \|\hat{s} - g\|_2$$

$$= \|(f - s)\|_2 \|g\|_2 + \|s\|_2 \|g - \hat{s}\|_2 + \epsilon$$

$$+ \|s - f\|_2 \|\hat{s}\|_2 + \|f\|_2 \|\hat{s} - g\|_2$$

$$\leq \epsilon \|g\|_2 + \epsilon \|s\|_2 + \epsilon + \epsilon \|\hat{s}\|_2 + \epsilon \|f\|_2,$$

where we have used Proposition 9.1 and the Cauchy-Schwarz inequality. Therefore (9.2) holds for all $f, g \in L^2(\mu)$.

(iv) Suppose that the limit (9.3) is true. Let $A \in \Sigma$ be such that $\mu(A) > 0$ and define the function $f_A(x) = \chi_A(x)/\mu(A)$ for all $x \in X$. Then $\lim_{n\to\infty} \|U_S^n f_A - 1\|_2 = 0$. Since the measure μ is S-invariant, by Proposition 8.1, $\int_X U_S^n \chi_{S^n(A)}\, d\mu = \int_X \chi_{S^n(A)} d\mu$. So,

$$\mu(S^n(A)) = \int_X \chi_{S^n(A)} d\mu$$

$$= \int_X f_A U_S^n \chi_{S^n(A)}\, d\mu - \int_X (f_A - 1)\chi_{S^n(A)}\, d\mu$$

$$= \int_{S^{-n}(S^n(A))} f_A d\mu - \int_{S^{-n}(S^n(A))} (f_A - 1)d\mu$$

$$= 1 - \int_{S^n(A)} U_S^n(f_A - 1)d\mu = 1 - \int_{S^n(A)} (U_S^n f_A - 1)d\mu,$$

and hence

$$\mu(S^n(A)) \geq 1 - \|U_S^n f_A - 1\|_1 \geq 1 - \|U_S^n f_A - 1\|_2 \to 1$$

as $n \to \infty$. Therefore, S is exact. $\qquad\square$

Remark 9.1. In the previous theorem, it is enough to require the convergence for f and g in a fundamental set of $L^2(\mu)$.

Based on the above result which characterizes ergodicity and mixing for general measure-preserving transformations on finite measure spaces, the following Theorem 9.5 characterizes the properties of ergodicity and mixing in the context of topological dynamics of continuous transformations. For its proof, a modification of Theorem 9.4 is needed here. Note that $C(K)$ is a subspace of $L^\infty(K)$ for any $\mu \in \mathcal{M}$.

Lemma 9.1. *Let* $S : K \to K$ *be a continuous transformation and let* $\mu \in \mathcal{M}$ *be an invariant probability measure of* S. *Then*
(i) μ *is ergodic if and only if for all* $f \in L^1(\mu)$ *and* $g \in C(K)$,

$$\lim_{n\to\infty} \frac{1}{n} \sum_{k=0}^{n-1} \int_K f U_S^k g \, d\mu = \int_K f d\mu \int_K g d\mu;$$

(ii) μ *is mixing if and only if for all* $f \in L^1(\mu)$ *and* $g \in C(K)$,

$$\lim_{n\to\infty} \int_K f U_S^n g \, d\mu = \int_K f d\mu \int_K g d\mu.$$

Proof (i) Suppose that the convergence condition holds. Let f, $g \in L^2(\mu)$. Then $f \in L^1(\mu)$ since $L^2(\mu) \subset L^1(\mu)$, so

$$\lim_{n\to\infty} \frac{1}{n} \sum_{k=0}^{n-1} \int_K f U_S^k h \, d\mu = \int_K f d\mu \int_K h d\mu, \ \forall \, h \in C(K).$$

Expressing g in $L^2(\mu)$ as a limit of a sequence $\{h_n\}$ of continuous functions in $C(K)$, we get from the above that

$$\lim_{n\to\infty} \frac{1}{n} \sum_{k=0}^{n-1} \int_K f U_S^k g \, d\mu = \int_K f d\mu \int_K g d\mu,$$

which together with Theorem 9.4(i) implies that μ is ergodic.

Conversely, suppose that μ is ergodic with respect to μ. Let $g \in C(K)$. Then $g \in L^2(\mu)$. So when $h \in L^2(\mu)$, we have

$$\lim_{n\to\infty} \frac{1}{n} \sum_{k=0}^{n-1} \int_K h U_S^k g \, d\mu = \int_K h d\mu \int_K g d\mu.$$

Approximating $f \in L^1(\mu)$ by $h \in L^2(\mu)$ in L^1-norm gives

$$\lim_{n\to\infty} \frac{1}{n} \sum_{k=0}^{n-1} \int_K f U_S^k g \, d\mu = \int_K f d\mu \int_K g d\mu.$$

The proof of (ii) is similar by using Theorem 9.4(ii). $\qquad\square$

We can employ the notions of Frobenius-Perron operators on measures to characterize the concepts of ergodicity and mixing for continuous transformations on compact metric spaces.

Theorem 9.5. *Let* $S : K \to K$ *be a continuous transformation,* P_S *the corresponding Frobenius-Perron operator on measures, and* $\mu \in \mathcal{M}$ *an* S-*invariant probability measure. Then,*

(i) μ *is ergodic if and only if for any* $\nu \in \mathcal{M}$ *with* $\nu \ll \mu$,

$$\lim_{n \to \infty} \frac{1}{n} \sum_{k=0}^{n-1} P_S^k \nu = \mu \tag{9.4}$$

under the w'-*topology of* $\mathcal{M}$;

(ii) μ *is mixing if and only if for any* $\nu \in \mathcal{M}$ *such that* $\nu \ll \mu$, *we have under the* w'-*topology of* $\mathcal{M}$

$$\lim_{n \to \infty} P_S^n \nu = \mu.$$

Proof (i) Suppose that the given invariant probability measure μ is ergodic. Let $\nu \in \mathcal{M}$ be such that $\nu \ll \mu$, so $g = d\nu/d\mu \in \mathcal{D}$. If $f \in C(K)$, then Lemma 9.1(i) implies that

$$\lim_{n \to \infty} \int_K f d \left(\frac{1}{n} \sum_{k=0}^{n-1} P_S^k \nu \right) = \lim_{n \to \infty} \frac{1}{n} \sum_{k=0}^{n-1} \int_K U_S^k f \, d\nu$$

$$= \lim_{n \to \infty} \frac{1}{n} \sum_{k=0}^{n-1} \int_K g U_S^k f \, d\mu = \int_K f d\mu \int_K g d\mu.$$

Therefore (9.4) is true.

Conversely, suppose that (9.4) is satisfied. Let $f \in L^1(\mu)$ be such that $f \geq 0$. Define $\nu \in \mathcal{M}$ by

$$\nu(A) = \frac{\int_K f \chi_A d\mu}{\int_K f d\mu}, \quad \forall A \in \mathcal{B}.$$

Then for any $g \in C(K)$, (9.4) ensures that

$$\lim_{n \to \infty} \frac{1}{n} \sum_{k=0}^{n-1} \int_K f U_S^k g \, d\mu = \int_K f d\mu \int_K g d\mu. \tag{9.5}$$

For $f \in L^1(\mu)$, write $f = f^+ - f^-$. Applying the above to $f^+ \geq 0$ and $f^- \geq 0$, we see that (9.5) is satisfied by all $f \in L^1(\mu)$ and $g \in C(K)$. Hence, μ is ergodic thanks to Lemma 9.1(i).

By using Lemma 9.1(ii) we can proves (ii). $\square$

The following result strengthens Birkhoff's pointwise ergodic theorem in the case of topological dynamical systems.

Theorem 9.6. *Let $S : K \to K$ be a continuous transformation. Suppose that $\mu \in \mathcal{M}$ is S-invariant and ergodic. Then there exists $A \in \mathcal{B}$ with $\mu(A) = 1$ such that*

$$\lim_{n\to\infty} \frac{1}{n} \sum_{k=0}^{n-1} U_S^k f(x) = \int_K f d\mu, \ \forall\, x \in A, \ f \in C(K). \tag{9.6}$$

Proof Pick up a sequence $\{f_j\}_{j=1}^{\infty}$ as a countable dense subset of $C(K)$. By Birkhoff's pointwise ergodic theorem, for every j there exists a set $A_j \in \mathcal{B}$ such that $\mu(A_j) = 1$ and

$$\lim_{n\to\infty} \frac{1}{n} \sum_{k=0}^{n-1} U_S^k f_j(x) = \int_K f_j d\mu, \ \forall\, x \in A_j.$$

Denote $A = \bigcap_{j=1}^{\infty} A_j$. Then $\mu(A) = 1$ and

$$\lim_{n\to\infty} \frac{1}{n} \sum_{k=0}^{n-1} U_S^k f_j(x) = \int_K f_j d\mu, \ \forall\, x \in A, \ j \geq 1.$$

Therefore, (9.6) follows since $\overline{\{f_j\}_{j=1}^{\infty}} = C(K)$. $\qquad\square$

The above limit theorem implies another characterization of ergodicity. Recall that $\delta_x \in \mathcal{M}$ is the Dirac measure centered at x for any $x \in K$. Clearly $P_S^k \delta_x = \delta_{S^k(x)}$ for any positive integer k.

Theorem 9.7. *Let $S : K \to K$ be a continuous transformation and let $\mu \in \mathcal{M}$ be S-invariant. Then μ is ergodic if and only if*

$$\lim_{n\to\infty} \frac{1}{n} \sum_{k=0}^{n-1} P_S^k \delta_x = \mu, \ \forall\, x \in K \ \mu - \text{a.e.}$$

under the w'-topology of $\mathcal{M}$.

Proof The necessity part is from Theorem 9.6 immediately. We now prove the sufficiency part. Suppose that there is $A \in \mathcal{B}$ with $\mu(A) = 1$, such that in the w'-topology of $\mathcal{M}$,

$$\lim_{n\to\infty} \frac{1}{n} \sum_{k=0}^{n-1} P_S^k \delta_x = \mu, \ \forall\, x \in A.$$

Then for all $x \in A$, $f \in L^1(\mu)$, and $g \in C(K)$, we have

$$\lim_{n \to \infty} \frac{1}{n} \sum_{k=0}^{n-1} f(x) U_S^k g(x) = f(x) \cdot \int_K g d\mu,$$

hence Lebesgue's dominated convergence theorem implies

$$\lim_{n \to \infty} \frac{1}{n} \sum_{k=0}^{n-1} \int_K f U_S^k g \, d\mu = \int_K f d\mu \int_K g d\mu.$$

This means that μ is ergodic. □

At the end of this section, we study the situation when there is only one invariant measure in $\mathcal{M}$ for the given S.

Definition 9.1. A continuous transformation $S : K \to K$ is said to be *uniquely ergodic* if there is only one S-invariant Borel probability measure.

Theorem 9.6 can be strengthened if S is uniquely ergodic.

Theorem 9.8. *Let $S : K \to K$ be a continuous transformation. Then the following are equivalent:*

(i) S is uniquely ergodic.

(ii) Given each $f \in C(K)$, the sequence $\left\{ n^{-1} \sum_{k=0}^{n-1} U_S^k f(x) \right\}$ converges to a constant function uniformly for all $x \in K$.

(iii) Given each $f \in C(K)$, the sequence $\left\{ n^{-1} \sum_{k=0}^{n-1} U_S^k f(x) \right\}$ converges to a constant function for all $x \in K$.

(iv) There exists an invariant probability measure μ such that for all functions $f \in C(K)$ and all points $x \in K$,

$$\lim_{n \to \infty} \frac{1}{n} \sum_{k=0}^{n-1} U_S^k f(x) = \int_K f d\mu.$$

Proof (i) $\Rightarrow$ (ii) Let μ be the unique invariant Borel probability measure of S. If (ii) is not satisfied, then there exist a function $f \in C(K)$ and a number $\epsilon_0 > 0$ such that for any positive integer n_0 there exist an integer $n > n_0$ and a point $x_n \in K$ satisfying

$$\left| \frac{1}{n} \sum_{k=0}^{n-1} U_S^k f(x_n) - \int_K f d\mu \right| \geq \epsilon_0. \tag{9.7}$$

In other words, $\left| \int_K f d\mu_n - \int_K f d\mu \right| \geq \epsilon_0$, where

$$\mu_n = \frac{1}{n} \sum_{k=0}^{n-1} \delta_{S^k(x_n)} = \frac{1}{n} \sum_{k=0}^{n-1} P_S^k \delta_{x_n}.$$

Since the sequence $\{\mu_n\}$ is pre-compact in the w'-topology, it has a subsequence $\{\mu_{n_i}\}$ that converges to some $\nu \in \mathcal{M}$. Moreover, the proof of Theorem 9.3 implies that ν is an invariant probability measure. Taking limit $n_i \to \infty$ in (9.7) gives $|\int_K f d\nu - \int_K f d\mu| \geq \epsilon_0$, so $\nu \neq \mu$. This contradicts the unique ergodicity of S.

(ii) $\Rightarrow$ (iii) It is obvious.

(iii) $\Rightarrow$ (iv) Note that for all positive integers n,

$$\left| \frac{1}{n} \sum_{k=0}^{n-1} U_S^k f(x) \right| \leq \|f\|_\infty.$$

Hence the given condition defines a continuous linear functional

$$\xi(f) = \lim_{n \to \infty} \frac{1}{n} \sum_{k=0}^{n-1} U_S^k f(x), \ \forall f \in C(K).$$

It is clear that $\xi(1) = 1$ and $\xi(f) \geq 0$ for $f \geq 0$. Thus by the Riesz representation theorem, there exists a Borel probability measure μ such that $\xi(f) = \int_K f d\mu$ for all functions $f \in C(K)$. Since $\xi(U_S f) = \xi(f)$ by a direct verification, $\int_K U_S f d\mu = \int_K f d\mu$ for all $f \in C(K)$. Therefore, μ is S-invariant by Corollary 9.2.

(iv) $\Rightarrow$ (i) Suppose that ν is an S-invariant Borel probability measure. Let $f \in C(K)$. Then (iv) says that

$$\lim_{n \to \infty} \frac{1}{n} \sum_{k=0}^{n-1} U_S^k f(x) = \int_K f d\mu, \ \forall x \in K.$$

Integrating the above equality with respect to ν and using Lebesgue's dominated convergence theorem give

$$\int_K f d\nu = \int_K \left(\int_K f d\mu \right) d\nu = \int_K f d\mu \int_K 1 d\nu = \int_K f d\mu.$$

Hence $\nu = \mu$ by Corollary 9.1, namely S is uniquely ergodic. $\qquad \square$

9.4 Topological Transitivity and Ergodicity

In this section we introduce several basic concepts of topological dynamical systems such as topological transitivity for continuous transformations and minimality for homeomorphisms of compact metric spaces, and we also explore some relations between the topological properties and the ergodic properties of such transformations. Let K be a compact metric space.

Definition 9.2. A homeomorphism $S : K \to K$ is *minimal* if the *two-sided* orbit $\{S^n(x)\}_{n=-\infty}^\infty$ of each $x \in K$ is dense in K.

Theorem 9.9. *Let $S : K \to K$ be a homeomorphism. Then the following statements are equivalent:*

(i) *S is minimal.*

(ii) *If $F \subset K$ is closed and $S(F) = F$, then $F = \emptyset$ or K.*

(iii) *$\bigcup_{n=-\infty}^{\infty} S^n(G) = K$ for any open subset $G \neq \emptyset$ of K.*

Proof (i) $\Rightarrow$ (ii) Suppose that S is minimal and $F \neq \emptyset$ is a closed subset of K such that $S(F) = F$. Let $x \in F$. Since $\{S^n(x)\}_{n=-\infty}^{\infty}$ is contained in F and is dense in K, $F = K$.

(ii) $\Rightarrow$ (iii) If G is a nonempty open subset of K, then the set $F = \left(\bigcup_{n=-\infty}^{\infty} S^n(G)\right)^c$ is closed and $S(F) = F$. Since $G \neq \emptyset$, $F \neq K$, so from (ii) $F = \emptyset$. Hence $\bigcup_{n=-\infty}^{\infty} S^n(G) = K$.

(iii) $\Rightarrow$ (i) Given a point $x \in K$. Let G be any nonempty open subset of K. By (iii), $x \in S^n(G)$ for some integer n. So $S^{-n}(x) \in G$, and hence the two-sided orbit of x is dense in K. $\qquad \square$

Theorem 9.9(ii) means that a minimal homeomorphism of a compact metric space is "indecomposable," a similar property to ergodicity for a measurable transformation of a measure space. The following theorem asserts that the only invariant functions of the composition operator $U_S : C(K) \to C(K)$ associated with a minimal homeomorphism $S : K \to K$ are constant ones.

Theorem 9.10. *Let $S : K \to K$ be a minimal homeomorphism. If $U_S f = f$ for some $f \in C(K)$, then f is a constant function.*

Proof The condition $f \circ S = f$ implies that $f \circ S^n = f$ for all integers n. Since S is a minimal homeomorphism, we know that f is constant on the dense two-sided orbit of any $x \in X$. The fact that f is continuous then ensures that it must be a constant function. $\qquad \square$

Remark 9.2. Although Theorem 9.10 is similar to Theorem 8.8 for ergodic transformations, the condition that the only invariant functions of S in $C(K)$ are constant ones does not imply minimality of S in general.

We now introduce topological transitivity for continuous transformations, a concept which is weaker than minimality.

Definition 9.3. A continuous transformation $S : K \to K$ is called *one-sided topologically transitive* if there exists some $x \in K$ whose *orbit* $\{S^n(x)\}_{n=0}^{\infty}$ is dense in K. A homeomorphism $S : K \to K$ is said to be *topologically transitive* if there is some $x \in K$ whose two-sided orbit $\{S^n(x)\}_{n=-\infty}^{\infty}$ is dense in K.

Theorem 9.11. *Let $S : K \to K$ be a homeomorphism. Then the following statements are equivalent:*

(i) *S is topologically transitive.*

(ii) *If F is a closed subset of K such that $S(F) = F$, then either $F = K$ or F is nowhere dense in K.*

(iii) *If G_1 and G_2 are nonempty open subsets of K, then there exists an integer n such that $S^n(G_1) \cap G_2 \neq \emptyset$.*

(iv) *The set of all points $x \in K$ whose two-sided orbit is dense in K is a dense intersection of countably many open sets.*

Proof (i) $\Rightarrow$ (ii) Suppose that the two-sided orbit $\{S^n(x_0)\}_{n=-\infty}^{\infty}$ of some $x_0 \in K$ is dense in K. Let $F \subset K$ be closed such that $S(F) = F$. If F is not nowhere dense in K, then there is a nonempty open set $G \subset F$. Thus, there is an integer n with $S^n(x_0) \in G$. Since $G \subset F$, the two-sided orbit is contained in F. We must have $F = K$ since $\{S^n(x_0)\}_{n=-\infty}^{\infty}$ is dense in K.

(ii) $\Rightarrow$ (iii) Suppose that G_1 and G_2 are any two nonempty open subsets of K. Then the union $\bigcup_{n=-\infty}^{\infty} S^n(G_1)$ is open in K and S-invariant, so it is dense by condition (ii). Thus $\bigcup_{n=-\infty}^{\infty} S^n(G_1) \cap G_2 \neq \emptyset$, which implies that $S^n(G_1) \cap G_2 \neq \emptyset$ for some integer n.

(iii) $\Rightarrow$ (iv) Let $G_1, G_2, \cdots, G_n, \cdots$ be a countable base for K. Then $\bigcap_{n=1}^{\infty} \bigcup_{k=-\infty}^{\infty} S^k(G_n)$ is the set of all points $x \in K$ whose two-sided orbit is dense in K. Since $\bigcup_{k=-\infty}^{\infty} S^k(G_n)$ is a dense subset of K for each n by condition (iii), we obtain (iv).

(iv) $\Rightarrow$ (i) This is obvious. $\qquad\square$

Similarly, we can prove the following theorem for one-sided topological transitivity for continuous and onto transformations.

Theorem 9.12. *Let $S : K \to K$ be a continuous transformation with $S(K) = K$. Then the following statements are equivalent:*

(i) *S is one-sided topologically transitive.*

(ii) *If F is a closed subset of K such that $S(F) \subset F$, then either $F = K$ or F is nowhere dense in K.*

(iii) *If G_1 and G_2 are nonempty open subsets of K, then there exists a positive integer $n \geq 1$ such that $S^{-n}(G_1) \cap G_2 \neq \emptyset$.*

(iv) *The set of all points $x \in K$ whose orbit is dense in K is a dense intersection of countably many open sets.*

In what follows we further explore some relationships between the topological properties and the ergodic properties for continuous transformations of compact metric spaces. The theorem below indicates that ergodicity implies topological transitivity in some sense.

Theorem 9.13. *Let $S : K \to K$ be a homeomorphism and let a measure $\mu \in \mathcal{M}$ satisfy the condition that $\mu(G) > 0$ for every nonempty open subset G of K. If S is an ergodic measure-preserving transformation with respect to μ, then the two-sided orbit of x is dense in K for $x \in K$ μ-a.e. Consequently, S is topologically transitive.*

Proof Suppose that $G_1, G_2, \cdots, G_n, \cdots$ is a countable base for K. Then $\bigcap_{n=1}^{\infty} \bigcup_{k=-\infty}^{\infty} S^k(G_n)$ is the set of all points $x \in K$ whose two-sided orbit is dense in K. For each n, the open set $\bigcup_{k=-\infty}^{\infty} S^k(G_n)$ is S-invariant, so by ergodicity it has measure 0 or 1. Since this set contains G_n and $\mu(G_n) > 0$ by assumption, $\mu\left(\bigcup_{k=-\infty}^{\infty} S^k(G_n)\right) = 1$. Therefore,

$$\mu\left(\bigcap_{n=1}^{\infty} \bigcup_{k=-\infty}^{\infty} S^k(G_n)\right) = 1. \square$$

The concept of minimality is closely related to that of unique ergodicity as the following theorem demonstrates.

Theorem 9.14. *Let a homeomorphism $S : K \to K$ be uniquely ergodic with the invariant probability measure μ. S is minimal if and only if $\mu(G) > 0$ for all open subsets $G \neq \emptyset$ of K.*

Proof Let S be minimal. If $G \neq \emptyset$ is open, then $K = \bigcap_{n=-\infty}^{\infty} S^n(G)$ by Theorem 9.9(iii). So $\mu(G) > 0$ since $\mu(G) = \mu(S^n(G))$ for all integers n.

Conversely, assume $\mu(G) > 0$ for all open sets $G \neq \emptyset$. If S is not minimal, then by Theorem 9.9(ii), there is a closed subset $F \neq \emptyset$ of K such that $S(F) = F$. By Theorem 9.3, $S : F \to F$ has an invariant Borel probability measure μ_F. Then the measure ν on K defined by $\nu(A) = \mu_F(F \cap A)$ for all Borel subsets A of K is S-invariant. Since F^c is nonempty and open, $\mu(F^c) > 0$ by the assumption. Thus $\nu \neq \mu$ since $\nu(F^c) = 0$. This contradicts the unique ergodicity of S. $\square$

9.5   Notes and Remarks

1. Topological Markov chains and the related shift transformations in Examples 9.1 and 9.2 are important mathematical models of dynamical systems studied in [Bowen (1975); Ruelle (1978)].

2. More relationships of topological properties and ergodic properties related to invariant measures for continuous transformations of compact metric spaces can be found in [Walters (1982)].

Exercises

9.1 Let $\{\mu_n\}$ be a sequence in $\mathcal{M}$. Prove:

(i) $\lim_{n\to\infty} \mu_n = \mu \in \mathcal{M}$ under the w'-topology if and only if $\limsup_{n\to\infty} \mu(F) \leq \mu_n(F)$ for each closed subset F of K.

(ii) $\lim_{n\to\infty} \mu_n = \mu \in \mathcal{M}$ under the w'-topology if and only if $\liminf_{n\to\infty} \mu_n(G) \geq \mu(G)$ for each open subset G of K.

(iii) $\lim_{n\to\infty} \mu_n = \mu \in \mathcal{M}$ under the w'-topology if and only if $\lim_{n\to\infty} \mu_n(A) = \mu(A)$ for each $A \in \mathcal{B}(K)$ with $\mu(\partial A) = 0$.

9.2 Consider the index set $\{1, 2, \cdots, k\}$ on which a measure μ is defined such that $\mu(\{i\}) = p_i$ for $i = 1, 2, \cdots, k$ and $\sum_{i=1}^{k} p_i = 1$. Denote $K = \prod_{n=-\infty}^{\infty}\{1, 2, \cdots, k\}$ with the product measure $\hat{\mu}$. Show that the two-sided $(p_1, \cdots, p_k)$-shift $S : K \to K$, defined by $S\left(\{x_n\}_{n=-\infty}^{\infty}\right) = \{y_n\}_{n=-\infty}^{\infty}$ with $y_n = x_{n+1}$ for all n, preserves $\hat{\mu}$.

9.3 Let $K = \prod_{n=0}^{\infty}\{1, 2, \cdots, k\}$ as above with the product measure $\hat{\mu}$. Let the one-sided $(p_1, \cdots, p_k)$-shift $S : K \to K$ be defined by $S(\{x_0, x_1, \cdots\}) = \{x_1, x_2, \cdots\}$. Show that $\hat{\mu}$ is S-invariant.

9.4 Show that the Frobenius-Perron operator on measures $P_S : \mathcal{M}(K) \to \mathcal{M}(K)$ is continuous under the w'-topology.

9.5 Show that the composition operator $U_S : C(K) \to C(K)$ is isometric if $S : K \to K$ is continuous and onto, and thus U_S is a multiplicative linear isometry from $C(K)$ onto $C(K)$ if S is a homeomorphism from K onto K.

9.6 Let $S : K \to K$ be a homeomorphism and let $F \subset K$ be a closed set such that $SF \subset F$. If the restriction $S|_F$ of S to F is minimal, then F is called a *minimal set* for S. Show that every homeomorphism $S : K \to K$ has a minimal set.

9.7 Show that the conclusion of Theorem 9.10 is still true for any topologically transitive homeomorphism or one-sided topologically transitive continuous transformation $S : K \to K$.

9.8 Show that the rotation map $S(z) = az$ in Example 8.6 is minimal if and only if $\{a^n\}_{n=-\infty}^{\infty}$ is dense in $\partial\mathbb{D}^2$.

9.9 The *ω-limit set* of a point $x \in K$ for a continuous transformation

$S : K \to K$ is defined to be the set $\omega(x) = \{ y \in K : \lim_{i \to \infty} S^{n_i}(x) = y$ for some $\{n_i\} \subset \{n\}\}$. Show that

    (i) $\omega(x)$ is a nonempty closed subset of K;

    (ii) $S(\omega(x)) = \omega(x)$.

9.10 Find a non-homeomorphism $S : K \to K$ for which $\omega(x)$ is a proper subset of $S^{-1}(\omega(x))$ for some $x \in K$.

9.11 Let $S : K \to K$ be a homeomorphism and let $F \subset K$ be a minimal set for S. Show that $\omega(x) = F$ for all $x \in F$. In particular, $\omega(x) = K$ for all $x \in K$ if S is minimal.

9.12 Let $S : K \to K$ be a topologically transitive homeomorphism which is also isometric with respect to the metric of K. Show that S is minimal.

9.13 Show that the set of all invariant probability measures of a continuous transformation $S : K \to K$ is a closed and convex subset of $\mathcal{M}(K)$.

9.14 Show that every extreme point of the set of all invariant Borel probability measures of a continuous transformation $S : K \to K$ is an ergodic invariant measure of S.

9.15 Let μ and ν be two invariant probability measures of $S : K \to K$ such that $\nu \ll \mu$ and S is ergodic with respect to μ. Show that $\nu = \mu$.

9.16 Let $S : K \to K$ be a continuous transformation and let $x \in K$ be a given point. Show that $S^n(x) = x$ if and only if $n^{-1} \sum_{i=0}^{n-1} \delta_{S^i(x)}$ is an invariant probability measure of S.

9.17 Show that $S : \partial \mathbb{D}^2 \to \partial \mathbb{D}^2$ defined by $S(\exp(2\pi i \theta)) = \exp(2\pi i \theta^2)$, $\theta \in [0, 1)$ is a uniquely ergodic homeomorphism which is not minimal. Find the unique invariant probability measure of S.

9.18 Show that the rotation map S introduced in Example 8.5 is uniquely ergodic if and only if S is minimal.

9.19 Let $P_S : \mathcal{M}(K) \to \mathcal{M}(K)$ be the Frobenius-Perron operator on measures and let $x \in K$. Show that $P_S^n \delta_{x_0} = \delta_{S^n(x_0)}$.

9.20 Let a Borel measurable transformation $S : \mathbb{R}_+ \to \mathbb{R}_+$ be defined via

$$S(x) = x \chi_{[1,\infty)}(x), \ \forall \, x \in \mathbb{R}_+.$$

Denote by $\mathcal{M}(\mathbb{R}_+)$ the set of all Borel probability measures on $\mathbb{R}_+$. Show that the corresponding Frobenius-Perron operator on measures $P_S : \mathcal{M}(\mathbb{R}_+) \to \mathcal{M}(\mathbb{R}_+)$ has the expression

$$P_S \mu(A) = \mu([0, 1)) \delta_0 + \mu([1, \infty)) \cap A), \ \forall \, \mu \in \mathcal{M}(\mathbb{R}_+), \ A \in \mathcal{B}.$$

9.21 Show that not every measurable transformation on a measure space has an invariant measure by providing an example.

Chapter 10

Approximations of Markov Operators

Approximations of infinite dimensional linear operators by finite dimensional ones are much needed in applications. Since a bounded linear operator normalized by its norm has norm 1, and since Markov operators constitute an important class of positive operators with norm 1, the present chapter is only concerned with the problem of approximating infinite dimensional Markov operators via finite dimensional Markov operators. In the next chapter we shall study the problem of approximating continuous functions by using finite dimensional positive operators.

Let (X, Σ, μ) be a σ-finite measure space and let $P : L^1(X) \to L^1(X)$ be a Markov operator. By Theorem 7.4, a stationary density of P describes the eventual behavior for the iteration of P. In particular, when P_S is a Frobenius-Perron operator associated with $S : X \to X$, a stationary density of P_S is the density of an absolutely continuous S-invariant probability measure. In applications of ergodic theory to physical sciences and engineering, however, the analytic expression of such a stationary density is often not available, even though its existence may be guaranteed theoretically. Therefore, it is important to be able to numerically compute the stationary density to a prescribed precision.

Since Markov operators are positive operators that preserve the integral of nonnegative functions, it is natural to compute their stationary densities with finite dimensional positive operators, preferably Markov operators, to preserve the intrinsic structure of the original operator. With this principle in mind, we introduce several structure-preserving finite approximations of Markov operators using piecewise polynomials. Such methods will be called *Markov finite approximations* since they use finite dimensional Markov operators to approximate the original Markov operator. Solving the resulting fixed point problem of the discretized operator is equivalent to solving the eigenvector problem of a stochastic matrix corresponding to eigenvalue 1. This approach originates from Ulam's famous book, A Collection of Mathematical Problems [Ulam (1960)], in which he proposed a piecewise constant numerical scheme to approximate a stationary density of the Frobenius-Perron operator associated with an interval mapping.

We investigate Ulam's method in Section 10.1 and extend it to the piecewise linear Markov approximation method in Section 10.2. Their convergence rate analysis will be performed in Sections 10.3 and 10.4, respectively. This chapter is an introduction to the modern research area of computational ergodic theory, and we list some recent works [Froyland (1995); Keane *et al.* (1998); Murray (1998); Dellnitz and Junge (1999); Liverani (2001); Jenkinson and Pollicott (2004); Terhesiu and Froyland (2008)] for reader's interest in further study.

10.1   Ulam's Piecewise Constant Approximations

In 1960, Ulam proposed [Ulam (1960)] a piecewise constant approximation scheme for computing a stationary density of the Frobenius-Perron operator P_S associated with an interval mapping. He conjectured that the resulting sequence of piecewise constant densities converge to a stationary density of P_S if P_S has a stationary density. In 1976, Li [Li (1976)] proved Ulam's conjecture for the class of piecewise C^2 and stretching mappings for which the existence of a stationary density was proved by Lasota and Yorke, as presented in Section 8.3.

Suppose that (X, Σ, μ) is a given finite measure space throughout this chapter. Let n be any chosen positive integer, and let $X_1, \cdots, X_n \in \Sigma$ be n mutually disjoint measurable subsets of X that form a nontrivial measurable partition of X, that is, $\mu(X_k) > 0$ for every k and $X = \bigcup_{k=1}^{n} X_k$. Denote by Δ_n the space of all the *piecewise constant functions* associated with the above partition of X. Then Δ_n is obviously an n-dimensional subspace of $L^1(X)$ which is spanned by the *canonical basis*

$$\chi_{X_1}, \ \chi_{X_2}, \ \cdots, \ \chi_{X_n}$$

of the characteristic functions of $X_1, \cdots, X_n$. The functions

$$1_k = \frac{1}{\mu(X_k)} \chi_{X_k}, \quad k = 1, \cdots, n$$

are all densities and constitute a *density basis* for Δ_n, which will be more convenient for us to use in the subsequent discussion.

We define a linear operator Q_n from $L^1(X)$ into itself by

$$Q_n f(x) = \sum_{k=1}^{n} \frac{1}{\mu(X_k)} \int_{X_k} f d\mu \cdot \chi_{X_k}(x), \quad \forall\, x \in X, \tag{10.1}$$

that is, $Q_n f$ is a piecewise constant function whose value on each subset X_k is the *average value* $\mu(X_k)^{-1} \int_{X_k} f d\mu$ of f on X_k.

Proposition 10.1. *The linear operator* $Q_n : L^1(X) \to L^1(X)$ *defined by (10.1) satisfies the following properties:*
(i) Q_n *is a Markov operator with range* $R(Q_n) = \Delta_n$.
(ii) Q_n *is a Galerkin projection from* L^1 *onto* Δ_n, *namely,*

$$\langle f - Q_n f, \chi_{X_k} \rangle = 0, \quad k = 1, \cdots, n, \quad \forall\, f \in L^1(X).$$

(iii) $\|Q_n g\|_\infty \leq \|g\|_\infty$ *for any* $g \in L^\infty(X)$.

Proof (i) Q_n is obviously positive. Let $f \in L^1(X)$. Then

$$\int_X Q_n f d\mu = \sum_{k=1}^{n} \frac{1}{\mu(X_k)} \int_{X_k} f d\mu \cdot \int_X \chi_{X_k} d\mu$$

$$= \sum_{k=1}^{n} \int_{X_k} f d\mu = \int_{\cup_{k=1}^{n} X_k} f d\mu = \int_X f d\mu,$$

thus, Q_n preserves the integral and so is a Markov operator.
(ii) Q_n is also a Galerkin projection onto Δ_n since

$$\langle f - Q_n f, \chi_{X_k} \rangle = \int_{X_k} f d\mu - \int_{X_k} f d\mu = 0, \quad \forall\, k, \ f \in L^1(X).$$

(iii) This is easily seen from the definition of Q_n. $\qquad\square$

Suppose that $P : L^1(X) \to L^1(X)$ is a Markov operator. Define the operator $P_n = Q_n P$ as an approximation of P since Q_n approximates the identity operator from Proposition 10.1(ii).

Proposition 10.2. P_n *satisfies the following properties:*
(i) $P_n : L^1(X) \to L^1(X)$ *is a Markov operator with range* Δ_n.
(ii) *Let* $\tilde{\Delta}_n = \{\sum_{k=1}^{n} a_k 1_k : a_k \geq 0, \sum_{k=1}^{n} a_k = 1\}$. *Then* $P_n(\tilde{\Delta}_n) \subset \tilde{\Delta}_n$.
Hence, P_n *has a stationary density* $f_n \in \Delta_n$.

Proof (i) Since Q_n and P are Markov operators, so is $Q_n P$.
(ii) Since P_n is a Markov operator and each function of $\tilde{\Delta}_n$ is a density, $P_n(\tilde{\Delta}_n) \subset \tilde{\Delta}_n$. Since $\tilde{\Delta}_n$ is a compact convex subset of Δ_n, by Lemma 2.1, $P_n f_n = f_n$ for some $f_n \in \tilde{\Delta}_n$. $\qquad\square$

Ulam's method for computing a stationary density of a Markov operator P is to approximate P by the above P_n and then solve the finite dimensional fixed point equation $P_n f_n = f_n$ in Δ_n. In the following, we focus on the case when P is a Frobenius-Perron operator.

To compute an approximate stationary density of the Frobenius-Perron operator $P_S : L^1(\mu) \to L^1(\mu)$ associated with a nonsingular transformation $S : X \to X$, corresponding to the partition $\{X_1, \cdots, X_n\}$ of X, we construct an $n \times n$ nonnegative matrix

$$\hat{P}_n = (p_{ij}), \quad p_{ij} = \frac{\mu(X_i \cap S^{-1}(X_j))}{\mu(X_i)}, \quad i, j = 1, \cdots, n, \qquad (10.2)$$

which can be viewed as the transition matrix of a finite state stationary Markov chain as Example 4.1 indicated, so that the (i, j)-entry p_{ij} of the matrix $\hat{P}$ can be interpreted as the "probability" that a point in the ith subset X_i is mapped into the jth subset X_j under the transformation S. The nonnegative matrix $\hat{P}_n$ defined by (10.2) is referred as the *Ulam matrix* associated with S and the partition $\{X_1, \cdots, X_n\}$ of X.

Proposition 10.3. *Let $\hat{P}_n$ be a Ulam matrix. Then*

(i) $\hat{P}_n$ *is a stochastic matrix;*

(ii) $\hat{P}_n$ *is the matrix representation of $P_n = Q_n P_S$ under the basis $\{1_1, \cdots, 1_n\}$ of densities for Δ_n.*

Proof (i) For each $i = 1, \cdots, n$,

$$\sum_{j=1}^{n} p_{ij} = \sum_{j=1}^{n} \frac{\mu(X_i \cap S^{-1}(X_j))}{\mu(X_i)} = \frac{\mu\left(X_i \cap \bigcup_{j=1}^{n} S^{-1}(X_j)\right)}{\mu(X_i)}$$

$$= \frac{\mu\left(X_i \cap S^{-1}\left(\bigcup_{j=1}^{n} X_j\right)\right)}{\mu(X_i)} = \frac{\mu(X_i \cap X)}{\mu(X_i)} = \frac{\mu(X_i)}{\mu(X_i)} = 1.$$

(ii) From the definitions (8.1) of P_S and (10.1) of Q_n,

$$P_n 1_i = Q_n P_S 1_i = \sum_{j=1}^{n} \left(\frac{1}{\mu(X_j)} \int_{X_j} P_S 1_i d\mu\right) \chi_{X_j}$$

$$= \sum_{j=1}^{n} \int_{X_j} P_S 1_i d\mu \cdot 1_j = \sum_{j=1}^{n} \int_{S^{-1}(X_j)} 1_i d\mu \cdot 1_j$$

$$= \sum_{j=1}^{n} \frac{\mu(X_i \cap S^{-1}(X_j))}{\mu(X_i)} 1_j = \sum_{j=1}^{n} p_{ij} 1_j, \quad \forall i = 1, \cdots, n.$$

Thus $\hat{P}_n$ is the matrix representation of the operator $P_n : \Delta_n \to \Delta_n$ under the density basis $\{1_1, \cdots, 1_n\}$ of Δ_n. □

The main numerical work in the performance of Ulam's method after forming the Ulam matrix $\hat{P}_n$ is to calculate a normalized nonnegative left eigenvector $v = (v_1, \cdots, v_n)^T$ of $\hat{P}_n$ associated with the maximal eigenvalue 1. Then the linear combination

$$f_n = \sum_{k=1}^{n} v_k 1_k$$

is a piecewise constant density which is a fixed point of P_n by Proposition 10.3(ii), and provides an approximation of a stationary density of the original Frobenius-Perron operator P_S.

Ulam's method was originally proposed for one dimensional mappings $S : [0, 1] \to [0, 1]$, in which the interval $I = [0, 1]$ is divided into n subintervals $I_k = [x_{k-1}, x_k]$ for $k = 1, 2, \cdots, n$ such that $h \equiv \max_k (x_k - x_{k-1}) \to 0$ as $n \to \infty$. In this particular case, the definition of Q_n in (10.1) becomes

$$Q_n f(x) = \sum_{k=1}^{n} \frac{1}{x_k - x_{k-1}} \int_{x_{k-1}}^{x_k} f(t)dt \cdot \chi_{[x_{k-1}, x_k]}(x), \qquad (10.3)$$

which has some more refined properties as follows.

Proposition 10.4. *Let Q_n be defined by (10.3). Then*
(i) $\lim_{n \to \infty} \|Q_n f - f\|_1 = 0$ *for $f \in L^1(0, 1)$;*
(ii) *for any $f \in BV(0, 1)$,*

$$\bigvee_0^1 Q_n f \le \bigvee_0^1 f;$$

(iii) $\|Q_n f - f\|_1 \le h \bigvee_0^1 f$ *for $f \in BV(0, 1)$;*
(iv) *if $f \in L^1(0, 1)$ is increasing (or decreasing), then so is $Q_n f$.*

Proof (i) It is enough to show $\|Q_n f - f\|_1 \to 0$ for any $f \in C[0, 1]$ since $C[0, 1]$ is dense in $L^1(0, 1)$ and $\|Q_n\|_1 \equiv 1$ for all n. Let $\epsilon > 0$. From the uniform continuity of f, for n large enough,

$$|\hat{f}_k - f(x)| < \epsilon$$

for all $x \in [x_{k-1}, x_k]$ and all k, where

$$\hat{f}_k = \frac{1}{x_k - x_{k-1}} \int_{x_{k-1}}^{x_k} f(t)dt$$

is the average value of f over $[x_{k-1}, x_k]$. Hence, from (10.3),

$$\|Q_n f - f\|_1 = \sum_{k=1}^{n} \int_{x_{k-1}}^{x_k} |\hat{f}_k - f(x)| dx < \sum_{k=1}^{n} \int_{x_{k-1}}^{x_k} \epsilon \, dx = \epsilon.$$

(ii) Let $Q_n f = \sum_{k=1}^{n} \hat{f}_k \chi_{[x_{k-1}, x_k]}$. Then

$$\bigvee_0^1 Q_n f = \sum_{k=1}^{n-1} |\hat{f}_k - \hat{f}_{k+1}|.$$

Choose s_k and t_k in $[x_{k-1}, x_k]$ for each k such that

$$f(s_k) \le \hat{f}_k \le f(t_k).$$

Let $\{\hat{f}_{k_j}\} \subset \{\hat{f}_k\}$ be the set of all local maximal values and local minimal values among $\hat{f}_k$, and we define

$$z_j = \begin{cases} s_{k_j}, & \text{if } \hat{f}_{k_j} \text{ is a local minimal value,} \\ t_{k_j}, & \text{if } \hat{f}_{k_j} \text{ is a local maximal value.} \end{cases}$$

Then it is easy to see that

$$\bigvee_0^1 Q_n f \le \sum_j |f(z_j) - f(z_{j+1})| \le \bigvee_0^1 f.$$

(iii) Since $|f(x) - \hat{f}_k| \le \bigvee_{x_{k-1}}^{x_k} f$ on each interval $[x_{k-1}, x_k]$,

$$\|f - Q_n f\|_1 = \sum_{k=1}^{n} \int_{x_{k-1}}^{x_k} |f(x) - \hat{f}_k| dx \le \sum_{k=1}^{n} \int_{x_{k-1}}^{x_k} \bigvee_{x_{k-1}}^{x_k} f \, dx$$

$$\le \sum_{k=1}^{n} (x_k - x_{k-1}) \bigvee_{x_{k-1}}^{x_k} f \le h \sum_{k=1}^{n} \bigvee_{x_{k-1}}^{x_k} f = h \bigvee_0^1 f.$$

(iv) Since $Q_n f = \sum_{k=1}^{n} \hat{f}_k \chi_{[x_{k-1}, x_k]}$, the finite sequence $\{\hat{f}_1, \cdots, \hat{f}_n\}$ is monotonically increasing or decreasing, depending on whether f is monotonically increasing or decreasing, respectively. $\qquad\square$

Although P_n does have stationary densities for any positive integer n by Proposition 10.2(ii), the original Frobenius-Perron operator P_S may not possess a stationary density. One example is $S(x) = x/2$ on $[0, 1]$. Let $S : [0, 1] \to [0, 1]$ and let $f_n \in \mathcal{D} \cap \Delta_n$ be a piecewise constant fixed point of P_n from Ulam's method for each n. Ulam *conjectured* in [Ulam (1960)] that if P_S has a stationary density, then $\{f_n\}$ converges to a stationary density of P_S. This conjecture was proved by Li [Li (1976)] for the Lasota-Yorke class of piecewise C^2 and stretching mappings for which the existence of stationary densities was proved by Theorem 8.5.

Theorem 10.1. (Li) *Let $S : [0,1] \to [0,1]$ satisfy the condition of Theorem 8.5 and the additional assumption $\inf_{x \in [0,1] \setminus \{a_0, \cdots, a_r\}} |S'(x)| > 2$, let P_S be the corresponding Frobenius-Perron operator, and let $\{P_n\}$ be the sequence of Ulam's approximations of P_S. Then for any sequence $\{f_n\}$ of stationary densities of P_n in Δ_n, there is a subsequence $\{f_{n_k}\}$ such that*

$$\lim_{k \to \infty} \|f_{n_k} - f^*\|_1 = 0,$$

where f^* is a stationary density of P_S. If in addition f^* is the unique stationary density of P_S, then

$$\lim_{n \to \infty} \|f_n - f^*\|_1 = 0.$$

Proof From the proof of Theorem 8.5, for $f \in \mathcal{D} \cap BV(0,1)$,

$$\bigvee_0^1 P_S f \leq \alpha \bigvee_0^1 f + C,$$

where $\alpha = 2/\inf |S'| < 1$ and $C > 0$. So, by Proposition 10.4(ii),

$$\bigvee_0^1 f_n = \bigvee_0^1 Q_n P_S f_n \leq \bigvee_0^1 P_S f_n \leq \alpha \bigvee_0^1 f_n + C,$$

which implies that

$$\bigvee_0^1 f_n \leq \frac{C}{1 - \alpha}, \quad \forall n.$$

Hence, by Lemma 8.2, a subsequence $\{f_{n_k}\}$ of $\{f_n\}$ exists such that $\lim_{k \to \infty} \|f_{n_k} - f^*\|_1 = 0$ for some $f^* \in \mathcal{D}$. Since $\|P_n\|_1 \equiv 1$,

$$\|f^* - P_S f^*\|_1 \leq \|f^* - f_{n_k}\|_1 + \|f_{n_k} - P_{n_k} f_{n_k}\|_1$$
$$+ \|P_{n_k} f_{n_k} - P_{n_k} f^*\|_1 + \|P_{n_k} f^* - P_S f^*\|_1$$
$$\leq \|f^* - f_{n_k}\|_1 + \|f_{n_k} - f^*\|_1 + \|P_{n_k} f^* - P_S f^*\|_1.$$

The right-hand side of the above inequality approaches 0 as $k \to \infty$, so $P_S f^* = f^*$. If in addition f^* is the unique stationary density of P_S, then necessarily $\lim_{n \to \infty} f_n = f^*$ since any convergent subsequence of $\{f_n\}$ converges to the same limit from the above. $\square$

Remark 10.1. Ulam's method is still a convergent one under the milder condition that $\inf_{x \in [0,1] \setminus \{a_0, \cdots, a_r\}} |S'(x)| > 1$ [Li (1976)].

The convergence of Ulam's method for the class of piecewise convex mappings from the interval $[0, 1]$ into itself with the strong repellor 0 was established by Miller in [Miller (1994)].

Theorem 10.2. (Miller) *Suppose that $S : [0,1] \rightarrow [0,1]$ is a piecewise convex mapping with a strong repellor as in Theorem 8.6. Then for any n, there is a stationary density of P_n which is a decreasing function, and for any sequence $\{f_n\}$ of decreasing stationary densities of P_n in Δ_n,*

$$\lim_{n \to \infty} \|f_n - f^*\|_1 = 0,$$

where f^* is the stationary density of P_S and is also decreasing.

Proof From the proof of Theorem 8.6, the density $P_S f$ is a decreasing function for any decreasing function $f \in \mathcal{D}$ and

$$P_S f(x) \leq \alpha f(0) + C, \quad \forall x \in [0,1],$$

where $\alpha = S'(0)^{-1} < 1$ and C is a positive constant. Let

$$\mathcal{D}_n = \{f \in \Delta_n \cap \mathcal{D} : f \text{ is decreasing on } [0,1]\}.$$

Then $\mathcal{D}_n$ is invariant under P_n by Proposition 10.4(iii). Since $\mathcal{D}_n$ is a compact convex set, Lemma 2.1 ensures that P_n has a fixed point $f_n \in \mathcal{D}_n$. Thus, from the fact that $\|Q_n\|_\infty \equiv 1$,

$$f_n(0) = Q_n P_S f_n(0) \leq \max_{x \in [0,1]} P_S f_n(x) \leq \alpha f_n(0) + C.$$

Hence, since $\alpha < 1$,

$$f_n(0) \leq \frac{C}{1 - \alpha}, \quad \forall n,$$

which implies that

$$\bigvee_0^1 f_n = f_n(0) - f_n(1) \leq \frac{C}{1 - \alpha}$$

uniformly, and the result follows from Lemma 8.2 and the uniqueness of the stationary density of P_S by Remark 8.7. $\square$

We present a convergence theorem of Ulam's method for multi-dimensional transformations. Let $\Omega = [0, 1]^N$ be the N-dimensional unit cube in $\mathbb{R}^N$, $S : \Omega \rightarrow \Omega$ a nonsingular transformation, and P_S the corresponding Frobenius-Perron operator. Divide Ω into $n = k^N$ equal sub-cubes $\{\Omega_1, \cdots, \Omega_n\}$ of side length $h = 1/k$. The corresponding $n \times n$ Ulam matrix, which is stochastic by Proposition 10.3(i), is

$$\hat{P}_h = (p_{ij}), \quad p_{ij} = \frac{m(\Omega_i \cap S^{-1}(\Omega_j))}{m(\Omega_i)}, \quad i, j = 1, \cdots, n.$$

In Ulam's method, a left nonnegative eigenvector of $\hat{P}_h$ corresponding to eigenvalue 1 is calculated to form a piecewise constant approximate invariant density f_h of P_S. We know that f_h is a stationary density of the Markov operator $P_h = Q_h P_S : \Delta_h \to \Delta_h$, where Δ_h is the space of all piecewise constant functions associated with the partition $\{\Omega_1, \cdots, \Omega_n\}$, and

$$Q_h f(x) = \sum_{k=1}^{n} \frac{1}{m(\Omega_k)} \int_{\Omega_k} f \, dm \cdot \chi_{\Omega_k}(x).$$

We list some consistency and stability results of the family $\{Q_h\}$; their proofs are referred to [Ding and Zhou (1996)].

Proposition 10.5. *There exists a constant C such that*

$$\|Q_h f - f\|_1 \leq ChV(f; \Omega), \quad \forall f \in BV(\Omega), \, h > 0.$$

Proposition 10.6. *There exists a constant C_{BV} such that*

$$V(Q_h f; \Omega) \leq C_{BV} V(f; \Omega), \quad \forall f \in BV(\Omega), \, h > 0.$$

Based on the above propositions, Ulam's conjecture was proved in [Ding and Zhou (1996)] for the Góra-Boyarsky class of piecewise C^2 and expanding transformations; see also [Miller (1997)].

Theorem 10.3. (Ding-Zhou) *Suppose that $S : \Omega \to \Omega$ is piecewise C^2 and b-expanding, and satisfies the condition of Theorem 8.7 with b large enough. Let $\{f_{h_n}\}$ be a sequence of the stationary densities of P_{h_n} in Δ_{h_n} obtained from Ulam's method such that $\lim_{n \to \infty} h_n = 0$. Then there is a subsequence $\{f_{h_{n_k}}\}$ of $\{f_{h_n}\}$ such that*

$$\lim_{k \to \infty} \|f_{h_{n_k}} - f^*\|_1 = 0,$$

where f^* is a stationary density of P_S. If in addition f^* is the unique stationary density of P_S, then

$$\lim_{n \to \infty} \|f_{h_n} - f^*\|_1 = 0.$$

10.2 Piecewise Linear Markov Approximations

Ulam's method preserves the *Markov property* of Markov operators, that is, the finite dimensional approximation P_n of P maps densities to densities, so its representation under any density basis is a stochastic matrix which must have a nonnegative left eigenvector corresponding to the maximal eigenvalue 1. However, since the method uses only piecewise constant functions to approximate L^1 functions, fast convergence is not expected. Indeed, [Keller (1982)] has shown that for the Lasota-Yorke class of interval mappings, Ulam's method applied to the corresponding Frobenius-Perron operator has only the L^1-norm convergence rate of $O(\ln n/n)$, where n is the number of subintervals for the partition of $[0,1]$. Moreover, one dimensional mappings have been constructed in [Bose and Murray (2001)] for which Ulam's method has exactly the same order as $\ln n/n$.

In this section, we present a higher order structure-preserving numerical scheme using piecewise linear continuous functions. This approach was first proposed in [Ding (1990); Ding and Li (1991)] for computing one dimensional absolutely continuous invariant measures, and was then generalized in [Ding and Zhou (1995, 2001, 2002)] for multi-dimensional transformations. This scheme is called the *piecewise linear Markov finite approximation method* since, like Ulam's method, it approximates infinite dimensional Markov operators by finite dimensional Markov operators. In the following we first introduce this method for Markov operators on $L^1(0,1)$, and then we construct Markov approximations for functions of multi-variable.

Let $P : L^1(0,1) \to L^1(0,1)$ be a Markov operator. Our purpose is to approximate P by a sequence of positive operators based on piecewise linear continuous functions. For this purpose it is sufficient to approximate all functions $f \in L^1(0,1)$ by such more special functions which preserve the integral of $f \in L^1(0,1)$ and positivity of $f \in L^1(0,1)_+$. As in Ulam's method, divide $[0,1]$ into n subintervals $I_k = [x_{k-1}, x_k]$ for $k = 1, \cdots, n$, which have the same length $h = 1/n$ for the sake of simplicity and so $x_k = k/n$ for $k = 0, 1, \cdots, n$. For the sake of convenience, we denote by Δ_n, the same notation that we have used for Ulam's piecewise constant method, the corresponding space of all piecewise linear continuous functions. Then Δ_n is an $(n+1)$-dimensional subspace of $C[0,1]$. The *canonical basis* for Δ_n consists of $n+1$ dilated and translated tent functions

$$\phi_k(x) = w\left(\frac{x - x_k}{h}\right), \quad k = 0, 1, \cdots, n$$

obtained from the basic tent function

$$w(x) = \begin{cases} 1 + x, & -1 \le x \le 0, \\ 1 - x, & 0 < x \le 1, \\ 0, & x \notin [-1, 1]. \end{cases}$$

It is easy to see that the sum of the basis functions is identical to 1 and that if $f = \sum_{k=0}^n q_k \phi_k$, then $f(x_k) = q_k$ for all k.

For $k = 1, \cdots, n$ let $\hat{x}_k = (x_{k-1} + x_k)/2 = (k - 1/2)h$ be the middle point of the interval I_k. Let $\hat{x}_0 = x_0 = 0$ and $\hat{x}_{n+1} = x_n = 1$. Denote by

$\hat{I}_k = [\hat{x}_k, \hat{x}_{k+1}]$ for $k = 0, 1, \cdots, n$. Then $\{\hat{I}_k\}_{k=0}^n$ form a partition of the interval $[0, 1]$. Furthermore,

$$\|\phi_k\|_1 = \int_0^1 \phi_k(x)dx = m(\hat{I}_k), \quad k = 0, 1, \cdots, n. \qquad (10.4)$$

We now define a linear operator $Q_n : L^1(0, 1) \to L^1(0, 1)$ by

$$Q_n f = \sum_{k=0}^n \left(\frac{1}{m(\hat{I}_k)} \int_{\hat{I}_k} f \, dm \right) \phi_k. \qquad (10.5)$$

Proposition 10.7. Q_n *is a Markov operator for each* n. *Moreover,* $\|Q_n g\|_\infty \le \|g\|_\infty$ *for every* $g \in L^\infty(0, 1)$.

Proof It is obvious that Q_n is a positive operator. Let $f \in L^1(0, 1)$. Then, from (10.5) and noting (10.4), we have

$$\int_0^1 Q_n f(x) dx = \sum_{k=0}^n \frac{1}{m(\hat{I}_k)} \int_{\hat{I}_k} f \, dm \cdot \int_0^1 \phi_k(x) dx$$

$$= \sum_{k=0}^n \int_{\hat{I}_k} f \, dm = \int_0^1 f(x) dx.$$

So Q_n is a Markov operator. The last conclusion is obvious. $\qquad \square$

Remark 10.2. The operator sequence $\{Q_n\}$ of piecewise linear Markov approximations defined by (10.5) was proposed in [Ding and Rhee (2006)], which improves the original Markov approximation scheme introduced in [Ding (1990)]; see also [Ding and Li (1991); Ding et al. (2007); Ye and Ding (2006)] for further discussions of this topic.

Remark 10.3. It has been proved by Ding, Eifler, and Rhee in [Ding et al. (2007)] that basically there are only two types of piecewise linear Markov approximations of $f \in L^1(0, 1)$, the one presented by (10.5) above and the original one constructed in [Ding (1990)]. The idea of a proof to this fact can be seen from the hint to Exercise 10.4.

As in Proposition 10.4(i), the operator sequence $\{Q_n\}$ associated with the piecewise linear Markov method satisfies the consistency condition $\lim_{n \to \infty} \|Q_n f - f\|_1 = 0$ under the L^1-norm for all functions $f \in L^1(0, 1)$. If the function f is second order continuously differentiable on the closed interval $[0, 1]$, one can estimate the rate explicitly for the above "local convergence." In order to give the best possible upper bound, we need two standard results from approximation theory.

Lemma 10.1. *Suppose that $f \in C^2[a, b]$. Let l be a linear polynomial function that satisfies $l(a) = f(a)$ and $l(b) = f(b)$. Then for any $x \in [a, b]$, there is a point $\zeta(x) \in (a, b)$ such that*

$$f(x) = l(x) + \frac{f''(\zeta(x))}{2}(x - a)(x - b).$$

Lemma 10.2. *If $f \in C^1[a, b]$, then*

$$\left| \int_a^b f(x)dx - f(c)(b - a) \right| \leq \frac{(b - a)^2}{2} \max_{x \in [a,b]} |f'(x)| \tag{10.6}$$

for $c = a$ or b, and if $f \in C^2[a, b]$, then for $c = (a + b)/2$,

$$\left| \int_a^b g(x)dx - g(c)(b - a) \right| \leq \frac{(b - a)^3}{24} \max_{x \in [a,b]} |f''(x)|. \tag{10.7}$$

Lemma 10.3. *Let $L_n f = \sum_{k=0}^n f(x_k)\phi_k$ be the piecewise linear Lagrange interpolation of $f \in C^2[0, 1]$. Then*

$$\|L_n f - f\|_1 \leq \frac{h^2}{12} \max_{x \in [0,1]} |f''(x)|. \tag{10.8}$$

Proof Given a function $f \in C^2[0, 1]$. Then, using Lemma 10.1 for each subinterval $[x_k, x_{k+1}]$, we obtain that

$$\|L_n f - f\|_1 = \sum_{k=0}^{n-1} \int_{x_k}^{x_{k+1}} |L_n f(x) - f(x)| \, dx$$

$$= \sum_{k=0}^{n-1} \int_{x_k}^{x_{k+1}} \left| \frac{f''(\zeta_k(x))}{2}(x - x_k)(x - x_{k+1}) \right| dx$$

$$\leq \sum_{k=0}^{n-1} \int_{x_k}^{x_{k+1}} \frac{1}{2} \max_{x \in [x_k, x_{k+1}]} |f''(x)| \, |(x - x_k)(x - x_{k+1})| dx$$

$$\leq \frac{1}{2} \max_{x \in [0,1]} |f''(x)| \sum_{k=0}^{n-1} \int_{x_k}^{x_{k+1}} (x - x_k)(x_{k+1} - x) dx$$

$$= \frac{1}{2} \max_{x \in [0,1]} |f''(x)| \sum_{k=0}^{n-1} \frac{(x_{k+1} - x_k)^3}{6} = \frac{h^2}{12} \max_{x \in [0,1]} |f''(x)|. \square$$

The usefulness of the auxiliary operator L_n and Lemma 10.3 is clear from the proof of the following L^1-norm estimate.

Proposition 10.8. *If $f \in C^2[0, 1]$, then*

$$\|Q_n f - f\|_1 \leq \frac{h^2}{8} \left(2 \max_{x \in \hat{I}_0 \cup \hat{I}_n} |f'(x)| + \max_{x \in [0,1]} |f''(x)| \right). \tag{10.9}$$

Proof The formula (10.5) for the definition of the operator Q_n can be written more explicitly as

$$Q_n f = \frac{2}{h} \int_0^{\frac{h}{2}} f \, dm \cdot \phi_0 + \sum_{k=1}^{n-1} \frac{1}{h} \int_{(k-\frac{1}{2})h}^{(k+\frac{1}{2})h} f \, dm \cdot \phi_k$$

$$+ \frac{2}{h} \int_{1-\frac{h}{2}}^1 f \, dm \cdot \phi_n.$$

Then, with the help of Lemma 10.2, it follows that

$$\|Q_n f - L_n f\|_1 \leq \left| \frac{h}{2} f(0) - \int_0^{\frac{h}{2}} f(t) \, dt \right| \cdot \frac{2}{h} \int_0^1 \phi_0(x) \, dx$$

$$+ \sum_{k=1}^{n-1} \left\{ \left| h f(kh) - \int_{(k-\frac{1}{2})h}^{(k+\frac{1}{2})h} f(t) \, dt \right| \cdot \frac{1}{h} \int_0^1 \phi_k(x) \, dx \right\}$$

$$+ \left| \frac{h}{2} f(1) - \int_{1-\frac{h}{2}}^1 f(t) \, dt \right| \cdot \frac{2}{h} \int_0^1 \phi_n(x) \, dx$$

$$= \left| \frac{h}{2} f(0) - \int_0^{\frac{h}{2}} f(t) \, dt \right| + \sum_{k=1}^{n-1} \left| h f(kh) - \int_{(k-\frac{1}{2})h}^{(k+\frac{1}{2})h} f(t) \, dt \right|$$

$$+ \left| \frac{h}{2} f(1) - \int_{1-\frac{h}{2}}^1 f(t) \, dt \right|$$

$$\leq \frac{(\frac{h}{2})^2}{2} \max_{x \in [0,\frac{h}{2}]} |f'(x)| + \sum_{i=1}^{n-1} \frac{h^3}{24} \max_{x \in [(k-\frac{1}{2})h, (k+\frac{1}{2})h]} |f''(x)|$$

$$+ \frac{(\frac{h}{2})^2}{2} \max_{x \in [1-\frac{h}{2},1]} |f'(x)|$$

$$\leq \frac{h^2}{8} \left(\max_{x \in [0,\frac{h}{2}]} |f'(x)| + \max_{x \in [1-\frac{h}{2},1]} |f'(x)| \right) + \frac{h^2}{24} \max_{x \in [0,1]} |f''(x)|$$

$$\leq \frac{h^2}{8} \left(2 \max_{x \in [0,\frac{h}{2}] \cup [1-\frac{h}{2},1]} |f'(x)| + \max_{x \in [0,1]} |f''(x)| \right).$$

Therefore, by means of (10.8), we obtain (10.9). □

Remark 10.4. Proposition 10.8 indicates that the piecewise linear Markov approximation method has the local convergence rate of order two as compared to the rate of order one for Ulam's piecewise constant method.

Like Ulam's method, the discretized operator Q_n for the Markov method does not increase the variation of a function.

Proposition 10.9. *For any $f \in BV(0,1)$,*

$$\bigvee_0^1 Q_n f \leq \bigvee_0^1 f.$$

Moreover, if f is increasing or decreasing, then so is $Q_n f$.

Proof If we write $Q_n f = \sum_{k=0}^n q_k \phi_k$, then $q_k = m(\hat{I}_k)^{-1} \int_{\hat{I}_k} f dm$ is the average value of f over $\hat{I}_k$ for $k = 0, \cdots, n$. Therefore,

$$\bigvee_0^1 Q_n f = \sum_{k=1}^n |q_k - q_{k-1}|$$

$$= \sum_{k=1}^n \left| \frac{1}{m(\hat{I}_k)} \int_{\hat{I}_k} f dm - \frac{1}{m(\hat{I}_{k-1})} \int_{\hat{I}_{k-1}} f dm \right| \leq \bigvee_0^1 f,$$

where the last inequality is from Proposition 10.4(ii). The second conclusion of the proposition is obvious. $\square$

We now show that $\lim_{n\to\infty} Q_n f = f$ under the variation norm $\| \; \|_{BV}$ by establishing an upper bound of $V_0^1(Q_n f - f)$ for $f \in BV(0,1)$, a property that is not shared by Ulam's method in general. For this purpose, we need the following two lemmas due to [Chiu *et al.* (1992)].

Lemma 10.4. *Suppose that $f \in C^2[a,b]$ and l is a linear polynomial such that $l(a) = f(a)$ and $l(b) = f(b)$. Then*

$$\bigvee_a^b (f - l) \leq \frac{(b-a)^2}{2} \max_{x \in [a,b]} |f''(x)|.$$

Proof Consider $g(x) \equiv f(x) - l(x)$ on $[a,b]$. Since $g(a) = g(b) = 0$, there is $c \in (a,b)$ such that $g'(c) = 0$. Since $g''(x) = f''(x)$, we have $|g'(x)| = |\int_c^x g''(t)dt| = |\int_c^x f''(t)dt|$. Thus,

$$\bigvee_a^b (f - l) = \bigvee_a^b g = \int_a^b |g'(x)| dx = \int_a^b \left| \int_c^x f''(t)dt \right| dx$$

$$\leq \max_{x \in [a,b]} |f''(x)| \int_a^b |x - c| dx \leq \frac{(b-a)^2}{2} \max_{x \in [a,b]} |f''(x)|. \square$$

Lemma 10.5. *Suppose that $L_n f = \sum_{k=0}^n f(x_k) \phi_k$ is the piecewise linear Lagrange interpolation of $f \in C^2[0,1]$. Then*

$$\bigvee_0^1 (L_n f - f) \leq \frac{h}{2} \max_{x \in [0,1]} |f''(x)|. \tag{10.10}$$

Proof By Lemma 10.4,

$$\bigvee_0^1 (L_n f - f) = \sum_{k=1}^n \bigvee_{x_{k-1}}^{x_k} (L_n f - f)$$

$$\leq \sum_{k=1}^n \frac{(x_k - x_{k-1})^2}{2} \max_{x \in [x_{k-1}, x_k]} |f''(x)| = \frac{h^2}{2} \sum_{k=1}^n \max_{x \in [x_{k-1}, x_k]} |f''(x)|$$

$$\leq \frac{h^2}{2} \sum_{k=1}^n \max_{x \in [0,1]} |f''(x)| = \frac{h}{2} \max_{x \in [0,1]} |f''(x)|. \qquad \square$$

Proposition 10.10. *Let* $f \in C^2[0,1]$. *Then*

$$\bigvee_0^1 (Q_n f - f) \leq \frac{h}{4} \left(2 \max_{x \in [0, \frac{h}{2}] \cup [1 - \frac{h}{2}, 1]} |f'(x)| + \frac{7}{3} \max_{x \in [0,1]} |f''(x)| \right).$$

Proof Because of (10.10), it is enough to estimate $\bigvee_0^1 (Q_n f - L_n f)$. Let $Q_n f = \sum_{k=0}^n q_k \phi_k$ and $L_n f = \sum_{k=0}^n f(x_k) \phi_k$. Since $\bigvee_0^1 \phi_0 = \bigvee_0^1 \phi_n = 1$ and $\bigvee_0^1 \phi_k = 2$ for $k = 1, \cdots, n-1$, we have

$$\bigvee_0^1 (Q_n f - L_n f) \leq \sum_{k=0}^n |q_k - f(x_k)| \bigvee_0^1 \phi_k$$

$$\leq |q_0 - f(0)| + 2 \sum_{k=1}^{n-1} |q_k - f(x_k)| + |q_n - f(1)|.$$

Since x_k is the middle point of $\hat{I}_k$, using (10.7), we find that

$$|q_k - f(x_k)| = \left| \frac{1}{m(\hat{I}_k)} \int_{\hat{I}_k} f dm - f(x_k) \right| \leq \frac{h^2}{24} \max_{x \in \hat{I}_k} |f''(x)|$$

for $k = 1, \cdots, n-1$. Similarly, using (10.6), we obtain

$$|q_0 - f(0)| \leq \frac{h}{4} \max_{x \in \hat{I}_0} |f'(x)| \quad \text{and} \quad |q_n - f(1)| \leq \frac{h}{4} \max_{x \in \hat{I}_n} |f'(x)|.$$

It follows that

$$\bigvee_0^1 (Q_n f - L_n f)$$

$$\leq \frac{h}{4} \max_{x \in \hat{I}_0} |f'(x)| + \sum_{k=1}^{n-1} \frac{h^2}{12} \max_{x \in \hat{I}_k} |f''(x)| + \frac{h}{4} \max_{x \in \hat{I}_n} |f'(x)|$$

$$\leq \frac{h}{4} \left(\max_{x \in \hat{I}_0} |f'(x)| + \max_{x \in \hat{I}_n} |f'(x)| + \frac{1}{3} \max_{x \in [0,1]} |f''(x)| \right)$$

$$\leq \frac{h}{4} \left(2 \max_{x \in \hat{I}_0 \cup \hat{I}_n} |f'(x)| + \frac{1}{3} \max_{x \in [0,1]} |f''(x)| \right). \qquad \square$$

Remark 10.5. In general, for Ulam's method,

$$\bigvee_0^1 (f - Q_n f) \geq \bigvee_0^1 f,$$

where Q_n is the Markov operator corresponding to Ulam's method. Thus, we do not expect the local convergence of Ulam's method under the BV-norm; see [Ding and Li (1998); Bose and Murray (2001)].

Propositions 10.8 and 10.10 immediately imply

Proposition 10.11. *If $f \in C^2[0,1]$, then*

$$\|Q_n f - f\|_{BV} = O\left(\frac{1}{n}\right) \to 0 \text{ as } n \to \infty.$$

Let $P_n = Q_n P$. Then P_n is a Markov operator with range Δ_n and $\lim_{n \to \infty} \|P_n f - Pf\|_1 = 0$ for $f \in L^1(0,1)$.

Proposition 10.12. *P_n has a stationary density $f_n \in \Delta_n$.*

Proposition 10.12 shows that the piecewise linear Markov method for computing a stationary density $f_n \in \Delta_n$ of a Markov operator P is well-posed. It is L^1-norm convergent whenever Ulam's method is. However, we have a stronger convergence result.

Theorem 10.4. *Let $S : [0,1] \to [0,1]$ satisfy the condition of Theorem 8.5 with $\inf_{x \in [0,1] \setminus \{a_0, \cdots, a_r\}} |S'(x)| > 2$, let P_S be the corresponding Frobenius-Perron operator which has a unique stationary density $f^* \in C^2[0,1]$, and let $\{P_n\}$ be the sequence of piecewise linear approximations of P_S. Then for any sequence $\{f_n\}$ of the stationary densities of P_n in Δ_n,*

$$\lim_{n \to \infty} \|f_n - f^*\|_{BV} = 0.$$

Proof The proof of Theorem 8.5 and the property that $\|Q_n\|_{BV} \leq 1$ for all n ensure the inequality

$$\|P_n f\|_{BV} \leq \alpha \|f\|_{BV} + C\|f\|_1, \ \forall \, f \in BV(0,1)$$

for some constants $0 < \alpha < 1$ and $C > 0$. Since

$$f^* - f_n = f^* - P_n f^* + P_n f^* - P_n f_n = P_n(f^* - f_n) + f^* - Q_n f^*,$$

$$\|f^* - f_n\|_{BV} = \|P_n(f^* - f_n) + f^* - Q_n f^*\|_{BV}$$
$$\leq \alpha \|f^* - f_n\|_{BV} + C\|f^* - f_n\|_1 + \|f^* - Q_n f^*\|_{BV}.$$

It follows that

$$\|f^* - f_n\|_{BV} \leq \frac{C\|f^* - f_n\|_1 + \|f^* - Q_n f^*\|_{BV}}{1 - \alpha}. \tag{10.11}$$

Since $\lim_{n\to\infty} \|f^* - f_n\|_1 = 0$, which can be proved with exactly the same argument as in the proof of Theorem 10.1, the theorem follows from the above inequality and Proposition 10.11. □

The following theorem from [Ding (1996)] can be proved by the same technique as was done for Theorem 10.2.

Theorem 10.5. *Let S be as in Theorem 10.2. Then for any n, there is a decreasing stationary density of P_n, and for any sequence $\{f_n\}$ of decreasing stationary densities of P_n in Δ_n,*

$$\lim_{n\to\infty} \|f_n - f^*\|_1 = 0,$$

where f^* is the unique stationary density of P_S.

We turn to Markov approximations of Markov operators on $L^1(\Omega)$, where Ω is an N-dimensional polygonal region in $\mathbb{R}^N$.

Let $h > 0$ and let $\mathcal{T}_h$ be a *quasi-uniform* simplicial triangulation of Ω with diam $\tau \leq h$ for all simplices τ of $\mathcal{T}_h$. Namely, there exists a constant $c > 0$ such that [Ciarlet (1978)]

$$\frac{\text{diam } \tau}{\text{diam } B_\tau} \leq c, \quad \forall \, \tau \in \mathcal{T}_h, \; h > 0,$$

where B_τ is the ball inscribed in τ.

Associated with $\mathcal{T}_h$ let $\Delta_h \subset L^1(\Omega)$ be the space of the corresponding piecewise linear continuous functions. Let $\{v_1, v_2, \cdots, v_n\}$ be all the vertices in $\mathcal{T}_h$. For each vertex v in $\mathcal{T}_h$ let n_v denote the number of the simplices of $\mathcal{T}_h$ with v as a vertex. Denote by ϕ_k the unique function in Δ_h such that

$$\phi_k(v_j) = \delta_{kj}, \quad k, j = 1, \cdots, n.$$

Then $\{\phi_k\}_{k=1}^n$ is the *canonical basis* for Δ_h. supp ϕ_k is the union of all n_{v_k} simplices of $\mathcal{T}_h$ with v_k as a vertex. Note that

$$\|\phi_k\|_1 = \frac{1}{N+1} m(\text{supp } \phi_k)$$

for each k. Moreover, $\sum_{k=1}^n \phi_k(x) \equiv 1$ for all $x \in \Omega$. Furthermore, every function $g \in \Delta_h$ can be represented as $g = \sum_{k=1}^n g(v_k)\phi_k$. Now we define a linear operator $Q_h : L^1(\Omega) \to L^1(\Omega)$ by

$$Q_h f = \sum_{k=1}^n \left(\frac{1}{m(\text{supp } \phi_k)} \int_{\text{supp } \phi_k} f \, dm \right) \phi_k. \tag{10.12}$$

Proposition 10.13. $Q_h : L^1(\Omega) \to L^1(\Omega)$ *is a Markov operator.*

Proof By its definition (10.12), Q_h is clearly a positive operator with range $R(Q_h) = \Delta_h$. Let $f \in L^1(\Omega)$. Then

$$\int_\Omega Q_h f \, dm = \sum_{k=1}^n \frac{1}{m(\text{supp } \phi_k)} \int_{\text{supp } \phi_k} f \, dm \int_\Omega \phi_k \, dm$$

$$= \frac{1}{N+1} \sum_{k=1}^n \int_{\text{supp } \phi_k} f \, dm = \int_\Omega f \, dm$$

since each simplex has exactly $N+1$ vertices. So, Q_h is integral-preserving, and thus it is a Markov operator. $\qquad\square$

The consistency and stability results for Ulam's method still hold for the Markov method, and Proposition 10.5 can be strengthened [Ding and Zhou (2002, 2009)] to the BV-norm provided that T_h is *symmetric*, namely all the simplices are symmetric with respect to each interior vertex of the region. Examples of symmetric triangulations are the uniform triangulation and the English flag triangulation of a region in $\mathbb{R}^2$.

Proposition 10.14. *If the triangulation T_h is symmetric, then*

$$\|Q_h f - f\|_{BV} = O(h), \ \ \forall f \in C^2(\overline{\Omega}).$$

Such consistency and stability results and the proof of Theorem 8.7 imply the following convergence result of the piecewise linear Markov method for the Góra-Boyarsky class of piecewise C^2 and expanding transformations from Ω into itself [Ding and Zhou (2002)].

Theorem 10.6. *Suppose that $S : \Omega \to \Omega$ satisfies the condition of Theorem 8.7 with sufficiently large b and the corresponding Frobenius-Perron operator P_S has only one stationary density f^*. Let $\{f_{h_n}\}$ be a sequence of stationary densities of P_{h_n} in Δ_{h_n} such that $\lim_{n\to\infty} h_n = 0$. Then*

$$\lim_{n\to\infty} \|f_{h_n} - f^*\|_1 = 0.$$

Furthermore, if the stationary density $f^* \in C^2(\overline{\Omega})$ and all of the triangulations T_{h_n} of Ω are symmetric, then

$$\|f_{h_n} - f^*\|_{BV} = O(h_n) \to 0 \ \text{as } n \to \infty.$$

10.3 Convergence Rate of Ulam's Method

We turn to investigate the convergence rate of Ulam's method, based on the concept of *stochastic stability* of Markov operators introduced in [Keller

(1982)]. Keller gave a perturbation analysis for a class of Markov operators and first obtained the L^1-norm convergence rate of order $O(\ln n/n)$ for the Lasota-Yorke class of piecewise C^2 and stretching mappings, as an immediate application of a simplified version of his general theory. We shall study his theory in the present section.

Throughout this section we assume that all the function spaces are complex ones. Let $\mathcal{P}$ be the class of Markov operators $P : L^1(0,1) \to L^1(0,1)$ satisfying the following conditions:

(i) $P(BV(0,1)) \subset BV(0,1)$ and $\|P\|_{BV} < \infty$;

(ii) There exist two numbers $0 < \alpha < 1$, $C > 0$ such that

$$\|Pf\|_{BV} \le \alpha\|f\|_{BV} + C\|f\|_1, \quad \forall\, f \in BV(0,1). \tag{10.13}$$

Let $\mathcal{P}(\alpha, C)$ denote the subclass of $\mathcal{P}$ with the given constants α and C. An operator sequence $\{P_n\}$ is called $\mathcal{P}$-*bounded* if $P_n \in \mathcal{P}(\alpha, C)$ for all n. It follows from the Ionescu-Tulcea and Marinescu theorem [Boyarsky and Góra (1997)] that each Markov operator $P \in \mathcal{P}$ is quasi-compact as a linear operator from $(BV(0,1), \|\ \|_{BV})$ into itself, and P is also constrictive on $L^1(0,1)$ from Remark 8.4. Therefore, there are only finitely many *peripheral eigenvalues* $1 = \lambda_1, \lambda_2, \cdots, \lambda_r$ and every *eigenspace* E_i corresponding to the eigenvalue λ_i is finite dimensional (see also Theorem VIII.8.3 of [Dunford and Schwartz (1957)]). Furthermore, P has the following spectral decomposition [Boyarsky and Góra (1997)]

$$P = \sum_{i=1}^{r} \lambda_i \Phi_i + Q, \tag{10.14}$$

where each Φ_i is a projection onto E_i with the properties that $\|\Phi_i\|_1 = 1$ for all i and $\Phi_i\Phi_j = 0$ for $i \ne j$, and $Q : L^1(0,1) \to L^1(0,1)$ is a bounded linear operator such that $\sup_{n\ge 0} \|Q^n\|_1 \le r + 1$, $Q\Phi_i = \Phi_iQ = 0$ for each $i = 1, \cdots, r$, $Q(BV(0,1)) \subset BV(0,1)$, and $\|Q^n\|_{BV} \le Mq^n$, $\forall\, n \ge 1$, for two constants $0 < q < 1$ and $M > 0$.

Let $\lambda \in \partial\mathbb{D}^2$. Define a linear operator $\Phi(\lambda, P)$ by

$$\Phi(\lambda, P)f = \lim_{n\to\infty} \frac{1}{n} \sum_{j=0}^{n-1} (\bar{\lambda}P)^j f, \quad \forall\, f \in L^1(0,1).$$

It can be shown (Exercise 10.7) that

$$\Phi(\lambda, P) = \begin{cases} \Phi_i, & \text{if } \lambda = \lambda_i, \\ 0, & \text{otherwise.} \end{cases}$$

In particular, $\Phi(1, P) = \Phi_1$.

Lemma 10.6. *If* $P, \hat{P} \in \mathcal{P}$ *with* $P = \sum_{i=1}^{r} \lambda_i \Phi_i + Q$ *and* $\|Q^n\|_{BV} \leq Mq^n$, $\forall\, n \geq 1$, *then for* $\lambda \in \partial \mathbb{D}^2$, *the operator*

$$T(\lambda) \equiv \sum_{i=1, \lambda_i \neq \lambda}^{r} \frac{\bar{\lambda}\lambda_i}{\lambda - \lambda_i} \Phi_i + (1 - \bar{\lambda})\Phi(\lambda, P) + (\lambda I - Q)^{-1}$$

defines a bounded linear operator on $BV(0,1)$. *Moreover,*

(i) $T(\lambda) = (\lambda I - (P - \Phi(\lambda, P)))^{-1}$;

(ii) $(\Phi(\lambda, P) - I)\Phi(\lambda, \hat{P}) = T(\lambda)(P - \hat{P})\Phi(\lambda, \hat{P})$.

Proof Since $0 < q < 1$, the fact that $\|Q^n\|_{BV} \leq Mq^n$ is satisfied for all n implies that $r(Q) < 1$, which shows that $(\lambda I - Q)^{-1}$ exists and equals $\lambda^{-1} \sum_{j=0}^{\infty} (\lambda^{-1}Q)^j = \bar{\lambda} \sum_{j=0}^{\infty} (\bar{\lambda}Q)^j$ as a bounded linear operator from $BV(0,1)$ into itself. This ensures that $T(\lambda)$ is well-defined as a bounded linear operator on $BV(0,1)$. Now (i) and (ii) can be easily verified by a direct computation. $\qquad\square$

For the sake of the stability analysis for the class of Markov operators in the following, we need to define another norm $\|\|P\|\|$ for a given Markov operator P by the formula

$$\|\|P\|\| = \sup \left\{ \|Pf\|_1 : f \in BV(0,1), \|f\|_{BV} = 1 \right\}.$$

That is, $\|\|P\|\|$ is precisely the operator norm of $P : (BV(0,1), \|\ \|_{BV}) \to (L^1(0,1), \|\ \|_1)$. It is obvious that $\|\|P\|\| \leq 1$ and $\|\|P\|\| \leq \|P\|_{BV}$.

Lemma 10.7. *Let* $P \in \mathcal{P}$ *as in Lemma 10.6. If* $\hat{P} \in \mathcal{P}(\alpha, C)$ *and* $\lambda \in \partial \mathbb{D}^2$, *then there are constants* a *independent of* λ,

$$b = \sum_{i=1, \lambda_i \neq \lambda}^{r} \frac{1}{\lambda - \lambda_i} + |1 - \lambda| \cdot \|\Phi(\lambda, P)\|_1,$$

and $c = C/(1 - \alpha)$ *such that if* $\|\|P - \hat{P}\|\| < 1$, *then*

$$\|(\Phi(\lambda, P) - I)\Phi(\lambda, \hat{P})\|_1$$

$$\leq (a + b)c\|\|P - \hat{P}\|\| \left(2 + \frac{\ln \|\|P - \hat{P}\|\|}{\ln q} \right).$$

Proof Inequality (10.13) implies that for each $f \in BV(0,1)$,

$$\|\Phi(\lambda, \hat{P})f\|_{BV} \leq \limsup_{n \to \infty} \frac{1}{n} \sum_{j=0}^{n-1} \|\hat{P}^j f\|_{BV}$$

$$\leq \limsup_{n \to \infty} \|\hat{P}^n f\|_{BV} \leq \frac{C}{1 - \alpha} \|f\|_1 = c\|f\|_1.$$

Now, for any integer $j_0 > 0$, the Neumann series expression $\bar{\lambda} \sum_{j=0}^{\infty} (\bar{\lambda} Q)^j$ of $(\lambda I - Q)^{-1}$ can be written as

$$(\lambda I - Q)^{-1} = \bar{\lambda} \sum_{j=0}^{j_0-1} (\bar{\lambda} Q)^j + (\bar{\lambda} Q)^J (\lambda I - Q)^{-1}.$$

Hence, for $f \in L^1(0,1)$, from Lemma 10.6(ii) and the spectral decomposition (10.14) of P, we obtain that

$$\|(\Phi(\lambda, P) - I)\Phi(\lambda, \hat{P})f\|_1 \leq \|T(\lambda)(P - \hat{P})\Phi(\lambda, \hat{P})f\|_1$$
$$\leq [b + (r+1)j_0] \|(P - \hat{P})\Phi(\lambda, \hat{P})f\|_1$$
$$+ \frac{Mq^{j_0}}{1-q} \|(P - \lambda I)\Phi(\lambda, \hat{P})f\|_{BV}$$
$$\leq c \left[(b + (r+1)j_0) \||P - \hat{P}\|| + \frac{Mq^{j_0}}{1-q} (\|P\|_{BV} + 1) \right] \|f\|_1,$$

which implies that

$$\|(\Phi(\lambda, P) - I)\Phi(\lambda, \hat{P})f\|_1$$
$$\leq (a+b)c(j_0\||P - \hat{P}\|| + q^{j_0})\|f\|_1$$

holds for some constant a depending on P only. If we choose

$$j_0 = \frac{\ln \||P - \hat{P}\||}{\ln q} + 1,$$

then the lemma follows. $\square$

Denote by $\omega_\lambda(P, \hat{P})$ the smallest right-hand side of the conclusion in Lemma 10.7 for the given P, $\hat{P}$, and λ. Lemma 10.7 implies that if $P \in \mathcal{P}$ and $\{P_n\}$ is a $\mathcal{P}$-bounded sequence, then

$$\omega_\lambda(P, P_n) = O\left(\||P - P_n\|| \cdot |\ln \||P - P_n\|| |\right).$$

Proposition 10.15. *Suppose that P, $\hat{P} \in \mathcal{P}$. If P has a unique stationary density, then there are the equality*

$$\Phi(1, P) - \Phi(1, \hat{P}) = (I - (P - \Phi(1, P)))^{-1}(P - \hat{P})\Phi(1, \hat{P})$$

and the inequality

$$\|\Phi(1, P) - \Phi(1, \hat{P})\|_1 \leq \omega_1(P, \hat{P}).$$

Proof Since P preserves integrals and dim $E_1 = 1$,

$$\Phi(1,P)f = \int_0^1 f(x)dx \cdot \Phi(1,P)1$$

for any $f \in L^1(0,1)$. Thus,

$$\Phi(1,P)\Phi(1,\hat{P})f = \int_0^1 \Phi(1,\hat{P})f(x)dx \cdot \Phi(1,P)1$$

$$= \int_0^1 f(x)dx \cdot \Phi(1,P)1 = \Phi(1,P)f,$$

which together with Lemmas 10.6 and 10.7 yields the result. □

Theorem 10.7. (Keller) *Let* $P \in \mathcal{P}$. *If a* $\mathcal{P}$-*bounded sequence* $\{P_n\}$ *is such that* $\lim_{n\to\infty} \|\!|P - P_n|\!\| = 0$, *then for* $\lambda \in \partial\mathbb{D}^2$,

$$\|(\Phi(\lambda,P) - I)\Phi(\lambda,P_n)\|_1 = O\left(\|\!|P - P_n|\!\| \cdot |\ln\|\!|P - P_n|\!\| |\right).$$

If in addition P *has a unique stationary density, then*

$$\|\Phi(1,P) - \Phi(1,P_n)\|_1 = O\left(\|\!|P - P_n|\!\| \cdot |\ln\|\!|P - P_n|\!\| |\right).$$

Proof It follows from Lemma 10.7 and Proposition 10.15. □

We shall apply Theorem 10.7 to a class of *stochastic perturbations* of the Frobenius-Perron operators P_S associated with a nonsingular transformation $S : [0,1] \to [0,1]$ in the following sense, so that we are able to estimate the L^1-norm convergence rate of Ulam's method for the Lasota-Yorke class of piecewise monotonic interval mappings. Let $w : [0,1]^2 \to \mathbb{R}_+$ be a *doubly stochastic kernel* in the sense that

$$\int_0^1 w(x,y)dy = 1, \ x \in [0,1] \text{ a.e.}, \quad \int_0^1 w(x,y)dx = 1, \ y \in [0,1] \text{ a.e.}$$

Consider the operator $P_w : L^1(0,1) \to L^1(0,1)$ defined by

$$P_w f(x) = \int_0^1 f(y)w(S(y),x)dy = \int_0^1 P_S f(y)w(y,x)dy.$$

P_w is obviously a Markov operator.

Denote $w_z(y) = \int_0^z w(x,y)dx$, $z \in [0,1]$,

$$B(z) = \{(x,y) : x \le z < y \text{ or } y \le z < x\},$$

$b(z) = \int_{B(z)} wdm$, and $c(w) = \sup_{z \in [0,1]} b(z)$. Then we have the following estimates (see [Keller (1982)] or [Ding and Zhou (2009)] for their proof).

Proposition 10.16. *With the above assumptions,*

(i) $\bigvee_{[0,1]} P_w f \le \sup_{z \in [0,1]} \bigvee_{[0,1]} w_z \cdot \bigvee_{[0,1]} P_S f$;

(ii) $\|\!|P_S - P_w|\!\| \le c(w)\|P_S\|_{BV}$.

The following corollary gives a sufficient condition for the variation of $P_w f$ to be less than or equal to that of $P_S f$.

Corollary 10.1. *Let $P_S \in \mathcal{P}$ be the Frobenius-Perron operator associated with $S : [0,1] \to [0,1]$ such that (10.13) is satisfied and let w be a doubly stochastic kernel such that*

$$\int_0^z w(x, y_1) dx \geq \int_0^z w(x, y_2) dx \qquad (10.15)$$

for all $y_1, y_2, z \in [0,1]$ with $y_1 \leq y_2$. Then the conclusion of Proposition 10.16 holds. Furthermore,

$$\bigvee_{[0,1]} P_w f \leq \bigvee_{[0,1]} P_S f.$$

Consequently, both P_S and P_w are in the same class $\mathcal{P}(\alpha, C)$.

Proof The inequality (10.15) implies that the function $w_z(y) = \int_0^z w(x, y) dx$ is monotonically decreasing. Since $0 \leq w_z(y) \leq 1$ for $0 \leq y \leq 1$, we easily see that $\bigvee_{[0,1]} w_z \leq 1$ for all $z \in [0,1]$. Therefore, the corollary follows from Proposition 10.16. □

We can apply Corollary 10.1 to estimating the convergence rate of Ulam's method for the Lasota-Yorke class of mappings. To fulfill our task, we need to estimate the quantity $\|\!|\!| P_S - P_n |\!|\!|$, where $P_n = Q_n P_S$ and Q_n is the finite dimensional Markov operator as defined by (10.1). For the sake of simplicity, we assume that the partition of $[0,1]$ is uniform, that is, we assume that $x_i = i/n$ for $i = 0, 1, \cdots, n$. If we define

$$w_n(x, y) = \begin{cases} n, & \text{if } \frac{i-1}{n} \leq x, \; y < \frac{i}{n} \text{ for some } i, \\ 0, & \text{otherwise,} \end{cases}$$

then it is easy to see that $P_n = P_{w_n}$. Namely,

$$Q_n P_S f(x) = P_{w_n} f(x) = \int_0^1 P_S f(y) w_n(y, x) dy, \quad \forall f \in L^1(0,1).$$

Under exactly the same condition of Theorem 10.1 that guarantees the convergence of Ulam's method as n goes to infinity, the above corollary together with Theorem 10.7 can be directly used to establish the following error estimate for Ulam's method.

Theorem 10.8. (Keller) *Under the condition of Theorem 10.1, if f^* is a unique stationary density of P_S and $\{f_n\}$ is a sequence of the stationary densities of P_n from Ulam's method, then*

$$\|f_n - f^*\|_1 = O\left(\frac{\ln n}{n}\right). \qquad (10.16)$$

Proof It is a routine task to verify the inequality $c(w_n) \leq 1/(2n)$. Hence the sequence $\{P_n\}$ is $\mathcal{P}$-bounded and $\||P_S - P_n\|| = O(1/n)$ by Corollary 10.1. Using Theorem 10.7 gives rise to (10.16).   □

Remark 10.6. A Markov operator $P \in \mathcal{P}$ is said to be *stochastically pre-stable* if $\lim_{n\to\infty} \||P_n - P\|| = 0$ implies that $\lim_{n\to\infty} \|(\Phi(1, P_n) - I)\Phi(1, P)\|_1 = 0$ for each $\mathcal{P}$-bounded sequence $\{P_n\}$ of Markov operators. Similarly, $P \in \mathcal{P}$ is said to be *stochastically stable* if $\lim_{n\to\infty} \||P_n - P\|| = 0$ implies that $\lim_{n\to\infty} \|\Phi(1, P_n) - \Phi(1, P)\|_1 = 0$ for any $\mathcal{P}$-bounded sequence $\{P_n\}$ of Markov operators. The results of this section show that each $P \in \mathcal{P}$ is stochastically pre-stable, and if P has only one stationary density, then $P \in \mathcal{P}$ is stochastically stable.

10.4   Convergence Rate of the Markov Method

In this section we give a convergence rate analysis of the piecewise linear Markov approximation method for solving the Frobenius-Perron operator equation associated with one dimensional mappings. The approach adopted here, initiated by [Chiu *et al.* (1992)], is to use the Dunford integral of linear operators studied in Section 6.1, following the presentations of [Chiu *et al.* (1992); Ding and Li (1998)]. We shall show that the BV-norm convergence rate of the Markov method is $O(1/n)$, where n is the number of the subintervals of the uniform partition of $[0, 1]$.

Let X be a complex Banach space, $T \in B(X)$, and $\lambda_0 \in \sigma_p(T)$ an isolated point of $\sigma(T)$. Then the resolvent $R(\lambda, T) = (\lambda I - T)^{-1}$ of T has the Laurent series expansion at λ_0:

$$R(\lambda, T) = \sum_{k=-\infty}^{\infty} (\lambda_0 - \lambda)^k A_k, \quad 0 < |\lambda - \lambda_0| < r \leq \infty$$

for some radius r of convergence, where the coefficient operators

$$A_{-k} = -\frac{1}{2\pi i} \int_\Gamma (\lambda_0 - \lambda)^{k-1} R(\lambda, T) d\lambda, \quad \forall \, k,$$

and the small circle $\Gamma \subset \rho(T)$ centers at λ_0 and encloses no other spectral points. When $k = 1$, the coefficient operator

$$A_{-1} = -\frac{1}{2\pi i} \int_\Gamma R(\lambda, T) d\lambda \tag{10.17}$$

is the projection $E(\lambda_0, T)$ onto the generalized eigenspace of T corresponding to the eigenvalue λ_0, and the restriction $T|_{R(A_{-1})}$ of the operator

T to the range $R(A_{-1})$ of A_{-1} maps the space $R(A_{-1})$ into itself and $\sigma(T|_{R(A_{-1})}) = \{\lambda_0\}$, as Theorem 6.6 shows.

We assume that a nonsingular transformation $S : [0,1] \to [0,1]$ is given such that the corresponding Frobenius-Perron operator $P_S : L^1(0,1) \to L^1(0,1)$ satisfies the variation inequality (10.13).

Remark 10.7. If S satisfies the condition of Theorem 8.5 with $\inf_{x \in [0,1] \setminus \{a_0, \cdots, a_r\}} |S'(x)| > 2$, then (10.13) is true.

For the spectral analysis of P_S, we assume that the space $L^1(0,1)$ is a complex one, and thus the Frobenius-Perron operator $P_S : L^1(0,1) \to L^1(0,1)$ is a complex bounded linear operator which still has the operator norm 1 as in the real L^1-space case (prove it!). We also assume that $f^* \in BV(0,1)$ is a unique stationary density of P_S in $L^1(0,1)$.

Let $\{P_n\}$ be the sequence of the finite dimensional approximations of P_S associated with the piecewise linear Markov method. Then the inequality (10.13) implies that for all $f \in BV(0,1)$,

$$\|P_n f\|_{BV} \le \alpha \|f\|_{BV} + C\|f\|_1. \tag{10.18}$$

By Theorem 10.4, the Markov method produces a sequence $\{f_n\}$ of piecewise linear densities converging to f^* in the BV-norm. We further investigate the rate of the convergence.

Lemma 10.8. *Let $\lambda \in \mathbb{C}$ and $\lambda \ne 1$. If $P_S f = \lambda f$, then $\int_0^1 f(x)dx = 0$. Similarly, $P_n f = \lambda f$ implies $\int_0^1 f(x)dx = 0$.*

Proof Integrating both sides of $P_S f = \lambda f$ over $[0,1]$ yields

$$\lambda \int_0^1 f(x)dx = \int_0^1 P_S f(x)dx = \int_0^1 f(x)dx,$$

so $\int_0^1 f(x)dx = 0$ since $\lambda \ne 1$. The proof for P_n is the same. $\square$

Lemma 10.9. *For n large enough, the eigenspace E_n of P_n corresponding to the eigenvalue $\lambda = 1$ is one dimensional.*

Proof $E_n \subset \Delta_n$ contains a density f_n for each n. If the lemma is not true, then there is a subsequence $\{f_{n_i}\}$ of $\{f_n\}$ such that for each f_{n_i}, there is a fixed point $g_{n_i} \in \Delta_{n_i}$ of P_{n_i} with L^1-norm 1 which is linearly independent of f_{n_i}. Set $c_{n_i} = \int_0^1 g_{n_i} dm$ and

$$h_{n_i} = \frac{c_{n_i} f_{n_i} - g_{n_i}}{\|c_{n_i} f_{n_i} - g_{n_i}\|_1}.$$

Then $\int_0^1 h_{n_i} dm = 0, P_{n_i} h_{n_i} = h_{n_i}$, and $\|h_{n_i}\|_1 = 1$. By (10.18),

$$\|h_{n_i}\|_{BV} = \|P_{n_i} h_{n_i}\|_{BV} \leq \alpha \|h_{n_i}\|_{BV} + C,$$

from which we have

$$\|h_{n_i}\|_{BV} \leq \frac{C}{1 - \alpha}, \quad \forall\, i.$$

From Lemma 8.2, the sequence $\{h_{n_i}\}$ has a convergent subsequence, which is still denoted as $\{h_{n_i}\}$ for the simplicity of notation, such that $\lim_{i \to \infty} \|h_{n_i} - h\|_1 = 0$ for some $h \in L^1(0,1)$. Moreover, we have $\|h\|_1 = \lim_{i \to \infty} \|h_{n_i}\|_1 = 1$ and $\int_0^1 h\, dm = \lim_{i \to \infty} \int_0^1 h_{n_i} dm = 0$. Since

$$\|h - P_S h\|_1 \leq \|h - h_{n_i}\|_1 + \|h_{n_i} - P_{n_i} h_{n_i}\|_1$$
$$+ \|P_{n_i} h_{n_i} - P_{n_i} h\|_1 + \|P_{n_i} h - P_S h\|_1 \to 0$$

as $i \to \infty$, it follows that $P_S h = h$. By Proposition 7.4, the assumption that f^* is a unique stationary density of P_S implies that $\dim N(I - P_S) = 1$. Hence $h = cf^*$, where c is a nonzero complex number. Thus $\int_0^1 h(x) dx \neq 0$, which is a contradiction to the equality $\int_0^1 h(x) dx = 0$. $\qquad\square$

Lemma 10.10. $N(I - P_n) = N((I - P_n)^k)$ *for all positive integers k and n. Therefore, when n is large enough, the generalized eigenspace of P_n associated with eigenvalue 1 is one dimensional.*

Proof Given any positive integer n, since

$$\lim_{j \to \infty} \frac{\|P_n^j\|_1}{j} \leq \lim_{j \to \infty} \frac{1}{j} = 0,$$

and since P_n is a compact operator, the lemma is from Theorem VII.4.5 and Lemma VIII.8.1 of [Dunford and Schwartz (1957)]. $\qquad\square$

Let two subsets of $\mathbb{C}$ be defined by

$$\Lambda = \{\lambda \in \mathbb{C} : P_n g = \lambda g,\ g \neq 0 \text{ for some } n\} \setminus \{1\}$$

and

$$\Lambda_0 = \{\lambda \in \mathbb{C} : P_S g = \lambda g,\ g \in BV(0,1),\ g \neq 0\} \setminus \{1\}.$$

Our next result will establish the fact that the maximal eigenvalue 1 is not a limit point of the whole other spectral points of P_S and all P_n for $n \geq 1$, which is a key result to achieve our purpose for the convergence rate analysis of the Markov method applied to the class of Frobenius-Perron operators satisfying the condition (10.13).

Lemma 10.11. $1 \notin \overline{\Lambda} \cup \overline{\Lambda_0}$.

Proof If $1 \in \overline{\Lambda}$, then there is a sequence $\{\lambda_{n_i}\}$ of complex numbers in Λ such that $\lim_{i\to\infty} \lambda_{n_i} = 1$. Let g_{n_i} be an eigenvector with L^1-norm 1 corresponding to the eigenvalue λ_{n_i} of P_{n_i}. Then, $\int_0^1 g_{n_i}(x)dx = 0$ by Lemma 10.8. By means of the inequality (10.18), we obtain

$$\|g_{n_i}\|_{BV} = \left\| \frac{1}{\lambda_{n_i}} P_{n_i} g_{n_i} \right\|_{BV} \leq \frac{1}{|\lambda_{n_i}|} \left(\alpha \|g_{n_i}\|_{BV} + C \right).$$

Since $|\lambda_{n_i}| > \alpha$ and

$$\|g_{n_i}\|_{BV} \leq \frac{C}{|\lambda_{n_i}| - \alpha}$$

for i large enough, the sequence $\{g_{n_i}\}$ contains a convergent subsequence, which is still denoted as $\{g_{n_i}\}$ for the simplicity of notation, such that $\lim_{i\to\infty} \|g_{n_i} - g\|_1 = 0$ for some $g \in L^1(0,1)$ of L^1-norm 1. Note that

$$\|g - P_S g\|_1 \leq \|g - g_{n_i}\|_1 + |1 - \lambda_{n_i}| \|g_{n_i}\|_1 + \|\lambda_{n_i} g_{n_i} - P_{n_i} g_{n_i}\|_1$$
$$+ \|P_{n_i} g_{n_i} - P_{n_i} g\|_1 + \|P_{n_i} g - P_S g\|_1 \to 0$$

as i goes to infinity, we therefore have the equality $P_S g = g$, and hence $g = cf^*$ with $|c| = 1$. This leads to a contradiction that $c = c \int_0^1 f^*(x)dx = \int_0^1 g(x)dx = \lim_{i\to\infty} \int_0^1 g_{n_i}(x)dx = 0$.

If $1 \in \overline{\Lambda_0}$, then $P_S g_n = \lambda_n g_n$ for a sequence $\{\lambda_n\}$ in Λ_0 such that $g_n \in BV(0,1)$, $\|g_n\|_1 = 1$, and $\lim_{n\to\infty} \lambda_n = 1$. The same contradiction appears from Lemma 10.8 and (10.13). $\qquad\square$

Lemma 10.12. $\sigma(P_n) \setminus \{0\} \subset \Lambda$.

Proof Let $\lambda \neq 0$. Since P_n is a compact linear operator, $\lambda \in \sigma(P_n)$ if and only if λ is an eigenvalue of P_n. $\qquad\square$

Thanks to Lemma 10.11 which guarantees a positive distance between 1 and the set $\Lambda \cup \Lambda_0$, we can define a positive number

$$\epsilon = \frac{1}{2} \min\{\text{dist}(1, \Lambda \cup \Lambda_0), 1 - \alpha\}.$$

Let a closed subset F of $\mathbb{C}$ be defined by

$$F = \left\{ \lambda \in \mathbb{C} : \frac{\epsilon}{2} \leq |\lambda - 1| \leq \epsilon \right\}.$$

Since $\epsilon > 0$, F is a nontrivial ring-shaped region centered at 1 with inner and outside radii $\epsilon/2$ and ϵ, respectively. By Lemma 10.12, as a linear operator on $(BV(0,1), \| \ \|_{BV})$, the inverse of $(P_n - \lambda I)$ exists for n large enough and the inverse operators

$$(P_n - \lambda I)^{-1} : (BV(0,1), \| \ \|_{BV}) \to (BV(0,1), \| \ \|_{BV})$$

are bounded for all $\lambda \in F$. From now on we fix a circle Γ centered at 1 with radius $\epsilon_0 \in [\epsilon/2, \epsilon]$ so that $\Gamma \subset F$.

Lemma 10.13. *There exists a constant $b > 0$ such that for any sequence $\{f_n\}$ of functions in $BV(0,1)$,*

$$\|(\lambda I - P_n)^{-1} f_n\|_1 \le b \|f_n\|_{BV}, \quad \forall \lambda \in \Gamma.$$

Proof If the lemma is not true, then there is a sequence $\{n_i\}$ of natural numbers and a sequence $\{\lambda_{n_i}\}$ in Γ such that

$$\lim_{i \to \infty} \frac{\|(\lambda_{n_i} I - P_{n_i})^{-1} f_{n_i}\|_1}{\|f_{n_i}\|_{BV}} = \infty.$$

Denote for each i

$$\tilde{g}_{n_i} = \frac{(\lambda_{n_i} I - P_{n_i})^{-1} f_{n_i}}{\|(\lambda_{n_i} I - P_{n_i})^{-1} f_{n_i}\|_1} \quad \text{and} \quad \tilde{f}_{n_i} = \frac{f_{n_i}}{\|(\lambda_{n_i} I - P_{n_i})^{-1} f_{n_i}\|_1}.$$

Then $\|\tilde{g}_{n_i}\|_1 \equiv 1$, $\tilde{f}_{n_i} = (\lambda_{n_i} I - P_{n_i}) \tilde{g}_{n_i}$ and

$$\lim_{i \to \infty} \|\tilde{f}_{n_i}\|_{BV} = \lim_{i \to \infty} \frac{\|f_{n_i}\|_{BV}}{\|(\lambda_{n_i} I - P_{n_i})^{-1} f_{n_i}\|_1} = 0.$$

Since the set $\{\lambda_{n_i}\}_{i=1}^{\infty} \subset \Gamma$ is pre-compact, without loss of generality, we assume that $\lim_{i \to \infty} \lambda_{n_i} = \lambda_0 \in \Gamma$. Then,

$$\|\tilde{g}_{n_i}\|_{BV} = \left\| \frac{1}{\lambda_{n_i}} (P_{n_i} \tilde{g}_{n_i} + \tilde{f}_{n_i}) \right\|_{BV} \le \frac{\|P_{n_i} \tilde{g}_{n_i}\|_{BV} + \|\tilde{f}_{n_i}\|_{BV}}{|\lambda_{n_i}|}$$

$$\le \frac{\alpha}{1 - \epsilon} \|\tilde{g}_{n_i}\|_{BV} + \frac{C}{1 - \epsilon} + 1$$

for i large enough. Hence

$$\|\tilde{g}_{n_i}\|_{BV} \le \frac{C + 1 - \epsilon}{1 - \epsilon - \alpha}, \quad \forall i \ge i_0$$

are true for some positive integer i_0. By Lemma 8.2, there is a subsequence of $\{\tilde{g}_{n_i}\}$, again denoted by $\{\tilde{g}_{n_i}\}$, such that

$$\lim_{i \to \infty} \|\tilde{g}_{n_i} - g\|_1 = 0$$

for some $g \in BV(0,1)$ and $\|g\|_1 = \lim_{i \to \infty} \|\tilde{g}_{n_i}\|_1 = 1$. From

$$\|\lambda_0 g - P_S g\|_1$$
$$\le |\lambda_0 - \lambda_{n_i}| \|g\|_1 + |\lambda_{n_i}| \|g - \tilde{g}_{n_i}\|_1 + \|(\lambda_{n_i} I - P_{n_i}) \tilde{g}_{n_i}\|_1$$
$$+ \|P_{n_i} (\tilde{g}_{n_i} - g)\|_1 + \|P_{n_i} g - P_S g\|_1 \to 0,$$

we have $\lambda_0 \in \Lambda_0$, which contradicts the fact that $F \cap \Lambda_0 = \emptyset$. $\qquad \square$

We are ready to use Dunford integrals of bounded linear operators to obtain our final result. For all positive integers n large enough, the resolvent $R(\lambda, P_n)$ of $P_n : (BV(0,1), \| \ \|_{BV}) \to (BV(0,1), \| \ \|_{BV})$ is well-defined and analytic for all numbers $\lambda \in F$ with the Laurent series expansion

$$R(\lambda, P_n) = \sum_{k=-\infty}^{\infty} A_k(P_n)(1 - \lambda)^k$$

at $\lambda = 1$, where $A_k(P_n)$ is given by the Dunford integral

$$A_{-k}(P_n) = -\frac{1}{2\pi i} \int_{\Gamma} (1 - \lambda)^{k-1} R(\lambda, P_n) d\lambda, \ \ \forall \, k.$$

By Lemma 10.10, $\lambda = 1$ is a simple pole of $R(\lambda, P_n)$, thus $A_{-k}(P_n) = 0$ for all $k \geq 2$. From (10.17), we have

Lemma 10.14. $A_{-1}(P_n) = -(2\pi i)^{-1} \int_{\Gamma} R(\lambda, P_n) d\lambda$ *is a projection from* $BV(0,1)$ *onto* $N(I - P_n)$ *for* n *large enough.*

Lemma 10.15. *For* n *large enough,*

$$A_{-1}(P_n)(f^* - f_n) = 0.$$

Proof Because dim $N(I - P_n) = 1$ for sufficiently large integers n, we have the equality $A_{-1}(P_n)(f^* - f_n) = c_n f_n$ for some constant c_n. Since $\int_0^1 f_n(x) dx = 1$, a direct calculation gives

$$\begin{aligned}
c_n &= \int_0^1 A_{-1}(P_n)(f^* - f_n)(x) dx \\
&= -\int_0^1 \frac{1}{2\pi i} \left(\int_{\Gamma} (\lambda I - P_n)^{-1} (f^* - f_n) d\lambda \right)(x) dx \\
&= -\frac{1}{2\pi i} \int_{\Gamma} \int_0^1 (\lambda I - P_n)^{-1} (f^* - f_n)(x) dx d\lambda.
\end{aligned}$$

Let $g_n(\lambda) = (\lambda I - P_n)^{-1}(f^* - f_n)$. Then $(\lambda I - P_n) g_n(\lambda) = f^* - f_n$. Thus, since P_n preserves integrals,

$$(\lambda - 1) \int_0^1 g_n(\lambda)(x) dx$$

$$= \int_0^1 [(\lambda I - P_n) g_n(\lambda)](x) dx = \int_0^1 (f^* - f_n)(x) dx = 0,$$

from which we have $\int_0^1 g_n(\lambda)(x) dx = 0$ for all $\lambda \in \Gamma$. Hence,

$$c_n = -\frac{1}{2\pi i} \int_{\Gamma} \int_0^1 g_n(\lambda)(x) dx d\lambda = 0. \qquad \square$$

The above preliminary results imply the following theorem.

Theorem 10.9. (Chiu-Du-Ding-Li) *Let f_n be the stationary densities of P_n from the Markov method to approximate the stationary density f^* of P_S. Then for n large enough,*

$$\|f^* - f_n\|_1 \le b\|f^* - Q_n f^*\|_{BV}, \tag{10.19}$$

$$\|f^* - f_n\|_{BV} \le \frac{Cb+1}{1-\alpha}\|f^* - Q_n f^*\|_{BV}. \tag{10.20}$$

Moreover, if $f^* \in C^2[0,1]$, then

$$\|f_n - f^*\|_{BV} = O\left(\frac{1}{n}\right). \tag{10.21}$$

Proof From (10.11), (10.19) implies (10.20), which and Proposition 10.10 lead to (10.21). So we just show (10.19). Since

$$(I - P_n)(f^* - f_n) = f^* - Q_n f^*,$$

$$\frac{f^* - f_n}{\lambda - 1} = (\lambda I - P_n)^{-1}(f^* - f_n) + \frac{(\lambda I - P_n)^{-1}(f^* - Q_n f^*)}{\lambda - 1}$$

for all $\lambda \in \Gamma$, Integrating the above equality over Γ, noting that $(2\pi i)^{-1} \int_\Gamma (\lambda - 1)^{-1} d\lambda = 1$, and using Lemma 10.15, we have

$$f^* - f_n = \frac{1}{2\pi i} \int_\Gamma \frac{1}{\lambda - 1}(\lambda I - P_n)^{-1}(f^* - Q_n f^*) d\lambda.$$

Thus, it follows from Lemma 10.13 that

$$\|f^* - f_n\|_1 \le \frac{1}{2\pi} \frac{2\pi\epsilon_0}{\epsilon_0} \max_{\lambda \in \Gamma} \|(\lambda I - P_n)^{-1}(f^* - Q_n f^*)\|_1$$
$$\le b\|f^* - Q_n f^*\|_{BV}. \qquad \square$$

Remark 10.8. While Ulam's method has the L^1-norm error bound no better than $O(\ln n/n)$, the Markov method does converge under the BV-norm with the convergence rate $O(1/n)$ for the class of Frobenius-Perron operators satisfying (10.13), which implies directly that the convergence rate of the piecewise linear Markov approximation method under the L^1-norm is at least $O(1/n)$. Indeed, a higher order of $O(\ln n/n^2)$ has been observed from numerical experiments [Ding (1990); Ding and Li (1991); Ding et al. (2003b); Ding and Rhee (2004)], but no theoretical justification has appeared in the literature. [Liverani (2001)] contains new ideas for the numerical analysis of Frobenius-Perron operators.

10.5   Notes and Remarks

1. A more detailed study of Ulam's method is contained in, for example, [Ding and Zhou (2009)] in which Murray's delicate analysis [Murray (1998)] on an explicit error estimate of Ulam's method was also presented.

2. It seems that the piecewise linear interpolation method proposed in [Ding and Rhee (2004)], which is also an approximation method via positive operators, is faster than the piecewise linear Markov method from numerical experiments, but it has not been successful to prove it rigorously.

Exercises

10.1 Prove Proposition 10.12.

10.2 Let $\{\phi_i\}_{i=0}^n$ be the canonical basis of Δ_n as introduced in Section 10.2. Let $Q_n : L^1(0,1) \to \Delta_n$ be a bounded linear operator so that it can be represented by the expression

$$Q_n f = \sum_{i=0}^n \int_0^1 f(x) w_i(x) dx \cdot \phi_i,$$

where $w_i \in L^\infty(0,1)$ for all i. Show that

(i) $Q_n 1 = 1$ if and only if $\int_0^1 w_i(x) dx = 1$ for $i = 0, 1, \cdots, n$;

(ii) Q_n is nonnegative if and only if $w_i(x) \geq 0$ a.e. for all i;

(iii) Q_n preserves integrals if and only if the equality $w_0(x) + 2 \sum_{i=1}^{n-1} w_i(x) + w_n(x) = 2n$ a.e. holds;

(iv) Q_n is a Markov operator if and only if $w_i(x) \geq 0$ a.e. for all i and $w_0(x) + 2 \sum_{i=1}^{n-1} w_i(x) + w_n(x) = 2n$ a.e.

10.3 Let Q be the Galerkin projection of $L^1(0,1)$ onto the space of linear polynomials. That is, for all $f \in L^1(0,1)$,

$$\int_0^1 (Qf(x) - f(x)) dx = 0, \quad \int_0^1 (Qf(x) - f(x)) x dx = 0.$$

Show that Q is not a Markov operator.

10.4 Let $Q_n : L^1(0,1) \to L^1(0,1)$ be defined by

$$Q_n f = \sum_{i=0}^n \frac{1}{m(F_i)} \int_{F_i} f(x) dx \cdot \phi_i, \quad \forall f \in L^1(0,1),$$

where each F_i is a closed subinterval of the closure of $\overline{\operatorname{supp} \phi_i}$. Show that if Q_n is a Markov operator, then Q_n is given by either the formula (10.5) or the one obtained in [Ding (1990)].

Hint: You may first show that if Q_n is integral-preserving and if B is a closed subinterval of $[x_k, x_{k+1}]$ and $0 \le k < n$, then

$$m(B) = \frac{m(B \cap F_k)}{m(F_k)} \frac{m(\text{supp } \phi_k)}{2} + \frac{m(B \cap F_{k+1})}{m(F_{k+1})} \frac{m(\text{supp } \phi_{k+1})}{2}.$$

Then show that if $m(F_i \cap F_{i+1}) = 0$ for $i = 0, 1, \cdots, n-1$, then Q_n is as given in (10.5), and if $F_k = \overline{\text{supp } \phi_k}$ for some $0 < k < n$, then Q_n is the same as the one given by [Ding (1990)].

10.5 Let (X, Σ, μ) be a probability measure space and let $\psi_1, \psi_2, \cdots, \psi_n$ be linearly independent and nonnegative measurable functions on X such that $\sum_{i=1}^n \psi_i = 1$. Define

$$T_n f = \sum_{i=1}^n \int_X f w_i d\mu \cdot \psi_i, \quad \forall \, f \in L^1(X),$$

where $w_1, w_2, \cdots, w_n \in L^\infty(X)$. Show that $T_n : L^1(X) \to L^1(X)$ preserves integrals if and only if $\sum_{i=1}^n \int_X \psi_i d\mu \cdot w_i = 1$.

10.6 Under the same assumption as in the above exercise, let $A_1, A_2, \cdots, A_n \in \Sigma$ be such that $\mu(A_i) > 0$ for each $i = 1, 2, \cdots, n$. Define a linear operator $Q_n : L^1(X) \to L^1(X)$ by

$$Q_n f = \sum_{i=1}^n \frac{1}{\mu(A_i)} \int_{A_i} f d\mu \cdot \psi_i.$$

Assume $\mu(A_i \cap A_j) = 0$ if $i \ne j$. Show that Q_n is a Markov operator if and only if $\int_X \psi_i d\mu = \mu(A_i)$ for each $i = 1, 2, \cdots, n$.

10.7 Let $\Phi(\lambda, P)$ be defined as in Section 10.3. Show that $\Phi(\lambda, P)f$ exists for all $f \in L^1(0, 1)$ and

$$\Phi(\lambda, P) = \begin{cases} \Phi_i, & \text{if } \lambda = \lambda_i, \\ 0, & \text{otherwise.} \end{cases}$$

10.8 Finish the proof of Lemma 10.6(i) and (ii).

10.9 Show that if a Markov operator $P \in \mathcal{P}$ has a unique stationary density, then for any function $f \in L^1(0, 1)$,

$$\Phi(1, P)f = \int_0^1 f(x)dx \cdot \Phi(1, P)1.$$

10.10 Prove Lemma 10.10 in more detail.

10.11 Construct a doubly stochastic symmetric kernel w for which w_z in Section 10.3 is not monotonically decreasing.

10.12 Construct a sequence of doubly stochastic kernels w_n corresponding to the sequence $\{P_n\}$ of the piecewise linear Markov approximation method introduced in Section 10.2, and investigate it to see whether Keller's approach can give a better L^1-norm convergence rate for the piecewise linear Markov method than Ulam's piecewise constant method.

Chapter 11

Applications of Positive Operators

Positive operators are often used in various areas of mathematical sciences. In this last chapter, we give one representative application of positive operators in each of the following areas: continuous function approximations via positive operators and in particular the generalized summation of Fourier series of continuous periodic functions; time dependent partial differential equations; computational molecular dynamics; and wireless communications. The first two of them are well-known in the literature and the last two are just recent active research areas.

11.1   Positive Approximations of Continuous Functions

We introduce a particular theory of approximating continuous real-valued functions by positive operators. Specifically the classic Bohman-Korovkin theorem will be studied. Then we treat the summation problem for the Fourier series associated with periodic continuous functions, following a classic approach. The general theory on the approximation of continuous functions by positive operators is a rich subject, and we refer to [DeVore (1972); Powell (1981)] for an extensive treatment of this topic.

In numerical analysis, especially numerical partial differential equations and solving operator equations such as the fixed point equation for Frobenius-Perron operators in the previous chapter, the given or unknown functions of some regularity are approximated by simpler functions such as polynomials, piecewise polynomials, or trigonometric polynomials. Naturally one faces the question of how to approximate continuous functions with good convergence properties. Here we present some classic schemes and results on approximating such functions via a sequence of finite dimensional positive operators. Such schemes are listed first before we study a fundamental approximation theorem. Recall that $C[a, b]$ is the M-space of all real-valued continuous functions on $[a, b]$ with the max-norm $\| \ \|_\infty$.

Example 11.1. Let $x_1, \cdots, x_n$ be n distinct points in the interval $[a, b]$ and let $h_1, \cdots, h_n$ be nonnegative functions in $C[a, b]$. The linear operator $H_n : C[a, b] \to C[a, b]$ defined by

$$H_n f(x) = \sum_{k=1}^{n} f(x_k) h_k(x), \quad \forall \, x \in [a, b]$$

is positive and is referred to as an *interpolation operator*. Note that Bernstein operators $B_n : C[0, 1] \to C[0, 1]$, which will be defined by (11.4) in the following, are special interpolation operators.

Example 11.2. Let $a = x_0 < x_1 < \cdots < x_{n-1} < x_n = b$ be $n + 1$ distinct points of the interval $[a, b]$, and let Δ_n be the space of all piecewise linear continuous functions corresponding to the partition of $[a, b]$. For any $f \in C[a, b]$ define $L_n f$ to be the unique function in Δ_n such that

$$L_n f(x_k) = f(x_k), \quad k = 0, 1, \cdots, n.$$

Then the *piecewise linear Lagrange interpolation operator* $L_n : C[a, b] \to C[a, b]$ is a positive operator with $R(L_n) = \Delta_n$, and it was used to study the Markov approximation in Section 10.2.

Example 11.3. Now we generalize Example 11.2 to interpolate L^1 functions in the following sense [Ding and Rhee (2004)]. Associated with the above partition of $[a, b]$, a *generalized piecewise linear Lagrange interpolation operator* $\hat{L}_n : L^1(a, b) \to \Delta_n$ is defined by letting $\hat{L}_n f$ be the unique function in Δ_n such that

$$\hat{L}_n f(x_k) = \frac{1}{m(\hat{I}_k)} \int_{\hat{I}_k} f(t) dt, \quad 0 \le k \le n$$

for any $f \in L^1(a, b)$, where each $\hat{I}_k$ is a small interval around x_k.

Example 11.4. Let $C[-\pi, \pi]$ denote the space of all real-valued continuous and periodic functions defined on the whole real line $\mathbb{R}$ with period 2π endowed with the max-norm $\| \ \|_\infty$, and let

$$s_n(f)(x) = \frac{a_0}{2} + \sum_{k=1}^{n} (a_k \cos kx + b_k \sin kx) \tag{11.1}$$

be the nth partial sum of the *Fourier series*

$$\frac{a_0}{2} + \sum_{n=1}^{\infty} (a_n \cos nx + b_n \sin nx) \tag{11.2}$$

of $f \in C[-\pi, \pi]$, where the *Fourier coefficients* are given by

$$a_n = \frac{1}{\pi} \int_{-\pi}^{\pi} f(t) \cos ntdt, \quad b_n = \frac{1}{\pi} \int_{-\pi}^{\pi} f(t) \sin ntdt, \quad \forall\, n.$$

The nth *Fejér sum* for the Fourier series of f is defined as

$$F_n f(x) = \frac{1}{n} \sum_{k=0}^{n-1} s_k(f)(x), \quad \forall\, x \in (-\infty, \infty),$$

namely $\{F_n f\}$ is just the Cesáro averages sequence of the partial sum sequence $s_n(f)$. Then for each n, $F_n : C[-\pi, \pi] \to C[-\pi, \pi]$ is a positive operator with range the space of trigonometric polynomials of degree at most n, which comes from the expression [Tolstov (1976)]

$$F_n f(x) = \frac{1}{2n\pi} \int_{-\pi}^{\pi} \left(\frac{\sin \frac{n(x-t)}{2}}{\sin \frac{x-t}{2}} \right)^2 f(t)dt. \tag{11.3}$$

We shall study the convergence of $F_n f$ later on.

Example 11.5. Let n be a positive integer. The positive operator $J_n : C[-\pi, \pi] \to C[-\pi, \pi]$ with range the space of all trigonometric polynomials of degree at most $2n$, defined by

$$J_n f(x) = \frac{a_n}{\pi} \int_{-\pi}^{\pi} \left(\frac{\sin \frac{n(x-t)}{2}}{\sin \frac{x-t}{2}} \right)^4 f(t)dt,$$

is called the nth *Jackson operator*. Here a_n is chosen so that

$$\frac{a_n}{\pi} \int_{-\pi}^{\pi} \left(\frac{\sin \frac{n(x-t)}{2}}{\sin \frac{x-t}{2}} \right)^4 dt = 1.$$

Before studying approximations of continuous functions via positive operators, we show that a general Lagrange interpolation does not preserve the positivity of nonnegative functions. Let $f \in C[a, b]$ and let $x_0 < x_1 < \cdots < x_{n-1} < x_n$ be $n + 1$ nodes in $[a, b]$. Then there is a unique polynomial p_n of degree at most n such that

$$p_n(x_k) = f(x_k), \quad k = 0, 1, \cdots, n.$$

Denote $\tilde{L}_n f = p_n$. Then $\tilde{L}_n : C[a, b] \to C[a, b]$ is a bounded linear operator. However, $\tilde{L}_n$ is not a positive operator for $n \geq 2$. For example, when $n = 2$, the quadratic polynomial that interpolates $f \in C[0, 1]_+$ with $f(x_0) = f(x_1) = 0$ and $f(x_2) > 0$ is negative on the open interval (x_0, x_1).

Even if the nodes x_k are evenly distributed on $[a, b]$ so that $\lim_{n \to \infty} \max_k (x_k - x_{k-1}) = 0$, the corresponding Lagrange interpolation polynomial sequence $\{p_n(x)\}$ does not converge to $f(x)$ uniformly on $[a, b]$ for *all* $f \in C[a, b]$, which is a consequence of the principle of uniform boundedness since the sequence $\{\|\tilde{L}_n\|_\infty\}$ is not bounded. In fact we have the following result [Powell (1981)].

Proposition 11.1. *Let $x_0^{(n)} < x_1^{(n)} < \cdots < x_n^{(n)}$ be $n + 1$ nodes of $[a, b]$ for all natural numbers n such that*

$$\lim_{n \to \infty} \max_{k=1}^{n} \left(x_k^{(n)} - x_{k-1}^{(n)} \right) = 0.$$

Then there exists a continuous function f defined on $[a, b]$ such that the corresponding interpolation polynomial sequence $\{p_n\}$ does not converge to f under the max-norm of $C[a, b]$.

Bernstein polynomials, however, provide a constructive proof of the celebrated Weierstrass approximation theorem. The nth *Bernstein polynomial* associated with $f \in C[0, 1]$ is defined as

$$B_n f(x) = \sum_{k=0}^{n} f\left(\frac{k}{n}\right) \binom{n}{k} x^k (1-x)^{n-k}. \tag{11.4}$$

It is easy to see that $B_n : C[0, 1] \to C[0, 1]$ is a positive operator which is called the nth *Bernstein operator*. The following theorem shows that the sequence $\{B_n\}$ of Bernstein operators strongly converges to the identity operator in the space $B(C[0, 1])$; its proof is referred to [Powell (1981)].

Theorem 11.1. *For any $f \in C[0, 1]$,*

$$\lim_{n \to \infty} B_n f(x) = f(x), \quad \forall\, x \in [0, 1] \; uniformly.$$

Although the interpolation operators L_n gives better local errors than Bernstein ones, the lacking of the positivity-preserving property explains why the strong convergence is missing. On the other hand, verification of the strong convergence of positive operators P_n to the identity operator in $B(C[a, b])$ can be reduced to the norm convergence of the sequence of the functions $P_n f$ in $C[a, b]$ for f to be one of the only three monomials $1, x$, and x^2, as the following Bohman-Korovkin theorem shows. For the sake of proving this classic result, we need a quantitative description for the degree of continuity of a function. Let

$$\omega(f, \delta) = \sup_{0 \le t \le \delta} \max_{x \in [a, b-t]} |f(x+t) - f(x)|, \quad 0 \le \delta \le b - a$$

be the *modulus of continuity* of $f \in C[a, b]$. Clearly, $\omega(f, 0) = 0$.

Lemma 11.1. $\omega(f, \delta)$ *has the following properties:*
(i) $\omega(f, \delta)$ *is a nondecreasing continuous function of* δ;
(ii) $\omega(f, \delta_1 + \delta_2) \leq \omega(f, \delta_1) + \omega(f, \delta_2)$;
(iii) $\omega(f, c\delta) \leq (c+1)\omega(f, \delta)$ *for any number* $c \geq 0$;
(iv) $\delta_2^{-1}\omega(f, \delta_2) \leq 2\delta_1^{-1}\omega(f, \delta_1)$.

Theorem 11.2. (The Bohman-Korovkin theorem) *Let* $\{P_n\}$ *be a sequence of positive operators from* $C[a, b]$ *into* $C[a, b]$. *Assume*

$$\lim_{n \to \infty} \|P_n f_k - f_k\|_\infty = 0,$$

where $f_k(x) = x^k$, $k = 0, 1, 2$. *Then for any* $f \in C[a, b]$,

$$\lim_{n \to \infty} \|P_n f - f\|_\infty = 0.$$

Proof Fix $x \in [a, b]$ and apply the operator P_n to the quadratic function q defined by $q(t) = (t - x)^2$ on $[a, b]$. Let

$$\beta_n(x) = (P_n q(x))^{\frac{1}{2}}.$$

Then $\lim_{n \to \infty} \beta_n(x) = 0$ for $x \in [a, b]$ uniformly. We first show

$$\begin{aligned}
|P_n f(x) - f(x)| &\leq |f(x)| \, |f_0(x) - P_n f_0(x)| \\
&+ \left[P_n f_0(x) + (P_n f_0(x))^{\frac{1}{2}} \right] \omega(f, \beta_n(x))
\end{aligned} \tag{11.5}$$

for the given $f \in C[a, b]$. From Lemma 11.1(iii),

$$|f(t) - f(x)| \leq \omega(f, |t - x|) \leq \left(1 + \frac{1}{\delta}|t - x|\right) \omega(f, \delta)$$

for $t \in [a, b]$ and $\delta > \beta_n(x)$, hence,

$$\begin{aligned}
&|P_n f(x) - f(x)| \\
&\leq (P_n |f(t) - f(x)|)(x) + |(P_n f(x) f_0)(x) - f(x)| \\
&\leq \omega(f, \delta) [P_n f_0(x) + (P_n |t - x|)(x)] + |f(x)| \, |P_n f_0(x) - f_0(x)|.
\end{aligned}$$

The Riesz representation theorem, Theorem 9.1, and the Cauchy-Schwarz inequality ensure the inequality

$$(P_n |t - x|)(x) \leq [(P_n (t - x)^2)(x)]^{\frac{1}{2}} (P_n f_0(x))^{\frac{1}{2}} = (P_n f_0(x))^{\frac{1}{2}} \beta_n(x).$$

It follows from the above that

$$\begin{aligned}
&|P_n f(x) - f(x)| \\
&\leq |f(x)| \, |f_0(x) - P_n f_0(x)| + \left[P_n f_0(x) + (P_n f_0(x))^{\frac{1}{2}} \right] \omega(f, \delta),
\end{aligned}$$

which implies (11.5) by letting $\delta \downarrow \beta_n(x)$.

Since $|f(x)| \le \|f\|_\infty$ uniformly on $[a, b]$, the right-hand side of (11.5) approaches zero uniformly on $[a, b]$ as $n \to \infty$ under the hypothesis of the theorem. This completes the proof. □

Theorem 11.2 implies the following corollary for continuous periodic functions, whose proof is left as an exercise.

Corollary 11.1. *Suppose that $\{P_n\}$ is a sequence of positive operators from $C[-\pi, \pi]$ into $C[-\pi, \pi]$. If*

$$\lim_{n\to\infty} \|P_n f - f\|_\infty = 0$$

for $f(x) = 1$, $\cos x$, and $\sin x$, then for any $f \in C[-\pi, \pi]$,

$$\lim_{n\to\infty} \|P_n f - f\|_\infty = 0.$$

We return to Example 11.4 on the Fejér summation of Fourier series (11.2) associated with a real-valued continuous and periodic function of period 2π defined on $\mathbb{R}$. Let $\{s_n\}$ be the partial sum sequence of the Fourier series by (11.1). Then

$$s_n(x) = \frac{1}{\pi} \int_{-\pi}^{\pi} f(x+u) \frac{\sin(n+\frac{1}{2})u}{2\sin\frac{u}{2}} du.$$

It is well-known that the Fourier series of f may not converge to $f(x)$ at a point x of continuity of f [Tolstov (1976)]. This can be explained by the fact that the integral kernel

$$w_n(x, t) = \frac{\sin(n+\frac{1}{2})(x-t)}{2\sin\frac{x-t}{2}},$$

which is usually called the *Dirichlet kernel* in the field of Fourier analysis, in the above representation of s_n is not positive. Because of the expression (11.3) of $F_n f$ in terms of a positive kernel, we may expect the convergence of $F_n f(x)$ to $f(x)$ as n approaches infinity. Indeed, Corollary 11.1 can be applied directly to show the uniform convergence of $F_n f$ to f for all $f \in C[-\pi, \pi]$, but in the following we adopt a direct and elementary approach to this problem from [Tolstov (1976)].

Lemma 11.2. *The Féjer operator $F_n : C[-\pi, \pi] \to C[-\pi, \pi]$ is an abstract composition operator. Consequently, for all n,*

$$1 = \frac{1}{2n\pi} \int_{-\pi}^{\pi} \left(\frac{\sin\frac{n(x-t)}{2}}{\sin\frac{x-t}{2}} \right)^2 dt. \tag{11.6}$$

Proof We first note that $C[-\pi, \pi]$ is an M-space with order unit 1. Let $f = 1$. Then $s_n 1 = 1$, so $F_n 1 = 1$. Thus F_n is an abstract composition operator by Definition 5.14. □

Theorem 11.3. *The Fourier series of a real-valued continuous and periodic function f of period 2π is uniformly convergent to f on the whole real axis $\mathbb{R}$ in the sense of the Fejér summation.*

Proof Let $x \in \mathbb{R}$. Then, using the formula (11.3) for F_n and the identity (11.6), and changing variables, we can write

$$F_n f(x) - f(x) = \frac{1}{2n\pi} \int_{-\pi}^{\pi} [f(x + u) - f(x)] \left(\frac{\sin \frac{nu}{2}}{\sin \frac{u}{2}} \right)^2 du.$$

Because of the uniform continuity of f on $\mathbb{R}$, for any real number $\epsilon > 0$, there is a number $\delta > 0$ such that

$$|f(x + u) - f(x)| < \frac{\epsilon}{2}, \quad \forall\, x \in \mathbb{R}$$

provided $|u| \leq \delta$. Thus, with $b = \|f\|_\infty$, we have

$$|F_n f(x) - f(x)|$$
$$\leq \frac{1}{2n\pi} \int_{-\pi}^{-\delta} |f(x + u) - f(x)| \left(\frac{\sin \frac{nu}{2}}{\sin \frac{u}{2}} \right)^2 du$$
$$+ \frac{1}{2n\pi} \int_{-\delta}^{\delta} |f(x + u) - f(x)| \left(\frac{\sin \frac{nu}{2}}{\sin \frac{u}{2}} \right)^2 du$$
$$+ \frac{1}{2n\pi} \int_{\delta}^{\pi} |f(x + u) - f(x)| \left(\frac{\sin \frac{nu}{2}}{\sin \frac{u}{2}} \right)^2 du$$
$$\leq \frac{1}{2n\pi \sin^2 \frac{\delta}{2}} \int_{-\pi}^{-\delta} |f(x + u) - f(x)| du + \frac{\epsilon}{2}$$
$$+ \frac{1}{2n\pi \sin^2 \frac{\delta}{2}} \int_{\delta}^{\pi} |f(x + u) - f(x)| du$$
$$\leq \frac{1}{n\pi \sin^2 \frac{\delta}{2}} \int_{-\pi}^{\pi} |f(x + u) - f(x)| du + \frac{\epsilon}{2}$$
$$\leq \frac{1}{n\pi \sin^2 \frac{\delta}{2}} \int_{-\pi}^{\pi} 2b\, du + \frac{\epsilon}{2} = \frac{4b}{n \sin^2 \frac{\delta}{2}} + \frac{\epsilon}{2}$$

for all $x \in \mathbb{R}$. Therefore we can find an integer $n_0 > 0$ such that

$$|F_n f(x) - f(x)| \leq \frac{\epsilon}{2} + \frac{\epsilon}{2} = \epsilon, \quad \forall\, x \in \mathbb{R}. \qquad \square$$

11.2   Evolution Differential Equations

A continuous dynamical system is often described by an ordinary differential equation defined in a region of $\mathbb{R}^N$. The solution of an evolution partial differential equation can be expressed in terms of a group or semigroup of linear operators P_t depending on the continuous time t. In this section we study some applications of Frobenius-Perron operators to solving initial value problems of ordinary differential equations.

For $k = 1, 2, \cdots, N$, let the function $F_k(x_1, \cdots, x_N)$ of N variables have continuous partial derivatives $\partial F_k / \partial x_j$ in its domain $\mathbb{R}^N$, $j = 1, \cdots, N$. Consider the initial value problem

$$\frac{dx}{dt} = F(x), \;\; x(0) = x^{(0)}, \tag{11.7}$$

where the point $x = (x_1, \cdots, x_N)^T$ and the vector function $F(x) = (F_1(x), \cdots, F_N(x))^T$. We assume that for every initial point $x^{(0)} \in \mathbb{R}^N$, the solution of (11.7) exists for all $t \in \mathbb{R}$. Let $x(t; x^{(0)})$ denote the unique solution of (11.7), that is,

$$\frac{d}{dt} x(t; x^{(0)}) \equiv F(x(t; x^{(0)}))$$

for all t and $x(0; x^{(0)}) = x^{(0)}$. For each $t \in \mathbb{R}$ let a transformation $S_t : \mathbb{R}^N \to \mathbb{R}^N$ be defined by

$$S_t(x^{(0)}) = x(t; x^{(0)}), \;\; \forall\, x^{(0)} \in \mathbb{R}^N. \tag{11.8}$$

Then, from the theory of ordinary differential equations and in particular the continuous dependence of solutions on initial conditions, $\{S_t\}_{t \in \mathbb{R}}$ is a *continuous group of nonsingular transformations*, namely $S_{r+t} = S_r \circ S_t$ and $S_{-t} = S_t^{-1}$ for all r and t. The group $\{S_t\}_{t \in \mathbb{R}}$ is also called the *flow* of the dynamical system (11.7). For each $t \in \mathbb{R}$ let $P_t \equiv P_{S_t} : L^1(\mathbb{R}^N) \to L^1(\mathbb{R}^N)$ be the Frobenius-Perron operator associated with S_t, defined by

$$\int_A P_t f(x)dx = \int_{S_t^{-1}(A)} f(x)dx, \;\; \forall\, A \in \mathcal{B}(\mathbb{R}^N), \; f \in L^1(\mathbb{R}^N),$$

and let $U_t \equiv U_{S_t} : L^\infty(\mathbb{R}^N) \to L^\infty(\mathbb{R}^N)$ be the corresponding Koopman operator, defined by

$$U_t g(x) = g(S_t(x)), \;\; \forall\, g \in L^\infty(\mathbb{R}^N).$$

By Proposition 8.3 and the multi-variable calculus, we have

$$P_t f(x) = f(S_{-t}(x))|\det J_{S_{-t}}(x)|, \tag{11.9}$$

where $J_{S_{-t}}$ is the Jacobian matrix of $S_{-t} : \mathbb{R}^N \to \mathbb{R}^N$.

Definition 11.1. Let (X, Σ, μ) be a measure space and let $1 \le p \le \infty$. A family of positive operators $T_t : L^p \to L^p$ for all parameter values $t \in \mathbb{R}$ is called a *group of weak contractions* if

    (i) $\|T_t f\|_p \le \|f\|_p$ for all $f \in L^p$ and $t \in \mathbb{R}$;
    (ii) $T_0 f = f$ for all $f \in L^p$;
    (iii) $T_{r+t} = T_r \circ T_t$ for all $r, t \in \mathbb{R}$;
    (iv) $T_{-t} = T_t^{-1}$ for all $t \in \mathbb{R}$.

If in addition

$$\lim_{t \to t_0} \|T_t f - T_{t_0} f\|_p = 0, \quad \forall\, t_0 \in \mathbb{R}, \ f \in L^p,$$

then the group $\{T_t\}_{t \in \mathbb{R}}$ is said to be *continuous*.

Using the expression (11.9) for P_t, one can prove that the family $\{P_t\}_{t \in \mathbb{R}}$ of the Frobenius-Perron operators corresponding to (11.8), the group $\{S_t\}_{t \in \mathbb{R}}$ generated by the initial value problem (11.7), is a continuous group of weak contractions on $L^1(\mathbb{R}^N)$.

Definition 11.2. Let $\{T_t\}_{t \ge 0}$ be a group of weak contractions on L^p. The operator $A : D \subset L^p \to L^p$ defined by

$$Af = \lim_{t \to 0} \frac{T_t f - f}{t}, \quad \forall\, f \in D$$

under the L^p-norm convergence, where the domain D of A is the set of all functions $f \in L^p$ such that the above limit exists, is called the *infinitesimal generator* of the group $\{T_t\}_{t \in \mathbb{R}}$.

It is easy to see that the domain D of the infinitesimal generator A is a subspace of L^p and A is linear on D. We now calculate the infinitesimal generator of the group $\{P_t\}_{t \in \mathbb{R}}$ of Frobenius-Perron operators associated with transformations (11.8). Let $t \in \mathbb{R}$. Since U_t is the dual operator of P_t, we have $\langle P_t f - f, g \rangle = \langle f, U_t g - g \rangle$ for all $f \in L^1$ and $g \in L^\infty$. So

$$\left\langle \frac{P_t f - f}{t}, g \right\rangle = \left\langle f, \frac{U_t g - g}{t} \right\rangle.$$

Let $A_P : D_P \subset L^1 \to L^1$ and $A_U : D_U \subset L^\infty \to L^\infty$ be the infinitesimal generators of $\{P_t\}_{t \in \mathbb{R}}$ and $\{U_t\}_{t \in \mathbb{R}}$, respectively. Then, taking the limit as $t \to 0$ in the above equality implies

$$\langle A_P f, g \rangle = \langle f, A_U g \rangle, \quad \forall\, f \in D_P, \ g \in D_U. \tag{11.10}$$

Let $g \in C_0^1(\mathbb{R}^N)$. Since $U_t g(x^{(0)}) = g(S_t(x^{(0)})) = g(x(t; x^{(0)}))$, by the mean value theorem, we have

$$\frac{U_t g(x^{(0)}) - g(x^{(0)})}{t} = \frac{g(x(t; x^{(0)})) - g(x^{(0)})}{t}$$

$$= \sum_{k=1}^{N} \frac{\partial g}{\partial x_k}(x(c_t; x^{(0)})) \frac{dx_k}{dt}(c_t; x^{(0)})$$

$$= \sum_{k=1}^{N} \frac{\partial g}{\partial x_k}(x(c_t; x^{(0)})) F_k(x(c_t; x^{(0)})),$$

where $0 \leq c_t \leq t$ or $t \leq c_t \leq 0$. By passing into limit $t \to 0$, since g is zero outside a compact subset of $\mathbb{R}^N$, we see that $g \in D_U$ and

$$A_U g(x) = \sum_{k=1}^{N} \frac{\partial g}{\partial x_k}(x) F_k(x), \quad \forall \, x \in \mathbb{R}^N. \tag{11.11}$$

Substituting (11.11) into (11.10) gives, since $g \in C_0^1(\mathbb{R}^N)$,

$$\langle A_P f, g \rangle = \left\langle f, \sum_{k=1}^{N} \frac{\partial g}{\partial x_k} F_k \right\rangle = \int_{\mathbb{R}^N} f \sum_{k=1}^{N} \frac{\partial g}{\partial x_k} F_k \, dm$$

$$= \sum_{k=1}^{N} \int_{\mathbb{R}^N} \left(\frac{\partial (fgF_k)}{\partial x_k} - g \frac{\partial (fF_k)}{\partial x_k} \right) dm$$

$$= -\sum_{k=1}^{N} \int_{\mathbb{R}^N} g \frac{\partial (fF_k)}{\partial x_k} \, dm = -\left\langle \sum_{k=1}^{N} \frac{\partial (fF_k)}{\partial x_k}, g \right\rangle,$$

where $f \in D_P \cap C^1(\mathbb{R}^N)$. Thus, for the same f,

$$A_P f(x) = -\sum_{k=1}^{N} \frac{\partial (fF_k)}{\partial x_k}(x), \quad \forall \, x \in \mathbb{R}^N. \tag{11.12}$$

Denote $u(t, x) \equiv u(t)(x) = P_t f(x)$ with $f \in L^1(\mathbb{R}^N)$. For the vector-valued function $u : \mathbb{R} \to L^1(\mathbb{R}^N)$, its *derivative* at a given point $t = t_0$ is the unique element $u'(t_0) \in L^1(\mathbb{R}^N)$ such that

$$\lim_{t \to t_0} \left\| \frac{u(t) - u(t_0)}{t - t_0} - u'(t_0) \right\|_1 = 0.$$

Now, for a given $f \in D_P$, we go to derive an evolution equation that u satisfies as follows. Fix $t_0 \in \mathbb{R}$. Then

$$u'(t_0) = \lim_{t \to t_0} \frac{u(t) - u(t_0)}{t - t_0} = \lim_{t \to t_0} \frac{P_t f(x) - P_{t_0} f(x)}{t - t_0}$$

$$= \lim_{t \to t_0} P_{t_0} \frac{P_{t-t_0} f(x) - f(x)}{t - t_0} = P_{t_0} A_P f$$

exists. On the other hand, since

$$\frac{u(t) - u(t_0)}{t - t_0} = \frac{P_{t-t_0}(P_{t_0}f) - (P_{t_0}f)}{t - t_0},$$

the existence of the derivative $u'(t)$ implies that the limit of the right-hand side of the above equality exists as $t \to t_0$, therefore $P_{t_0}f \in D_P$ and $u'(t_0) = A_P P_{t_0} f = A_P u(t_0)$. Thus, it follows from (11.12) that $u(t, x)$ satisfies the partial differential equation

$$\frac{\partial u}{\partial t} + \sum_{k=1}^{N} \frac{\partial (uF_k)}{\partial x_k} = 0, \ \ t \in \mathbb{R}, \ (x_1, \cdots, x_N) \in \mathbb{R}^N \qquad (11.13)$$

and the initial value condition $u(0, x) = f(x)$. The equation (11.13) is usually called the *continuity equation*.

Let $f \in L^1(\mathbb{R}^N) \cap \mathcal{D}(\mathbb{R}^N, \mathcal{L}(\mathbb{R}^N), m)$ and let a measure μ_f be defined by

$$\mu_f(A) = \int_A f(x)dx, \ \ \forall A \in \mathcal{L}(\mathbb{R}^N).$$

Then μ_f is an invariant probability measure of $\{S_t\}_{t \in \mathbb{R}^N}$ (in other words, the transformation S_t preserves μ_f for *all* values $t \in \mathbb{R}$) if and only if the function f is a stationary density of $\{P_t\}_{t \in \mathbb{R}^N}$, that is,

$$P_t f = f, \ \ \forall t \in \mathbb{R}^N.$$

From the definition of the infinitesimal generator A_P, if f is a stationary density of $\{P_t\}_{t \in \mathbb{R}^N}$, then $A_P f = 0$. That the converse is also true is due to the classic *Hille-Yosida theorem* [Dunford and Schwartz (1957)]. Hence, we have the following *Liouville's theorem*:

Theorem 11.4. (Liouville) *Let $\{P_t\}_{t \in \mathbb{R}^N}$ be the continuous group of Frobenius-Perron operators associated with the initial value problem (11.7). Then μ_f is a stationary density of $\{P_t\}_{t \in \mathbb{R}^N}$ if and only if f satisfies*

$$\sum_{k=1}^{N} \frac{\partial (fF_k)}{\partial x_k} = 0.$$

Corollary 11.2. *Under the assumption of the above theorem, the Lebesgue measure of $\mathbb{R}^N$ is invariant under the flow $\{S_t\}_{t \in \mathbb{R}}$ of the initial value problem (11.7) if and only if* div $F = 0$.

As a special case of the above system of ordinary differential equations, consider the following *Hamiltonian system*

$$\frac{dq_k}{dt} = \frac{\partial H}{\partial p_k}, \ \ \frac{dp_k}{dt} = -\frac{\partial H}{\partial q_k}, \ \ k = 1, \cdots, N, \qquad (11.14)$$

where the *Hamiltonian* $H(p, q)$ is the energy of the system with q and p the *generalized positions* and *generalized momenta*, respectively. The flow determined by (11.14) is called the *Hamiltonian flow* of the Hamiltonian dynamical system, and is denoted as $\{\Phi^t\}$. Since the divergence of the right-hand side vector field of (11.14) is zero everywhere, the Lebesgue measure of $\mathbb{R}^N \times \mathbb{R}^N$ is invariant under the flow of (11.14) by Corollary 11.2, and the corresponding continuity equation (11.13) takes the form

$$\frac{\partial u}{\partial t} + [u, H] = 0,$$

where the *Poisson bracket*

$$[u, H] = \sum_{k=1}^{N} \left(\frac{\partial u}{\partial q_k} \frac{\partial H}{\partial p_k} - \frac{\partial u}{\partial p_k} \frac{\partial H}{\partial q_k} \right).$$

For Hamiltonian systems, the change with time of an arbitrary function g of the variables $q_1, \cdots, q_N, p_1, \cdots, p_N$ is given by

$$\frac{dg}{dt} = \sum_{k=1}^{N} \left(\frac{\partial g}{\partial q_k} \frac{\partial H}{\partial p_k} - \frac{\partial g}{\partial p_k} \frac{\partial H}{\partial q_k} \right) = [g, H].$$

In particular, if we take g to be a function of the energy H, then

$$\frac{dg}{dt} = \frac{dg}{dH} \frac{dH}{dt} = \frac{dg}{dH}[H, H] = 0, \tag{11.15}$$

so any function of the energy H is a *constant of the motion*. We shall meet the Hamiltonian system again in the next section.

11.3 Conformational Molecular Dynamics

In the past decade, the research on computational molecular dynamics has attracted much attention in life science and in particular the drug design. This section is a brief introduction to the application of Markov operators in this fast developing area, based on the presentations of the survey papers [Schütte (1999); Deuflhard and Schütte (2004)].

The molecular dynamics is based on Hamiltonian mechanics. It is well-known that numerical integration of Hamiltonian differential equations of molecular systems to approximate individual solution curves is ill-conditioned when the time is relatively large, due to the fact that the solutions of the Hamiltonian initial value problem to be computed are asymptotically chaotic. In the bio-molecular design, real times of pharmaceutical interest are in the region of milliseconds to minutes, whereas simulation times are presently in the range of nanoseconds (1 nanosecond = 10^{-9} second) with time steps of less than 5 femtoseconds (1 femtosecond = 10^{-15} second). Consequently, the numerical simulation can only give the information about short-term solutions because of the limitation of available computational facilities. On the other hand, the long-term prediction and computational results are required in the drug design industry.

Because of the above observation, *conformational dynamics* has been proposed to model the long-term dynamics of molecular systems, based on the fundamental ideas from ergodic theory and statistical physics. In the following we present the basic idea and methodology in this new approach.

Assume that the dynamics of a molecular system with k atoms under consideration is governed by the Hamiltonian

$$H(q,p) = \frac{1}{2}p^T D p + V(q),$$

which is the sum of the *kinetic energy* depending on $p = (p_1, \cdots, p_k)^T \in \mathbb{R}^{3k}$ of generalized momenta via the inverse matrix D of a diagonal mass matrix and the *potential energy* depending only on $q = (q_1, \cdots, q_k)^T \in \mathbb{R}^{3k}$ of generalized positions. The value of the Hamiltonian $H(q,p)$ represents the *internal energy* of the system in the state $x = (q,p) \in \mathbb{R}^{6k}$, and the resulting Hamiltonian system (11.14) of the motion is

$$\dot{q} = Dp, \quad \dot{p} = -\text{grad } V(q). \tag{11.16}$$

For a given initial value $x(0) = x^{(0)} = (q^{(0)}, p^{(0)}) \in \mathbb{R}^{6k}$, the unique solution of (11.16) can be written in terms of the Hamiltonian flow Φ^t as

$$x(t) = (q(t), p(t)) = \Phi^t x^{(0)}. \tag{11.17}$$

In the new approach of conformational dynamics, the concept of *transfer operators* has been introduced to incorporate the stochastic elements of the deterministic system. This operator is a Markov one which arises from the Frobenius-Perron operator associated with the transformation Φ^t. In what follows we shall denote $N = 3k$.

Let $X = \Omega \times \mathbb{R}^N$ be the *phase space* of the Hamiltonian flow (11.17) in which Ω is the *position space* of the generalized position vector q. For each time moment t, let $P_t \equiv P_{\Phi^t}$ be the Frobenius-Perron operator associated with the diffeomorphism $\Phi^t : X \to X$. Because of the identity $\det J_{\Phi^t}(x) \equiv 1$, the formula (11.9) is reduced to

$$P_t f(x) = f(\Phi^{-t}(x)), \quad \forall f \in L^1.$$

By its definition in Section 8.2, the Koopman operator $U_t \equiv U_{\Phi^t}$ corresponding to Φ^t has the following expression

$$U_t g(x) = g(\Phi^t(x)), \ \forall \, g \in L^\infty.$$

From (11.15), the Hamiltonian flow Φ^t has the *conservation of energy* $H(\Phi^t(x)) \equiv H(x)$. Corollary 11.2 implies the *conservation of volume*, which is equivalent to $\det J_{\Phi^t}(x) \equiv 1$. Because of the energy and volume conservation properties of the Hamiltonian flow, the composition of any scalar function with the Hamiltonian function H is a fixed point of both the Frobenius-Perron operator P_t and the Koopman operator U_t.

If the distribution of the initial states of the molecular system is described by a probability density $f_0(x)$, then

$$f(x,t) \equiv P_t f_0(x) = f_0(\Phi^{-t}(x))$$

is the time-dependent probability density transported along the solution curves $\Phi^t(x)$ of the system. By far most experiments of the molecular dynamics are performed using *equilibrium ensembles*, that is, ensembles which are described by stationary densities f of the continuous time group of the Frobenius-Perron operators P_t. This means that f satisfies the equality $f(x) = f(\Phi^{-t}(x))$ for all t and $x \in X$. From (11.15), given any smooth function $\varphi : \mathbb{R} \to \mathbb{R}_+$ such that $\int_X \varphi(H(x))dx = 1$, the resulting density $f = \varphi \circ H$ is a stationary density of P_t, that is,

$$f = f \circ \Phi^{-t}.$$

Note that the flow Φ^t has the *reversibility* in the sense that $h \circ \Phi^{-t} \circ h = \Phi^t$, where the *momentum reversion h* is defined by $h(q,p) = (q, -p)$. The above "energy prepared" density is also *p-symmetric*, namely,

$$f = f \circ h. \tag{11.18}$$

Most experiments on molecular systems are performed under the equilibrium conditions of constant temperature, particle number, and volume. The corresponding stationary density is the *canonical density* associated with the Hamiltonian H:

$$f_{\mathrm{can}}(q,p) = \frac{1}{Z} \exp\left(-\frac{\eta}{2} p^T D p - \eta V(q) \right), \tag{11.19}$$

where

$$Z = \int_{\mathbb{R}^N} \int_\Omega \exp\left(-\frac{\eta}{2} p^T D p - \eta V(q) \right) dq dp$$

is a normalization factor, $\eta = 1/(\kappa_B T)$, T is system's temperature, and κ_B is *Boltzmann's constant*. f_{can} can be factorized as

$$f_{\text{can}}(q, p) = \mathcal{P}(p)\mathcal{Q}(q),$$

where

$$\mathcal{P}(p) = \frac{1}{Z_p} \exp\left(-\frac{\eta}{2} p^T D p\right) \quad \text{and} \quad \mathcal{Q}(q) = \frac{1}{Z_q} \exp(-\eta V(q)),$$

so that $\int_{\mathbb{R}^N} \mathcal{P}(p) dp = 1$ and $\int_\Omega \mathcal{Q}(q) dq = 1$.

Given an arbitrary stationary density f_0. We would like to define a transition probability from one region, $A \subset X$, of the phase space to another one, $B \subset X$, after a time scale $\hat{t}$. The new ensemble of all such systems with states $x \in A$ selected from the ensemble f_0 at $t = 0$ has the density

$$f_A(x) = \frac{\chi_A(x) f_0(x)}{\int_A f_0(x) dx}.$$

Since all systems evolve due to Φ^t, after time $\hat{t}$, the relative frequency of systems in the ensemble f_A with states in B equals

$$\int_X \chi_B(\Phi^{\hat{t}}(x)) f_A(x) dx.$$

Thus, the *transition probability* should be defined as

$$\delta(A, B, \hat{t}) = \frac{\int_A \chi_B(\Phi^{\hat{t}}(x)) f_0(x) dx}{\int_A f_0(x) dx}. \tag{11.20}$$

It follows from (11.20) that a subset $B \subset X$ is *invariant* under the flow Φ^t if and only if $\delta(A, A, t) = 1$ for all $t \in \mathbb{R}$, which is *independent* of the stationary density f_0. In conformational dynamics of molecular systems, we are interested in *almost invariant* subsets A in the sense that $\delta(A, A, \hat{t})$ is sufficiently close to 1. Therefore,

$$A \subset X \text{ is almost invariant} \quad \Leftrightarrow \quad \delta(A, A, \hat{t}) \approx 1. \tag{11.21}$$

Remark 11.1. The concept of "almost invariant sets" was first introduced by Dellnitz and Junge [Dellnitz and Junge (1999)] to the computational problem of dynamical systems.

In the conformational dynamics of bio-molecules, any conformation is an almost invariant set of the ensemble in the sense of (11.21). But chemical intuition indicates that the phrase "conformation" does not refer to the momentum information. So *spatial* subsets of Ω will be used in the definition of transfer operators in the following investigation. The transition probability between A, $B \subset \Omega$ is defined as the transition probability between the associated phase space *cylinders* $X(A) = \{(q, p) \in X : q \in A\}$ and $X(B) = \{(q, p) \in X : q \in B\}$, namely

$$\delta(A, B, \hat{t}) = \delta(X(A), X(B), \hat{t}).   \qquad (11.22)$$

Naturally, $A \subset \Omega$ is said to be *almost invariant* if $\delta(A, A, \hat{t}) \approx 1$.

Let $f_0 \in L^1(X)$ be an arbitrary stationary density satisfying (11.18) such that the reduced density

$$\varphi(q) = \int_{\mathbb{R}^N} f_0(q, p) dp$$

is a finite-valued positive and smooth function on Ω. We define the spatial *transfer operator* $T \equiv T(\hat{t})$ by the formula

$$Tf(q) = \frac{1}{\varphi(q)} \int_{\mathbb{R}^N} f(\pi \Phi^{-\hat{t}}(q, p)) f_0(q, p) dp,   \qquad (11.23)$$

where f is a function of q defined on Ω and π is the *coordinate projection* $\pi(q, p) = q$. The operator T can be interpreted as a suitable *weighted average* of the Frobenius-Perron operator $P_{\hat{t}}$ over the momenta in each cross section $X(q) = \{q\} \times \mathbb{R}^N$ with weight f_0.

If $f_0 = f_{\text{can}}$ and $\varphi(q) = \mathcal{Q}(q)$, then (11.23) becomes

$$Tf(q) = \int_{\mathbb{R}^N} f(\pi \Phi^{-\hat{t}}(q, p)) \mathcal{P}(p) dp.   \qquad (11.24)$$

Consequently, in this special case, the spatial transfer operator T describes the momentum weighted fluctuations inside the canonical ensemble with respect to the chosen time scale $\hat{t}$.

We shall prove below that the positive operator T is a weak contraction on the weighted L^r-space

$$L_\varphi^r(\Omega) = \left\{ f : \int_\Omega |f(q)|^r \varphi(q) dq < \infty \right\}, \quad r = 1, 2.$$

The inner product of the Hilbert space $L_\varphi^2(\Omega)$ is defined by

$$\langle f, g \rangle_\varphi = \int_\Omega f(q) g(q) \varphi(q) dq,$$

which induces the L^2-norm $\|f\|_{2,\varphi} = \sqrt{\langle f, f \rangle_\varphi}$ on $L_\varphi^2(\Omega)$. The L^1-norm on $L_\varphi^1(\Omega)$ is $\|f\|_{1,\varphi} = \int_\Omega |f(q)| \varphi(q) dq$.

Proposition 11.2. $T : L_\varphi^1(\Omega) \to L_\varphi^1(\Omega)$ *is a Markov operator.*

Proof Let $f \in L^1_\varphi(\Omega)$ be nonnegative. Then so is T. From (11.23),

$$\|Tf\|_{1,\varphi} = \int_\Omega \frac{1}{\varphi(q)} \left| \int_{\mathbb{R}^N} f(\pi\Phi^{-\hat{t}}(q,p)) f_0(q,p) dp \right| \varphi(q) dq$$

$$= \int_\Omega \int_{\mathbb{R}^N} f(\pi\Phi^{-\hat{t}}(q,p)) f_0(q,p) dp dq = \int_X f(\pi\Phi^{-\hat{t}}(x)) f_0(x) dx$$

$$= \int_X f(\pi x) f_0(x) dx = \int_\Omega f(q) \int_{\mathbb{R}^N} f_0(q,p) dp dq,$$

where the fourth equality is from the invariance of f_0 with respect to the substitution $x \to \Phi^{\hat{t}}(x)$ and the volume conservation property of the flow. As a consequence,

$$\|Tf\|_{1,\varphi} = \int_\Omega f(q) \varphi(q) dq = \|f\|_{1,\varphi},$$

which means that T is a Markov operator. $\square$

Proposition 11.3. $T : L^2_\varphi(\Omega) \to L^2_\varphi(\Omega)$ *is well-defined and satisfies that* $\|Tf\|_{2,\varphi} \le \|f\|_{2,\varphi}$ *for all* $f \in L^2_\varphi(\Omega)$.

Proof Given any function $f \in L^2_\varphi(\Omega)$,

$$\|Tf\|^2_{2,\varphi} = \int_\Omega \frac{1}{\varphi(q)^2} \left(\int_{\mathbb{R}^N} f(\pi\Phi^{-\hat{t}}(q,p)) f_0(q,p) dp \right)^2 \varphi(q) dq$$

$$= \int_\Omega \frac{1}{\varphi(q)} \left(\int_{\mathbb{R}^N} f(\pi\Phi^{-\hat{t}}(q,p)) f_0(q,p) dp \right)^2 dq. \qquad (11.25)$$

For any point $q \in \Omega$, let $f_q(p) = f(\pi\Phi^{-\hat{t}}(q,p))$. Then, similar to the proof of Proposition 11.2, we obtain

$$\int_\Omega \int_{\mathbb{R}^N} [f_q(p)]^2 f_0(q,p) dp dq = \|f\|^2_{2,\varphi} < \infty. \qquad (11.26)$$

It follows from the Cauchy-Schwarz inequality that

$$\left(\int_{\mathbb{R}^N} f(\pi\Phi^{-\hat{t}}(q,p)) f_0(q,p) dp \right)^2 = \left(\int_{\mathbb{R}^N} f_q(p) f_0(q,p) dp \right)^2$$

$$\le \int_{\mathbb{R}^N} [f_q(p)]^2 f_0(q,p) dp \cdot \int_{\mathbb{R}^N} f_0(q,p) dp$$

$$= \int_{\mathbb{R}^N} [f_q(p)]^2 f_0(q,p) dp \cdot \varphi(q).$$

Combining the above inequality with (11.25) and (11.26) gives

$$\|Tf\|^2_{2,\varphi} \le \int_\Omega \int_{\mathbb{R}^N} |f_q(p)|^2 f_0(q,p) dp dq = \|f\|^2_{2,\varphi}. \qquad \square$$

We now use the transfer operator $T : L^2_\varphi(\Omega) \to L^2_\varphi(\Omega)$ to express the transition probability (11.22) as follows. Since the characteristic functions χ_A and χ_B belong to the space $L^2_\varphi(\Omega)$, we have

$$\langle T\chi_A, \chi_B \rangle_\varphi = \int_\Omega \frac{1}{\varphi(q)} \int_{\mathbb{R}^N} \chi_A(\pi\Phi^{-\hat{t}}(q,p))f_0(q,p)dp\chi_B(q)\varphi(q)dq$$

$$= \int_X \chi_A(\pi\Phi^{-\hat{t}}(x))f_0(x)\chi_{X(B)}(x)dx,$$

which together with the invariance of f_0, the volume conservation property of the flow, and the substitution $x = \Phi^\tau(y)$ yields

$$\langle T\chi_A, \chi_B \rangle_\varphi = \int_X \chi_{X(A)}(x)\chi_{X(B)}(\Phi^{\hat{t}}(x))f_0(x)dx. \qquad (11.27)$$

Since

$$\langle \chi_A, \chi_B \rangle_\varphi = \int_A \varphi(q)dq = \int_{X(A)} f_0(x)dx,$$

$$\delta(A, B, \hat{t}) = \frac{\int_{X(A)} \chi_{X(B)}(\Phi^{\hat{t}}(x))f_0(x)dx}{\int_{X(A)} f_0(x)dx} = \frac{\langle T\chi_A, \chi_B \rangle_\varphi}{\langle \chi_A, \chi_B \rangle_\varphi}.$$

Proposition 11.4. $T : L^2_\varphi(\Omega) \to L^2_\varphi(\Omega)$ *is self-adjoint, that is,* $T^* = T$. *Hence, its spectrum* $\sigma(T) \subset [-1, 1]$.

Proof Given measurable subsets A and B of Ω, the reversibility property of Φ^t and (11.27) imply that

$$\langle T\chi_A, \chi_B \rangle_\varphi = \int_X \chi_{X(A)}(x)\chi_{X(B)}(h(\Phi^{-\hat{t}}(h(x)))f_0(x)dx.$$

Since f_0 is chosen to satisfy (11.18) and the sets $X(A)$ and $X(B)$ include all possible momenta, the transformation $y = h(x)$ yields

$$\langle T\chi_A, \chi_B \rangle_\varphi = \int_X \chi_{X(A)}(y)\chi_{X(B)}(h(\Phi^{-\hat{t}}(y)))f_0(y)dy$$

$$= \int_X \chi_{X(A)}(\Phi^{\hat{t}}(x))\chi_{X(B)}(h(x))f_0(x)dx,$$

in which the second equality is from the volume conservation property of the Hamiltonian flow. Hence,

$$\langle T\chi_A, \chi_B \rangle_\varphi = \int_X \chi_{X(B)}(x)\chi_{X(A)}(\Phi^{\hat{t}}(x))f_0(x)dx = \langle \chi_A, T\chi_B \rangle_\varphi.$$

Since the simple functions are dense in $L^2_\varphi(\Omega)$, it follows that $\langle Tf, g \rangle_\varphi = \langle f, Tg \rangle_\varphi$ for all $f, g \in L^2_\varphi(\Omega)$. $\qquad \square$

We give an example with $f_0 = f_{\text{can}}$ to have a taste of the transfer operator in the simplest case [Schütte (1999)].

Example 11.6. Consider the one dimensional harmonic oscillator with $H(q,p) = (q^2 + p^2)/2$ and $\Omega = \mathbb{R}$. Then $\mathcal{P}(p) = \sqrt{\eta/(\beta\pi)}\exp(-\eta p^2/2)$. From the formula (11.24) we obtain

$$Tf(q) = \int_{-\infty}^{\infty} f(q\cos\hat{t} - p\sin\hat{t})\sqrt{\frac{\eta}{\beta\pi}}\exp\left(-\frac{\eta}{2}p^2\right)dp.$$

We distinguish two different cases. First assume that $\hat{t} = 2j\pi$ for some integer $j > 0$. Then $\cos\hat{t} = 1$ and $\sin\hat{t} = 0$, so

$$Tf(q) = \int_{-\infty}^{\infty} f(q)\sqrt{\frac{\eta}{\beta\pi}}\exp\left(-\frac{\eta}{2}p^2\right)dp = f(q), \ \ \forall\, f \in L_{\mathcal{Q}}^2.$$

Namely, the transfer operator T is the identity operator on $L_{\mathcal{Q}}^2$, therefore $\sigma(T) = \{1\}$ and every subset A of $\mathbb{R}$ is invariant. Similarly, if $\hat{t} = (2j+1)\pi$ for some integer $j \geq 0$, then T is the identity operator on the subspace of $L_{\mathcal{Q}}^2$ consisting of all the even functions.

We then turn to consider the case of all the other time scales $\hat{t}$. Then $|\cos\hat{t}| < 1$. Suppose that $f \in L_{\mathcal{Q}}^2$ is a smooth eigenvector of T associated with eigenvalue λ. Differentiating $Tf(q) = \lambda f(q)$ with respect to q yields

$$(Tf)'(q) = \cos\hat{t}\, Tf'(q) = \lambda f'(q).$$

Thus, $\lambda/\cos\hat{t}$ is also an eigenvalue of T with eigenvector f'. Since $T1 = 1$, by using mathematical induction we can find a sequence of eigenvectors which are polynomials $f_n \in L_{\mathcal{Q}}^2$ such that

$$f_1'(q) \equiv 1, \ \ f_n'(q) \equiv f_{n-1}(q), \ \ \forall\, n \geq 1.$$

With an additional condition that such polynomials are pairwise orthogonal under $\langle\,,\,\rangle_{\mathcal{Q}}$, they are uniquely determined as

$$f_1(q) = q, \ f_2(q) = q^2 - \frac{1}{\eta^2}, \ f_3(q) = q^3 - \frac{3}{\eta^2}q, \cdots$$

with the corresponding eigenvalues $\lambda_n = \cos^n\hat{t}$ for all $n \geq 1$. This means that, for $\hat{t} \neq j\pi$ with any positive integer j, the operator T has a purely discrete spectrum with 0 as its limit point.

Remark 11.2. Under some mild conditions the transfer operator T : $L_{\varphi}^2(\Omega) \to L_{\varphi}^2(\Omega)$ is quasi-compact [Schütte (1999)].

Next we give a stochastic interpretation [Lasota and Mackey (1994)] of the transfer operator in terms of finite state stationary Markov chains studied in Section 4.1, for which we shall need the concept of Foias operators on measures whose definition is taken from [Lasota and Mackey (1994)].

Definition 11.3. Given two positive integers i and j, let a vector-valued function $S : W \times Y \subset \mathbb{R}^i \times \mathbb{R}^j \to W$ be such that $S(\cdot, y)$ is continuous for each $y \in Y$ and $S(x, \cdot)$ is Borel measurable for each $x \subset W$. The dynamical system with random perturbations

$$x^{(n+1)} = S(x^{(n)}, y^{(n)}), \quad n = 0, 1, \cdots, \tag{11.28}$$

where $y^{(n)}$ are independent random vectors with measure

$$\nu(A) = \text{prob}(y^{(n)} \in A), \quad \forall A \in \mathcal{B}(Y)$$

the same for all n and the random vectors $x^{(0)}, y^{(0)}, y^{(1)}, \cdots$ are independent, is called a *regular stochastic dynamical system*. The linear operator $P : \mathcal{M}_{\text{fin}} \to \mathcal{M}_{\text{fin}}$ defined by

$$P\mu(A) = \int_W \int_Y \chi_A(S(x, y))\nu(dy)\mu(dx), \quad \forall \mu \in \mathcal{M}_{\text{fin}}, \ A \in \mathcal{B}(W),$$

where $\mathcal{M}_{\text{fin}}$ is the set of all real (namely signed) measures on W, is called the *Foias operator* corresponding to (11.28).

We only consider the case of the canonical ensemble $f_0(q, p) = f_{\text{can}}(q, p) = \mathcal{P}(p)\mathcal{Q}(q)$ as given by (11.19). Define the regular stochastic dynamical system

$$q^{(n+1)} = \pi\Phi^{\hat{t}}(q^{(n)}, p^{(n)}), \quad p^{(n)} \text{ is } \mathcal{P} - \text{distributed} \tag{11.29}$$

for any given $q^{(0)} \in \Omega$, which gives rise to a sequence of the probability measures μ_n for finding the probabilities of the position vectors q_n that lie in a set $A \in \mathcal{B}(\Omega)$, that is,

$$\mu_n(A) = \text{prob}(q^{(n)} \in A), \quad \forall A \in \mathcal{B}(\Omega), \ n \geq 0.$$

Actually the sequence $\{\mu_n\}$ is also given by the iterates of the following Foias operator P that maps $\mu \in \mathcal{M}_{\text{fin}}$ to $P\mu$ via

$$P\mu(A) = \int_\Omega \int_{\mathbb{R}^N} \chi_A(\pi\Phi^{\hat{t}}(q, p))\mathcal{P}(p)dp\mu(dq). \tag{11.30}$$

Namely, $\mu_n = P^n\mu_0$ if $\mu_0 \in \mathcal{M}_{\text{fin}}$ is the probability measure according to which the initial random position $q^{(0)}$ is distributed.

If we only consider absolutely continuous finite measures $\mu_f \in \mathcal{M}_{\text{fin}}$ with its Radon-Nikodym derivative $f = d\mu_f/(\mathcal{Q}dq) \in L^1_{\mathcal{Q}}$ with respect to $\mathcal{Q}dq$, then the definition (11.30) implies that

$$
\begin{aligned}
P\mu_f(A) &= \int_X \chi_A(\pi\Phi^{\hat{t}}(q,p))f(\pi x)f_{\text{can}}(x)dx \\
&= \int_{X(A)} f(\pi\Phi^{-\hat{t}}(x))f_{\text{can}}(x)dx,
\end{aligned}
$$

or equivalently

$$
P\mu_f(A) = \int_A \int_{\mathbb{R}^N} f(\pi\Phi^{-\hat{t}}(q,p))\mathcal{P}(p)dp\mathcal{Q}(q)dq.
$$

Therefore, the Radon-Nikodym derivative $d(P\mu_f)/(\mathcal{Q}dq)$ of $P\mu_f$ with respect to the measure $\mathcal{Q}dq$ has the expression

$$
Tf(q) = \int_{\mathbb{R}^N} f(\pi\Phi^{-\hat{t}}(q,p))\mathcal{P}(p)dp,
$$

which is exactly the same expression as (11.24) given by the transfer operator T. Hence, the transfer operator $T : L^1_{\mathcal{Q}}(\Omega) \to L^1_{\mathcal{Q}}(\Omega)$ is equivalent to the Foias operator $P : \mathcal{M}_{\text{fin}} \to \mathcal{M}_{\text{fin}}$ restricted to the space of absolutely continuous real measures, and the stochastic dynamical system (11.29) is a realization of the Markov chain induced by the Markov operator T, which combines a short term deterministic model, characterized by the Hamiltonian flow $\Phi^{\hat{t}}$, with a statistical model, characterized by the $\mathcal{P}$-distribution, the momentum part of the canonical distribution. Consequently, if the initial position $q^{(0)}$ of the chain (11.29) is distributed according to the probability density $f \in L^1_{\mathcal{Q}}(\Omega)$, then the probability density $f_n \in L^1_{\mathcal{Q}}(\Omega)$ of finding $q^{(n)} = q$ is given by $u_f(q) = T^n f(q)$.

A dynamical system is said to have a *meta-stable decomposition*, if its state space can be decomposed into a finite number of disjoint sets such that the probability of exit from each of them is extremely small. In the context of the molecular dynamics discussed here, the meta-stability is related to the maximal eigenvectors of the transfer operator. More specifically, meta-stable decompositions can be detected via the discrete eigenvalues of the transfer operator T close to its maximal eigenvalue 1. They can be identified from the structure of the corresponding eigenfunctions using the concept of almost invariant sets.

Since the meta-stability of any set A may be measured by the number $\delta(A, A, \hat{t})$ in (11.20), for any finite decomposition $\{A_1, \cdots, A_n\}$ of Ω into n

disjoint subsets $A_1, \cdots, A_n$, the *meta-stability* of the decomposition can be defined via the quantity

$$\delta_n(\hat{t}) = \sum_{i=1}^{n} \delta(A_i, A_i, \hat{t}).$$

The possibility of identifying an almost optimal meta-stable decomposition of the state space is related to the so-called *Perron cluster analysis* after the transfer operator T is approximated by finite dimensional operators (see [Deuflhard and Schütte (2004); Deuflhard and Weber (2003)]).

Note that the definition of the transfer operator is based on the Hamiltonian flow at time $\hat{t}$. One can then numerically solve the initial value problem of the Hamiltonian equation of the motion (11.16) with time step size $\Delta t = \hat{t}/s$ to yield a discrete flow $\Psi^{\Delta t}$ such that $\Phi^{\hat{t}} x^{(0)}$ is approximated by x_s via the iteration

$$x^{(j+1)} = \Psi^{\Delta t} x^{(j)}, \ j = 0, 1, \cdots, s - 1.$$

Then the chain

$$q^{(n+1)} = \pi(\Psi^{\Delta t})^s (q^{(n)}, p^{(n)}), \ p^{(n)} \text{ is } \mathcal{P} - \text{distributed}, \ n = 0, 1, \cdots$$

approximates the Markov chain (11.29) after using some statistical method such as the Metropolis acceptance procedure to yield a hybrid Monte Carlo (HMC) chain which contains good approximations of solution curves of the Hamiltonian system.

Finally, we can apply the idea of Ulam's method from Section 10.1 to approximate the transfer operator T in solving the eigenvalue problem $Tf = \lambda f$. In the context of the $L^2_{\mathcal{Q}}$ space here, Ulam's method is the same as the *orthogonal projection method* or the Galerkin method. Given a finite rectangular partition of Ω in terms of the N-dimensional rectangles or simplicies $\{A_1, \cdots, A_n\}$, let $\Delta_n = \text{span}\{\chi_{A_1}, \cdots, \chi_{A_n}\}$. Let $Q_n : L^2_{\mathcal{Q}} \to L^2_{\mathcal{Q}}$ be the operator that projects each $f \in L^2_{\mathcal{Q}}$ orthogonally onto Δ_n. Then

$$Q_n f = \sum_{i=1}^{n} \frac{1}{\langle \chi_{A_i}, \chi_{A_i} \rangle_{\mathcal{Q}}} \langle \chi_{A_i}, f \rangle_{\mathcal{Q}} \cdot \chi_{A_i}$$

$$= \sum_{i=1}^{n} \frac{1}{\int_{A_i} \mathcal{Q}(q) dq} \langle \chi_{A_i}, f \rangle_{\mathcal{Q}} \cdot \chi_{A_i}.$$

The resulting discretized eigenvalue problem is given by $T_n f_n = \lambda f_n$, where $T_n = Q_n T$. If we write $f_n = \sum_{i=1}^{n} v_i \chi_{A_i}$ as a linear combination of $\chi_{A_1}, \cdots, \chi_{A_n}$, then the resulting matrix eigenvalue problem is

$$\sum_{j=1}^{n} \frac{\langle T\chi_{A_i}, \chi_{A_j} \rangle_{\mathcal{Q}}}{\langle \chi_{A_i}, \chi_{A_i} \rangle_{\mathcal{Q}}} v_j = \lambda v_i, \ \forall \, i = 1, \cdots, n.$$

Thus, the (i, j)-entry of the $n \times n$ stochastic matrix $\hat{T}_n$ is

$$t_{ij} = \frac{\langle T\chi_{A_i}, \chi_{A_j}\rangle_Q}{\langle \chi_{A_i}, \chi_{A_i}\rangle_Q} = \delta(A_i, A_j, \tau). \tag{11.31}$$

For the numerical evaluation of the t_{ij}'s in (11.31), using a realization $\{q^{(1)}, q^{(2)}, \cdots, q^{(l)}\}$ of the HMC chain from above, the (i, j)-entry of the transition matrix $\hat{T}_n$ can be computed via

$$t_{ij} = \frac{\#\{q^{(n+1)} \in A_j, \ q^{(n)} \in A_i\}}{\#\{q^{(n)} \in A_i\}}, \quad i, j = 1, \cdots, n.$$

More details on the theoretical and computational issues of the computational molecular dynamics can be found in [Schütte (1999); Deuflhard and Schütte (2004); Deuflhard and Weber (2003)] and the references therein.

11.4 Third Generation Wireless Communications

The third generation wireless communication technique uses the *direct sequence code division multiple access* (DS-CDMA) in which the statistical properties of the *spreading sequences* generated from iterating a suitably chosen nonsingular transformations play a key role. With the increasing demand of wireless communications, research on wireless communication technologies has been extensive in electrical engineering. We give in this section a short introduction to the DS-CDMA and consider a simplified baseband equivalent of an asynchronous system model as an application; details are referred to [Setti *et al.* (2002)] and the references therein.

Suppose that there are N users of an asynchronous DS-CDMA system with user i's information signal $S^i(t) = \sum_{k=-\infty}^{\infty} s_k^i g_{t_0}(t - kt_0)$, where $s_k^i = \pm 1$ are the information symbols and g_{t_0} is the *rectangular pulse function* which is 1 on $[0, t_0]$ and 0 elsewhere. The spreading signal depends on sequences $x^i = \{x_k^i\}$ of finite length l (called the *spreading factor*) from the alphabet set X which are mapped into the set Y of all the dth roots of 1 by a complex quantization function $Q : X \rightarrow Y$. The combined signal $Q^i(t) = \sum_{k=-\infty}^{\infty} Q(x_k^i) g_{t_0/l}(t - kt_0/l)$, of which $\overline{Q^i(t)}$ is the complex conjugate function, is then multiplied by $S^i(t)$ and transmitted along the channel together with spread-spectrum signals from the other users. At the receiver of the useful jth user, the global signal is multiplied by $\overline{Q^j(t)}$ and fed into an integrate-and-dump stage.

The performance of a DS-CDMA system tends to be optimal if the codes for different users are almost *orthogonal*, namely the spreading sequences are generated with cross correlations almost zero. However, for the asynchronous communication in which a CDMA link is from the mobile transmitters to a fixed base station in a cellular system, the sequence design problem is much more complicated than the determination of the sequences with vanishing cross correlations. As signals of different links are not synchronized and the spreading sequences are not perfectly uncorrelated, orthogonality cannot be achieved and contributions from undesired signals appear at the output of the integrate-and-dump as the co-channel interference c_k^j to s_k^j for each integer k. A usual approach to the co-channel interference estimation is to consider each co-channel interference c_k^j as a Gaussian random variable whose variance can be estimated from the cross-correlation characteristics of the spreading sequences.

We proceed to introduce this approach briefly. If the *partial correlation function* Λ_τ between the ith spreading sequence x^i and the jth spreading sequence x^j is defined as

$$\Lambda_r(x^i, x^j) = \begin{cases} \sum_{k=0}^{l-r-1} Q(x_k^i)\overline{Q}(x_{k+r}^j), & 0 \le r < l, \\ \overline{\Lambda}_r(x^j, x^i), & -l < r < 0, \\ 0, & |r| \ge l, \end{cases}$$

then the *bit-error probability* P_{err} of the communication link has the following expression:

$$P_{\text{err}} \approx \frac{1}{2}\text{erfc}\left(\frac{1}{\sqrt{(N-1)\Pi}}\right), \tag{11.32}$$

where the function $\text{erfc}(t) = 2\pi^{-1/2}\int_t^\infty \exp(-x^2)dx$ and the quantity Π can be represented as

$$\Pi = \frac{1}{3l|\Lambda_0(x^j, x^j)|^2} \sum_{r=-l+1}^{l-1} E_{x^i, x^j}^{i \ne j}[2|\Lambda_r(x^i, x^j)|^2$$
$$+ \text{Re}\left(\Lambda_r(x^i, x^j)\overline{\Lambda}_{r+1}(x^i, x^j)\right)], \tag{11.33}$$

which assumes, using conditional expectations, the significance of an expected *signal-to-interference ratio* (SIR) per interfering user. The number Π is the expected degradation in performance if a new user is added to the communication system.

To get an optimal performance of communication, one needs to optimize Π, the global SIR, which is the *performance merit figure*. The value of Π depends on the spreading sequences. Traditionally they are chosen to be random-like such as the Gold or maximum-length sequences. In recent years suitable chaotic mappings have been used to deliver spreading sequences which are better than the random-like methods in terms of the merit figure

II. Thus, an optimal performance can be achieved if we can design spreading sequences characterized by the desired statistical features. So, it is the statistical study of chaotic deterministic systems that is essential in the performance analysis of DS-CDMA wireless communications.

Next we turn to analyze the spreading sequence performance. Note that the decay property of the correlations for chaotic transformations plays an important role in the statistical analysis of the spreading sequences. From its formula (11.32), it is clear to see that P_{err} will be decreased while Π is decreased when the number N of the users is fixed as a constant. As a consequence, the purpose of the performance improvement is to minimize the quantity Π. Thanks to its expression (11.33), the number Π can be evaluated by means of the expectations of the partial correlation functions Λ_r and their products as random variables with respect to all the spreading sequences. Since the spreading sequences x^i are generated by chaotic mappings starting from randomly chosen initial points, and any initial point x_0^i corresponds to the resulting iteration sequence $x^i = \{x_k^i\}$, expectations of $\Lambda_r(x^i, x^j)$ and their products taken over all the spreading sequences x^i and x^j can be translated into expectations over randomly chosen initial points x_0^i and x_0^j which determine the sequences themselves. Let E be the conditional expectation operator and let x_0^i, x_0^j be the random variables with a density f_0. Then the formula (11.33) of Π can be expressed to be

$$
\begin{aligned}
\Pi &= \frac{1}{3l|\Lambda_0(x^j, x^j)|^2} \sum_{r=-l+1}^{l-1} E_{x_0^i, x_0^j}^{i \neq j} \big[\, 2|\Lambda_r(x^i, x^j)|^2 \\
&\quad + \operatorname{Re}\big(\Lambda_r(x^i, x^j)\overline{\Lambda}_{r+1}(x^i, x^j)\big)\big] \\
&= \frac{1}{3l^3} \Big\{ 2E_{x_0^i, x_0^j}^{i \neq j}\big[|\Lambda_0(x^i, x^j)|^2\big] + \operatorname{Re} E_{x_0^i, x_0^j}^{i \neq j}\big[\Lambda_0(x^i, x^j)\overline{\Lambda}_1(x^i, x^j)\big] \\
&\quad + 2\sum_{r=1}^{l-1} E_{x_0^i, x_0^j}^{i \neq j}\big[|\Lambda_r(x^i, x^j)|^2 + |\Lambda_r(x^j, x^i)|^2\big] \\
&\quad + \operatorname{Re}\sum_{r=1}^{l-1} E_{x_0^i, x_0^j}^{i \neq j}\big[\Lambda_r(x^i, x^j)\overline{\Lambda}_{r+1}(x^i, x^j) + \Lambda_{r-1}(x^j, x^i)\overline{\Lambda}_r(x^j, x^i)\big] \Big\} \\
&= \frac{1}{3l^3} \Big\{ 2E_{x_0^i, x_0^j}^{i \neq j}\big[|\Lambda_0(x^i, x^j)|^2\big] + 4\sum_{r=1}^{l-1} E_{x_0^i, x_0^j}^{i \neq j}\big[|\Lambda_r(x^i, x^j)|^2\big] \\
&\quad + \operatorname{Re} E_{x_0^i, x_0^j}^{i \neq j}\big[\Lambda_0(x^i, x^j)\overline{\Lambda}_1(x^i, x^j)\big] \\
&\quad + \operatorname{Re}\sum_{r=1}^{l-1} E_{x_0^i, x_0^j}^{i \neq j}\big[\Lambda_r(x^i, x^j)\overline{\Lambda}_{r+1}(x^i, x^j) + \Lambda_{r-1}(x^i, x^j)\overline{\Lambda}_r(x^i, x^j)\big] \Big\}.
\end{aligned}
$$

For the spreading sequences generated by a chaotic mapping $S : [0, 1] \rightarrow [0, 1]$, the spreading sequences x^i and x^j are generated via $x_k^i = S^k(x_0^i)$ and $x_{k+r}^j = S^{k+r}(x_0^j)$ for all integers $k \geq 0$. Therefore, after carrying out the evaluation of the various conditional expectations, we see that

$$|\Lambda_0(x^j, x^j)|^2 = \left(\sum_{k=0}^{l-1} |Q(S^k(x_0^j))|^2\right)^2,$$

$$E_{x_0^i, x_0^j}^{i \neq j} \left[|\Lambda_r(x^i, x^j)|^2\right]$$

$$= \sum_{k=0}^{l-r-1} \sum_{n=0}^{l-r-1} E_{x_0^i, x_0^j}^{i \neq j} \left[Q(S^k(x_0^i))\overline{Q}(S^{k+r}(x_0^j)) \right.$$
$$\left. \times \overline{Q}(S^n(x_0^i))Q(S^{n+r}(x_0^j))\right]$$

$$= \sum_{k=0}^{l-r-1} \sum_{n=0}^{l-r-1} \int_0^1 \int_0^1 Q(S^k(x_0^i))\overline{Q}(S^{k+r}(x_0^j))\overline{Q}(S^n(x_0^i))$$
$$\times Q(S^{n+r}(x_0^j))f_0(x_0^i)f_0(x_0^j)dx_0^i dx_0^j$$

$$= \sum_{k=0}^{l-r-1} \sum_{n=0}^{l-r-1} \int_0^1 Q(S^k(x))\overline{Q}(S^n(x))f_0(x)dx$$
$$\times \int_0^1 \overline{Q}(S^{k+r}(x))Q(S^{n+r}(x))f_0(x)dx$$

for $0 < r < l$, and for $0 \leq r < l - 1$,

$$E_{x_0^i, x_0^j}^{i \neq j} \left[\Lambda_r(x^i, x^j)\overline{\Lambda}_{r+1}(x^i, x^j)\right]$$

$$= \sum_{k=0}^{l-r-1} \sum_{n=0}^{l-r-2} E_{x_0^i, x_0^j}^{i \neq j} \left[Q(S^k(x_0^i))\overline{Q}(S^{k+r}(x_0^j)) \right.$$
$$\left. \times \overline{Q}(S^n(x_0^i))Q(S^{n+r+1}(x_0^j))\right]$$

$$= \sum_{k=0}^{l-r-1} \sum_{n=0}^{l-r-2} \int_0^1 \int_0^1 Q(S^k(x_0^i))\overline{Q}(S^{k+r}(x_0^j))\overline{Q}(S^n(x_0^i))$$
$$\times Q(S^{n+r+1}(x_0^j))f_0(x_0^i)f_0(x_0^j)dx_0^i dx_0^j$$

$$= \sum_{k=0}^{l-r-1} \sum_{n=0}^{l-r-2} \int_0^1 Q(S^k(x))\overline{Q}(S^n(x))f_0(x)dx$$
$$\times \int_0^1 \overline{Q}(S^{k+r}(x))Q(S^{n+r+1}(x))f_0(x)dx.$$

So by (11.34), the computation of Π involves such integrals as

$$\int_0^1 (f \circ S^k)(g \circ S^n) f_0 dm, \tag{11.34}$$

which is closely related to the rate of decay of correlations for S that can be studied with Frobenius-Perron and Koopman operators [Baladi (2000)].

As an illustration how to proceed along this direction, we let $k = 0$ in (11.34). The discussion in Section 8.4 implies that if S is piecewise C^2 with $\inf_{x \in [0,1] \setminus \{a_1, \cdots, a_{r-1}\}} |S'(x)| > 1$, and if the invariant probability measure μ^* with density $f^* \in C^1[0,1]$ is mixing with respect to S, then it is also *exact* with respect to μ^* (see [Boyarsky and Góra (1997)]). Furthermore, there is a constant M such that for any function $f \in BV(0,1)$ and $n \geq 0$,

$$\left\| P_S^n f - \left(\int_0^1 f(x) dx \right) f^* \right\|_{BV} \leq M \|f\|_{BV} r_{\text{mix}}^n, \tag{11.35}$$

where $r_{\text{mix}} \in (0,1)$ is the *rate of mixing* (Theorem 9.1.2 of [Ding and Zhou (2009)]). The estimate (11.35) implies that the *correlation coefficient*

$$\left| \int_0^1 f(g \circ S^n) f^* dm - \int_0^1 f f^* dm \cdot \int_0^1 g f^* dm \right|$$

can be bounded by (Theorem 9.1.3 of [Ding and Zhou (2009)])

$$M \|f f^*\|_{BV} \|g\|_\infty r_{\text{min}} \leq C \|f\|_\infty \|g\|_\infty r_{\text{mix}}^n, \tag{11.36}$$

where C is a constant independent of the functions f and g from Lemma 8.3. In particular, if f and g satisfy the equality $\int_0^1 f(x) f^*(x) dx = 0$ or the equality $\int_0^1 g(x) f^*(x) dx = 0$, then

$$\left| \int_0^1 f(x) g(S^n(x)) f^*(x) dx \right| \leq C \|f\|_\infty \|g\|_\infty r_{\text{mix}}^n, \quad \forall\, n.$$

The above brief introductory analysis implies that the performance of the spreading sequences generated by a one dimensional chaotic mapping $S : [0,1] \rightarrow [0,1]$ depends on the correlation property of the underlying deterministic discrete dynamical system S, which is determined by the rate of mixing $r_{\text{mix}} \in (0,1)$. The corresponding Frobenius-Perron operator and its finite approximations represented by stochastic matrices provide useful analytic and computational tools for the design and evaluation of efficient wireless communication facilities. We refer the reader to the survey paper [Setti *et al.* (2002)] for a comprehensive description for the state of the art of this active research field.

11.5   Notes and Remarks

Since the pioneering works of Cesáro, Fejér, and other analysts, general-ized summations using sequences of Cesáro averages or more general convex combinations have often been used in analysis. The following example was brought to us by Chuanxi Qian.

Let $f : [0,1] \to [0,1]$ be continuous. By Lemma 2.1, f has a fixed point, but the direct iteration $x_{n+1} = f(x_n)$, $\forall\, n \geq 0$, $x_0 \subset [0,1]$ may not converge. However, if we define the sequence $\{x_n\}$ by

$$x_{n+1} = f(a_n), \quad \text{where } a_n = \frac{1}{n}\sum_{k=0}^{n-1} x_k, \quad \forall\, n = 0,1,\cdots, \ x_0 \in [0,1],$$

then $\lim_{n\to\infty} x_n = x^*$, where x^* is a fixed point of f in $[0,1]$. See the well-cited paper [Mann (1953)] for extensions to continuous transformations of compact subsets of Banach spaces and arbitrary convex combinations.

Exercises

11.1 Prove Lemma 11.1.

11.2 Using Corollary 11.1 to prove Theorem 11.3.

11.3 This problem is related to Exercise 8.10 and can be applied to do the next exercise. Let μ be a unique absolutely continuous invariant probability measure of $S : [0,1] \to [0,1]$ with the density f^*. Let $P_S : L^1(0,1) \to L^1(0,1)$ be the corresponding Frobenius-Perron operator with respect to the Lebesgue measure m. Define an operator P_μ on $L^1(\mu)$ into itself by

$$P_\mu f = \frac{P_S(f \cdot f^*)}{f^*}.$$

Show that P_μ is the Frobenius-Perron operator from $L^1((0,1), \mathcal{B}, \mu)$ into itself associated with the mapping S.

11.4 Let $X = [0,1]$ and let μ be a unique absolutely continuous invariant probability measure of $S : [0,1] \to [0,1]$. For any $f \in L^1(0,1)$ and $g \in L^\infty(0,1)$ the quantity

$$\text{Cor}(f,g,n) = \left| \int_0^1 f(g \circ S^n)\,dm - \int_0^1 f\,dm \int_0^1 g\,dm \right|$$

is called the nth *correlation coefficient*. Express the correlation coefficient $\text{Cor}(f,g,n)$ in terms of the Frobenius-Perron operator with respect to the Lebesgue measure m and the invariant measure μ, respectively.

11.5 Let

$$\frac{a_0}{2} + \sum_{n=1}^{\infty}(a_n \cos nx + b_n \sin nx)$$

be the Fourier series corresponding to a continuous periodic function f with period 2π. Show that its nth partial sum has the expression

$$s_n(x) = \frac{1}{\pi}\int_{-\pi}^{\pi} f(x+u)\frac{\sin(n+\frac{1}{2})u}{2\sin\frac{u}{2}}du.$$

11.6 Show that the nth Fejér sum of the Fourier series in the above exercise has the expression

$$F_n f(x) = \frac{1}{2n\pi}\int_{-\pi}^{\pi}\left(\frac{\sin\frac{n(x-t)}{2}}{\sin\frac{x-t}{2}}\right)^2 f(t)dt.$$

11.7 Let n be any positive integer and let the *Chebyshev polynomial T_n* of degree n be defined by the formula

$$T_n(x) = \cos(n\arccos x).$$

(i) Show that all the roots of T_n are given by

$$x_{k,n} = \cos\frac{(2k-1)\pi}{2n} \in (-1,1), \quad k = 1, 2, \cdots, n.$$

(ii) Let $H_n(f)$ be the polynomial of degree at most $2n-1$ which interpolates $f \in C[-1,1]$ at each $x_{k,n}$ and whose derivative is zero at these points. Show that $H_n : C[-1,1] \to C[-1,1]$ is actually a positive operator by finding its explicit expression.

11.8 Let $\{P_n\}$ be a sequence of positive operators from $C[-\pi,\pi]$ into $C[-\pi,\pi]$ and let $f \in C[-\pi,\pi]$. Prove Corollary 11.1 by establishing the following inequality

$$|P_n f(x) - f(x)| \le |f(x)|\,|f_0(x) - P_n f_0(x)|$$
$$+ \left[P_n f_0(x) + \pi(P_n f_0(x))^{\frac{1}{2}}\right]\omega(f,\beta_n(x))$$

for $x \in [-\pi,\pi]$, where $f_0(x) = 1$ and

$$\beta_n(x) = \sqrt{P_n\left(\sin^2\frac{t-x}{2},x\right)}.$$

11.9 Let $\{P_n\}$ be a sequence of positive operators from $C[-\pi,\pi]$ into $C[-\pi,\pi]$ defined by

$$P_n f(x) = \frac{1}{\pi}\int_{-\pi}^{\pi} w_n(x-t)f(t)dt,$$

where w_n is an even and nonnegative function on $[-\pi, \pi]$ and satisfies

$$\frac{1}{\pi} \int_{-\pi}^{\pi} w_n(t)dt = 1, \quad \forall \, n.$$

Define

$$\beta_n(x) = \sqrt{\frac{1}{\pi} \int_{-\pi}^{\pi} \sin^2 \frac{t}{2} w_n(t)dt}.$$

Show that

$$\|P_n f - f\|_\infty \le (1 + \pi)\omega(f, \beta_n).$$

11.10 Use the conclusion of Exercise 11.9 to show that for any $f \in C[-\pi, \pi]$ and all positive integers n,

$$\|F_n f - f\|_\infty \le (1 + \pi)\omega\left(f, \frac{1}{\sqrt{n}}\right).$$

11.11 Give a function $f \in C[-\pi, \pi]$ such that its Fourier series does not converge at a point of continuity of f.

11.12 Show that the partial sum sequence of the Fourier series associated with any function $f \in C[-\pi, \pi]$ converges to f under the L^2-norm. How about $f \in L^2(-\pi, \pi)$? Do you think the Fourier series associated with any function $f \in C[-\pi, \pi]$ converges to f under the L^1-norm?

11.13 Let $\{T_t\}_{t \in \mathbb{R}}$ be a group of positive operators on $L^2(\mathbb{R})$ defined by

$$T_t f(x) = f(x - ct),$$

where $c > 0$ is a constant. Show that

(i) $T_t : L^2 \to L^2$ is a linear isometry for each t;

(ii) $\{T_t\}_{t \in \mathbb{R}}$ is a continuous group of isometries;

(iii) the infinitesimal generator of $\{T_t\}_{t \in \mathbb{R}}$ is given by

$$Af = -cf', \quad \forall \, f \in D;$$

(iv) $u(t, x) = T_t f(x)$ satisfies the following initial value problem

$$\frac{\partial u}{\partial t} + c\frac{\partial u}{\partial x} = 0, \quad u(0, x) = f(x).$$

11.14 Use Liouville's theorem to show that the Lebesgue measure of $\mathbb{R}^N \times \mathbb{R}^N$ is invariant under the Hamiltonian flow of the system of the Hamiltonian differential equations (11.14).

11.15 Give a lower bound of the constant C in (11.35) that is independent of f and g.

Appendix A

Foundations of Matrix Theory

The following two appendices review the foundations of matrix theory and functional analysis. In Appendix A we go over basic concepts and fundamental results of linear algebra. While most of the materials are presented in standard linear algebra textbooks and so will only be stated without proof, some advanced results will be proved more analytically so that they can be extended to or compared with related results for bounded linear operators. Section A.1 is about elementary concepts and facts concerning matrices. In Section A.2 we present spectral properties of matrices and an analytic approach to the Jordan canonical form of a matrix. More advanced spectral results of matrices are studied in Section A.3. In the last section we shall review the concept of norms and several matrix convergence results for general square matrices. See [Horn and Johnson (1985); Bernstein (2005)] for more materials on matrix theory.

A.1 Basic Concepts and Results

First we introduce some standard notations. $\mathbb{R}^n$ will denote the real n-dimensional vector space of column vectors x with components of real numbers $x_1, \cdots, x_n \in \mathbb{R} \equiv \mathbb{R}^1$, and $\mathbb{C}^n$ is the complex n-dimensional vector space of column vectors z with components of complex numbers $z_1, \cdots, z_n \in \mathbb{C} \equiv \mathbb{C}^1$. For $x \in \mathbb{R}^n$, x^T denotes the *transpose* of x, and for $z \in \mathbb{C}^n$, z^H is the *conjugate transpose* of z, that is, $z^H = \bar{z}^T$. The vector of all components 0 is called the *zero vector*, denoted as 0, and the vector of all components 1 will be denoted by e. The vectors $e_1, \cdots, e_n$ are said to constitute the *canonical basis* of $\mathbb{R}^n$ and $\mathbb{C}^n$, where e_j is the jth coordinate vector whose components are 0 except for the jth component 1 for each j.

For the simplicity of presentation below, we mainly deal with real vectors and matrices, and most concepts and results can be easily extended to complex ones with obvious modifications.

An $m \times n$ real *matrix* $A = (a_{ij})$ is an array of mn real numbers in m rows and n columns with a_{ij} its entry at row i and column j. The *zero matrix*, denoted 0, is a matrix with all entries 0. If $m = n$, then A is called a *square matrix* of order n. A square matrix A of order n is called an *identity matrix*, which is denoted by I, if $a_{ij} = \delta_{ij}$, $\forall\ i, j = 1, 2, \cdots, n$, where the symbol δ_{ij} is the *Kronecker notation*: $\delta_{ij} = 0$ if $i \neq j$ and 1 if $i = j$. A square matrix A is said to be *upper triangular* if $a_{ij} = 0$ for all $i > j$ and *lower triangular* if $a_{ij} = 0$ whenever $i < j$. A matrix A is called *diagonal* if it is both upper and lower triangular.

Let $A = (a_{ij})$ and $B = (b_{ij})$ be two $m \times n$ matrices. Then their *sum* $A + B$ is the $m \times n$ matrix $C = (c_{ij})$ with $c_{ij} = a_{ij} + b_{ij}$ for all i, j. The *scalar product* cA of a number c with A is the $m \times n$ matrix $B = (b_{ij})$ with $b_{ij} = ca_{ij}$. The set of all $m \times n$ real (or complex) matrices is denoted by $\mathbb{R}^{m \times n}$ (or $\mathbb{C}^{m \times n}$), which becomes an mn-dimensional vector space under the matrix addition and scalar multiplication operations over the real (or complex) field. We identify $\mathbb{R}^n$ and $\mathbb{C}^n$ with $\mathbb{R}^{n \times 1}$ and $\mathbb{C}^{n \times 1}$, respectively, so that each n-dimensional vector can be viewed as an $n \times 1$ matrix.

For an $m \times n$ matrix A and an $n \times k$ matrix B, their *product* AB is defined to be an $m \times k$ matrix $C = (c_{ij})$ such that

$$c_{ij} = \sum_{l=1}^{k} a_{il} b_{lj}, \quad \forall\ i = 1, \cdots, m; \ j = 1, \cdots, k.$$

Matrix multiplication is an *associative* operation but not a commutative one, that is, $A(BC) = (AB)C$ when both sides are well-defined, but $AB \neq BA$ in general even if the multiplication operations of both sides are meaningful. Identity matrices I of suitable orders satisfy $AI = A$ and $IA = A$ for any rectangular matrix A.

An $m \times n$ matrix A defines a linear operator from $\mathbb{R}^n$ into $\mathbb{R}^m$ by $y = Ax \in \mathbb{R}^m$ for all $x \in \mathbb{R}^n$. Thus, we view $A \in \mathbb{R}^{m \times n}$ as either a matrix or a linear operator. The matrix A itself is the matrix representation of the operator A under the canonical basis $e_1, \cdots, e_n$ of $\mathbb{R}^n$ and the canonical basis $e_1, \cdots, e_m$ of $\mathbb{R}^m$.

Let $A : \mathbb{R}^n \to \mathbb{R}^m$ be a linear operator. The collection of all vectors Ax with $x \in \mathbb{R}^n$ is a vector subspace of $\mathbb{R}^m$, is called the *range* of A, and is denoted as $R(A)$. The collection of all vectors $x \in \mathbb{R}^n$ such that $Ax = 0$ is a vector subspace of $\mathbb{R}^n$, is called the *null space* of A, and is denoted as $N(A)$. If $R(A) = \mathbb{R}^m$, then $A : \mathbb{R}^n \to \mathbb{R}^m$ is said to be *onto*, and if $N(A) = \{0\}$, then $A : \mathbb{R}^n \to \mathbb{R}^m$ is said to be *one-to-one*. If $A : \mathbb{R}^n \to \mathbb{R}^m$ is both onto and one-to-one, then A is called *invertible*. Invertible matrices must be square ones. The *rank* of a matrix A, denoted $\text{rank}(A)$, is defined as the dimension of $R(A)$. It is well-known (also see Theorem A.2 below) that the rank of A is the same as the dimension of the range of the *transpose* A^T (called the *row space* of A), and their common number is also the largest order of nonzero sub-determinants of A.

A matrix $A \in \mathbb{R}^{n \times n}$ is *nonsingular* if its determinant $\det A \neq 0$. There are many other equivalent conditions for the non-singularity of A, which are listed in the following theorem.

Theorem A.1. *Let* $A \in \mathbb{R}^{n \times n}$. *The following are equivalent:*
 (i) *A is nonsingular.*
 (ii) *A^T is nonsingular.*
 (iii) *$A : \mathbb{R}^n \to \mathbb{R}^n$ is one-to-one.*
 (iv) *$A : \mathbb{R}^n \to \mathbb{R}^n$ is onto.*
 (v) *$A : \mathbb{R}^n \to \mathbb{R}^n$ is invertible.*
 (vi) *The equation $Ax = b$ has a solution for all $b \in \mathbb{R}^n$.*
 (vii) *The equation $Ax = 0$ has no nonzero solutions.*
 (viii) *The rank of A is n.*

The *Euclidean inner product* of $x,\ y \in \mathbb{R}^n$ is defined to be

$$(x, y) = y^T x = \sum_{k=1}^{n} x_k y_k,$$

and the *inner product* of $z,\ w \in \mathbb{C}^n$ is defined to be

$$(z, w) = w^H z = \sum_{k=1}^{n} z_k \bar{w}_k.$$

Under the inner product operation, $\mathbb{R}^n$ (or $\mathbb{C}^n$) becomes an *inner product space* and is called the n-dimensional *Euclidean space* (or *unitary space*). The induced *Euclidean 2-norm*

$$\|x\|_2 = \sqrt{(x, x)}, \quad \forall\, x \in \mathbb{R}^n\ (\mathbb{C}^n)$$

makes $\mathbb{R}^n$ (or $\mathbb{C}^n$) a *complete normed space*.

Two vectors $x,\ y \in \mathbb{R}^n$ are said to be *orthogonal* to each other if $x^T y = 0$. Similarly $z,\ w \in \mathbb{C}^n$ are orthogonal if $z^H w = 0$. A set of vectors $v_1, \cdots, v_k$ are said to be *orthonormal* if $(v_i, v_j) = \delta_{ij}, \ \forall\, i, j = 1, \cdots, k$. A matrix $A \in \mathbb{R}^{n \times n}$ is called an *orthogonal matrix* if it satisfies $A^T A = I$, that is, if the columns of A constitute an *orthonormal basis* of $\mathbb{R}^n$. Similarly,

$A \in \mathbb{C}^{n \times n}$ is said to be a *unitary matrix* if $A^H A = I$, where $A^H = \overline{A}^T$ is the *conjugate transpose* of A.

The *orthogonal complement* of a subset M of $\mathbb{R}^n$ is defined as

$$M^\perp = \{x \in \mathbb{R}^n : x^T y = 0, \ \forall \, y \in M\}.$$

$M^\perp$ is a subspace of $\mathbb{R}^n$. If M is a subspace of $\mathbb{R}^n$, then

$$\mathbb{R}^n = M \oplus M^\perp,$$

which is the *direct sum decomposition* of $\mathbb{R}^n$ in terms of M and $M^\perp$. As a consequence, $\dim M + \dim M^\perp = n$. The equality $(M^\perp)^\perp = M$ is obvious when M is a subspace of $\mathbb{R}^n$.

Theorem A.2. *Let $A \in \mathbb{R}^{m \times n}$. Then $\dim R(A) = \dim R(A^T)$.*

Remark A.1. We can define a generalized inverse $A^\dagger : \mathbb{R}^m \to \mathbb{R}^n$ of a matrix $A : \mathbb{R}^n \to \mathbb{R}^m$ as follows. On $R(A)$, $A^\dagger$ is just the inverse of A restricted to $N(A)^\perp$, and $A^\dagger$ is zero on $R(A)^\perp$. This inverse is referred to as the *Moore-Penrose inverse* of a general $m \times n$ matrix since Moore first defined it in 1920 and Penrose re-discovered it in 1955. Generalized inverses of matrices and linear operators are very important in many branches of mathematics and engineering [Campbell and Meyer (1979)].

A.2 The Spectral Theory of Matrices

We now review the fundamentals of the spectral theory for square matrices. Let $A \in \mathbb{C}^{n \times n}$. A complex number λ is called an *eigenvalue* of A if there is a nonzero vector $x \in \mathbb{C}^n$ such that

$$Ax = \lambda x.$$

The vector x is called an *eigenvector* of A corresponding to eigenvalue λ. A nonzero vector $y \in \mathbb{C}^n$ is called a *left eigenvector* of A associated with an eigenvalue λ if $y^T A = \lambda y^T$, which is exactly an eigenvector of A^T associated with λ. Since the homogeneous equation $(A - \lambda I)x = 0$ has a nonzero solution if and only if the matrix $A - \lambda I$ is singular by Theorem A.1, the eigenvalues of A are precisely the n complex roots (counting algebraic multiplicity) of the *characteristic polynomial*

$$p_A(\lambda) \equiv \det(\lambda I - A).$$

Let $p_A(\lambda)$ be factored as

$$p_A(\lambda) = (\lambda - \lambda_1)^{m_1} (\lambda - \lambda_2)^{m_2} \cdots (\lambda - \lambda_k)^{m_k}$$

in the field of complex numbers, where $\lambda_1, \cdots, \lambda_k$ are the distinct eigenvalues of A. For each $i = 1, \cdots, k$, the positive integer of power index m_i is called the *algebraic multiplicity* of λ_i, and the dimension of the null space $N(A - \lambda_i I)$ is called the *geometric multiplicity* of λ_i. If $m_i = 1$, then the eigenvalue λ_i is said to be *simple*. The geometric multiplicity of each eigenvalue is less than or equal to its algebraic multiplicity. Eigenvectors corresponding to distinct eigenvalues are linearly independent. The collection of all eigenvalues of A is called the *spectrum* of A and is denoted by $\sigma(A)$, its complement $\rho(A) = \mathbb{C} \setminus \sigma(A)$ in $\mathbb{C}$ is called the *resolvent set* of A, and the nonnegative number

$$r(A) = \max\{|\lambda| : \lambda \in \sigma(A)\}$$

is called the *spectral radius* of A.

Let M be a subspace of $\mathbb{C}^n$. Two vectors x, $y \in \mathbb{C}^n$ are said to be *equivalent* with respect to M if $x - y \in M$, and the equivalent relation between x and y is denoted as $x \sim y$, which is an *equivalent relation*. That is, the relation is *reflexive* ($x \sim x$), *symmetric* ($x \sim y$ implies $y \sim x$), and *transitive* ($x \sim y$ and $y \sim z$ imply $x \sim z$). The resulting *equivalent classes* of $\mathbb{C}^n$ are $[x] \equiv x + M = \{x + y : y \in M\}$ for all $x \in \mathbb{C}^n$.

Denote by $\mathbb{C}^n/M$ the corresponding *quotient space* which is defined as a vector space consisting of all equivalent classes of $\mathbb{C}^n$ with the addition $[x] + [y] = [x + y]$ and scalar multiplication $a[x] = [ax]$. If M is a vector subspace of $\mathbb{C}^n$ which is *invariant* under a matrix A, that is, $AM \subset M$, then we can define the *induced linear operator* $A_M : \mathbb{C}^n/M \to \mathbb{C}^n/M$ by

$$A_M[x] = [Ax], \quad \forall\, [x] \in \mathbb{C}^n/M.$$

It can be shown that

$$\sigma(A) = \sigma(A|_M) \cup \sigma(A_M), \tag{A.1}$$

where $A|_M : M \to M$ is the restriction of A to M.

Definition A.1. Let $A \in \mathbb{C}^{n \times n}$. The set of all eigenvalues of modulus $r(A)$ is called the *peripheral spectrum* of A. The peripheral spectrum of A is said to be *cyclic* if the condition $r(A)a \in \sigma(A), |a| = 1$ implies that $r(A)a^k \in \sigma(A)$ for all integers k; it is said to be *fully cyclic* if whenever $Ax = r(A)ax$ with $|a| = 1$ and $x \neq 0$, then

$$\left(|x_1|(\operatorname{sgn} x_1)^k, |x_2|(\operatorname{sgn} x_2)^k, \cdots, |x_n|(\operatorname{sgn} x_n)^k\right)^T$$

is an eigenvector corresponding to eigenvalue $r(A)a^k$ for all integers k, where $\operatorname{sgn} c = c/|c|$ for $c \neq 0$ and $\operatorname{sgn} c = 1$ for $c = 0$.

The *resolvent* $R(\lambda) \equiv R(\lambda, A)$ of a matrix $A \in \mathbb{C}^{n \times n}$ is defined to be $R(\lambda) = (\lambda I - A)^{-1}$ for all $\lambda \in \rho(A)$. The inverse formula of a matrix tells us that the resolvent $R(\lambda)$ is a rational function of its variable λ and its *poles* are just the eigenvalues of A.

If $p(\lambda) = a_0 + a_1\lambda + a_2\lambda^2 + \cdots + a_m\lambda^m$ is any polynomial, then the corresponding *matrix polynomial* $p(A) \in \mathbb{C}^{n \times n}$ is defined by

$$p(A) = a_0 I + a_1 A + a_2 A^2 + \cdots + a_m A^m$$

for any $A \in \mathbb{C}^{n \times n}$. If $\lambda_1, \cdots, \lambda_n$ are the eigenvalues of A with corresponding eigenvectors $x_1, \cdots, x_n$, then $x_1, \cdots, x_n$ are eigenvectors of $p(A)$ corresponding to the eigenvalues $p(\lambda_1), \cdots, p(\lambda_n)$. Similarly, if A is invertible, then the eigenvalues of A^{-1} are $\lambda_1^{-1}, \cdots, \lambda_n^{-1}$ with the same eigenvectors. The following Cayley-Hamilton theorem asserts that any square matrix satisfies its own characteristic equation.

Theorem A.3. (Cayley-Hamilton) *Let $A \in \mathbb{R}^{n \times n}$ and let $p_A(\lambda)$ be the characteristic polynomial of A. Then $p_A(A) = 0$.*

A polynomial $q(\lambda)$ with the leading coefficient 1 such that $q(A) = 0$ is called an *annihilating polynomial* for A. The Cayley-Hamilton theorem implies that there is a unique annihilating polynomial $p_m(\lambda)$ of minimal degree for A, which is called the *minimal polynomial* for A. The minimal polynomial $p_m(\lambda)$ for A is a factor of the characteristic polynomial $p_A(\lambda)$ of A with the same set of zeros as $p_A(\lambda)$. Thus, the multiplicity of each zero λ_0 of the minimal polynomial, which is precisely the same number as the multiplicity of λ_0 as a pole of the resolvent $R(\lambda)$, is less than or equal to the algebraic multiplicity of λ_0.

Two matrices A and B are said to be *similar*, written as $A \sim B$, if there is a nonsingular matrix P such that $P^{-1}AP = B$. If in addition P is a unitary matrix, then A and B are called *unitarily similar*. The Similarity relation among square matrices of the same order is an equivalent relation. If A and B are similar, then they have the same characteristic polynomial, and so the same eigenvalues. The following Schur's theorem is a fundamental result in matrix theory.

Theorem A.4. (Schur) *Let $A \in \mathbb{C}^{n \times n}$. Then there is a unitary matrix U such that $U^H A U = T$ is upper triangular.*

Since $A \sim T$ in the above theorem, $p_A(\lambda) = \prod_{i=1}^{n}(\lambda - t_{ii})$. A matrix $A \in \mathbb{C}^{n \times n}$ is said to be *normal* if $A^H A = A A^H$, *Hermitian* if $A^H = A$, and *symmetric* if $A^T = A$. Eigenvalues of normal matrices are all real numbers. Unitary, orthogonal, Hermitian, and real symmetric matrices are all normal. Theorem A.4 implies that if A is a normal matrix, then there is a unitary matrix U such that

$$U^H A U = D, \qquad (A.2)$$

where D is a diagonal matrix whose diagonal entries are the eigenvalues of A. That is, A is *unitarily diagonalizable*. In particular, if A is a real and symmetric matrix, then U in (A.2) is orthogonal. In other words, a real symmetric matrix is *orthogonally similar to* a real diagonal matrix. Furthermore, the columns of U are necessarily eigenvectors of A, so that a real symmetric matrix has n orthogonal eigenvectors.

We have known that a normal matrix is similar to a diagonal matrix. A square matrix A is said to be *diagonalizable* if it is similar to a diagonal matrix. The next theorem lists necessary and sufficient conditions for a matrix to be diagonalizable.

Theorem A.5. *Let $A \in \mathbb{C}^{n \times n}$. The following are equivalent:*

(i) *A is diagonalizable.*

(ii) *A has n linearly independent eigenvectors.*

(iii) *Each root of the minimal polynomial for A is simple.*

(iv) *For every eigenvalue of A, its algebraic multiplicity is equal to its geometric multiplicity.*

If a matrix A is similar to a diagonal matrix D, then D is the simplest "canonical form" for the given matrix A since D shares many spectral properties of A. Although any $A \in \mathbb{C}^{n \times n}$ is unitarily similar to an upper triangular matrix by Schur's theorem, the triangular matrix is not as simple as a special block diagonal matrix called the Jordan canonical form which is composed of Jordan blocks.

Definition A.2. Let λ be a complex number and let j be a positive integer. The $j \times j$ upper triangular matrix

$$J_\lambda = \begin{bmatrix} \lambda & 1 & & & \\ & \lambda & 1 & & \\ & & \cdot & \cdot & \\ & & & \cdot & \cdot \\ & & & & \lambda & 1 \\ & & & & & \lambda \end{bmatrix}$$

is called a *Jordan block* of order j. If $j = 1$, then the corresponding Jordan block is the 1×1 matrix $[\lambda]$.

Theorem A.6. (Jordan's canonical form) *Let $A \in \mathbb{C}^{n \times n}$. Then there is a nonsingular matrix P such that*

$$P^{-1}AP = J,$$

where J is called the Jordan canonical form *of A and it has the following structure: J is a block diagonal matrix*

$$J = \begin{bmatrix} J_{\lambda_1} & & \\ & \ddots & \\ & & J_{\lambda_r} \end{bmatrix}, \tag{A.3}$$

in which each J_{λ_i} is a Jordan block and λ_i is an eigenvalue of A of algebraic multiplicity at least equal to the order of J_{λ_i}.

Remark A.2. The eigenvalues $\lambda_1, \cdots, \lambda_r$ may not be distinct in the Jordan canonical form. In other words, an eigenvalue of A may be associated with more than one Jordan block. For example, all Jordan blocks of an identity matrix are [1].

Let j_i be the order of the ith Jordan block in the Jordan canonical form (A.3) for $i = 1, \cdots, r$, and let p_l be the lth column of the matrix P in Theorem A.6 for $l = 1, \cdots, n$. Then

$$p_1, p_{j_1+1}, \cdots, p_{j_1+\cdots+j_{r-1}+1}$$

are the eigenvectors of A associated with the eigenvalues $\lambda_1, \cdots, \lambda_r$, respectively. Other columns of P are often called *generalized eigenvectors*. In particular, if all the eigenvalues of A are distinct, then the Jordan canonical form of A is a diagonal matrix and all the columns of the matrix P are n linearly independent eigenvectors of A.

The concept of the index of an eigenvalue of A defined below will play an important role in the next section on the analytic theory of matrix functions, which is related to Jordan's canonical form.

Definition A.3. Let $A \in \mathbb{C}^{n \times n}$. The *index* $\nu(\lambda) \equiv \nu_A(\lambda)$ of a complex number λ with respect to A is the smallest nonnegative integer j such that $N((A - \lambda I)^{j+1}) = N((A - \lambda I)^j)$.

The following properties of the index can be easily verified.

Proposition A.1. *Let $\nu(\lambda)$ be the index of a complex number $\lambda \in \mathbb{C}$ with respect to $A \in \mathbb{C}^{n \times n}$. Then the following are true:*

(i) $\nu(\lambda) \leq n$;

(ii) $N((A - \lambda I)^j) = N((A - \lambda I)^{\nu(\lambda)})$ *for all* $j \geq \nu(\lambda)$;

(iii) *every inclusion relation in the following chain*

$$\{0\} \subset N(A - \lambda I) \subset N((A - \lambda I)^2) \subset \cdots \subset N((A - \lambda I)^{\nu(\lambda)})$$

is a strict one;

(iv) $\lambda \in \sigma(A)$ *if and only if* $\nu(\lambda) > 0$.

Eigenvectors associated with an eigenvalue λ of A are nonzero vectors of $N(A - \lambda I)$, and generalized eigenvectors of A corresponding to λ are those nonzero vectors of $N((A - \lambda I)^{\nu(\lambda)})$ which are not eigenvectors of A, where the index $\nu(\lambda)$ of λ is exactly the maximal order of all the Jordan blocks associated with the same eigenvalue λ.

The usefulness of the index concept associated with each spectral point of a square matrix is reflected in the following result, which can be extended to the Dunford integral of matrix-valued analytic functions in the next section and the Dunford integral of more general operator-valued analytic functions in Chapter 6.

Theorem A.7. *Suppose that* $A \in \mathbb{C}^{n \times n}$, *and* p *and* q *are two polynomials. Then* $p(A) = q(A)$ *if and only if every* $\lambda \in \sigma(A)$ *is a root of the polynomial* $p - q$ *with multiplicity at least* $\nu(\lambda)$.

Proof It is enough to assume $q = 0$. Suppose that p has a root of multiplicity $\nu(\lambda)$ at each $\lambda \in \sigma(A)$. By the Cayley-Hamilton theorem, $p_A(A) = 0$, where $p_A(\lambda) = \prod_{i=1}^{k}(\lambda - \lambda_i)^{r_i}$ is the characteristic polynomial of A and $\lambda_1, \cdots, \lambda_k$ are all the distinct eigenvalues of A. Let $p_k(\lambda) = \prod_{i=1}^{k}(\lambda - \lambda_i)^{s_i}$, where $s_i = \min\{r_i, \nu(\lambda_i)\}$. Then $p_k(A) = 0$. Since p_k is a factor of p, we see that p is divisible by p_m, hence $p(A) = 0$.

Conversely, suppose that $p(A) = 0$, where $p(\lambda) = \prod_{i=1}^{k}(\lambda - \lambda_i)^{t_i}$. If $\lambda_i \notin \sigma(A)$, then $(A - \lambda_i I)x = 0$ implies $x = 0$. So, without loss of generality, assume each $\lambda_i \in \sigma(A)$. If $\lambda \in \sigma(A)$, then $Ax = \lambda x$ for some $x \neq 0$, from which $0 = p(A)x = p(\lambda)x$. So $p(\lambda) = 0$, thus $\lambda = \lambda_i$ for some i. This shows that $\sigma(A) = \{\lambda_1, \cdots, \lambda_k\}$. If $t_1 < \nu(\lambda_1)$, then there is $x_1 \neq 0$ such that $(A - \lambda_1 I)^{t_1+1}x_1 = 0$ and $y_1 \equiv (A - \lambda_1 I)^{t_1}x_1 \neq 0$. Let $p(\lambda) = (\lambda - \lambda_1)^{t_1}q(\lambda)$. Then $q(\lambda_1) \neq 0$. Since $Ay_1 = \lambda_1 y_1$, we have

$$0 = p(A)x_1 = q(A)y_1 = q(\lambda_1)y_1 \neq 0.$$

This contradiction shows $t_1 \geq \nu(\lambda_1)$. Similarly, $t_i \geq \nu(\lambda_i), i = 2, \cdots, k$, so p has a root of multiplicity at least $\nu(\lambda_i)$ for each λ_i. $\square$

We end this section by presenting some results on locating the eigenvalues of a general complex square matrix. The eigenvalue location problem has the origin in the classic *Gersgörin disk theorem*, which is a fundamental result for the localization of eigenvalues of a matrix.

Theorem A.8. (Gersgörin) *Let $A = (a_{ij}) \in \mathbb{C}^{n \times n}$. Then*

$$\sigma(A) \subset \bigcup_{i=1}^{n} \left\{ \lambda \in \mathbb{C} : |\lambda - a_{ii}| \leq \sum_{j=1, j \neq i}^{n} |a_{ij}| \right\} \equiv G(A). \qquad (A.4)$$

Remark A.3. The set $G(A)$ in (A.4) is called the *Gersgörin region* which is the union of n *Gersgörin disks*. With the help of the intermediate value theorem of calculus and a perturbation argument, one can further show that if a union of k of these disks forms a connected region that is disjoint from the union of all the remaining $n - k$ disks, then there are precisely k eigenvalues of A in this region.

Since $\sigma(A^T) = \sigma(A)$, applying Theorem A.8 to A^T gives

Corollary A.1. *Let $A = (a_{ij}) \in \mathbb{C}^{n \times n}$. Then*

$$\sigma(A) \subset \bigcup_{j=1}^{n} \left\{ \lambda \in \mathbb{C} : |\lambda - a_{jj}| \leq \sum_{i=1, i \neq j}^{n} |a_{ij}| \right\} = G(A^T).$$

Since $P^{-1}AP$ has the same eigenvalues as A whenever P is a nonsingular matrix, we can apply the Gersgörin theorem to $P^{-1}AP$ with a particular choice of P. For example, the selection of $P = D \equiv \text{diag}(d_1, d_2, \cdots, d_n)$ with all $d_i > 0$ leads to the following result.

Corollary A.2. *Let $A = (a_{ij}) \in \mathbb{C}^{n \times n}$, and let $d_1, d_2, \cdots, d_n$ be n positive numbers. Then*

$$\sigma(A) \subset \bigcup_{i=1}^{n} \left\{ \lambda \in \mathbb{C} : |\lambda - a_{ii}| \leq \frac{1}{d_i} \sum_{j=1, j \neq i}^{n} d_j |a_{ij}| \right\} = G(D^{-1}AD)$$

as well as

$$\sigma(A) \subset \bigcup_{j=1}^{n} \left\{ \lambda \in \mathbb{C} : |\lambda - a_{jj}| \leq d_j \sum_{i=1, j \neq j}^{n} \frac{1}{d_i} |a_{ij}| \right\}.$$

A.3   Analytic Theory of Matrices

Theorem A.7 enables us to define the matrix $f(A)$ for f in a wider class of functions than polynomials. Any open set of $\mathbb{C}$ that contains $\sigma(A)$ is called a *neighborhood* of $\sigma(A)$. Let $\mathcal{H}(A)$ be the class of complex-valued functions f of a complex variable λ which are *analytic* in a neighborhood Ω of $\sigma(A)$, namely the derivative $f'(\lambda)$ exists at every point $\lambda \in \Omega$ which depends on the choice of $f \in \mathcal{H}(A)$ and need not be connected.

Definition A.4. Let $A \in \mathbb{C}^{n \times n}$ and let $f \in \mathcal{H}(A)$. We define $f(A) = p(A)$, where p is any polynomial such that $f^{(j)}(\lambda) = p^{(j)}(\lambda)$, $\forall \, j = 0, 1, \cdots, \nu(\lambda) - 1$, for each $\lambda \in \sigma(A)$.

From Theorem A.7, the matrix $f(A)$ does not depend on the choice of p. The following proposition follows immediately from the corresponding results for polynomial functions.

Proposition A.2. *Let $f, g \in \mathcal{H}(A)$ and let $a, b \in \mathbb{C}$. Then*

(i) $af + bg \in \mathcal{H}(A)$ *and* $(af + bg)(A) = af(A) + bg(A)$;

(ii) $fg \in \mathcal{H}(A)$ *and* $(fg)(A) = f(A)g(A)$;

(iii) *if* $f(\lambda) = \sum_{i=0}^{m} a_i \lambda^i$, *then* $f(A) = \sum_{i=0}^{m} a_i A^i$;

(iv) *if f has a power series representation* $f(\lambda) = \sum_{i=0}^{\infty} a_i \lambda^i$, *valid in an open disk that contains $\sigma(A)$, then* $f(A) = \sum_{i=0}^{\infty} a_i A^i$;

(v) $f(A) = 0$ *if and only if*

$$f^{(j)}(\lambda) = 0, \ \forall \, j = 0, 1, \cdots, \nu(\lambda) - 1, \ \lambda \in \sigma(A);$$

(vi) $f \in \mathcal{H}(A^H)$ *and* $f(A^H) = f(A)^H$.

For $\lambda_0 \in \mathbb{C}$, let $h_{\lambda_0} \in \mathcal{H}(A)$ be such that $h_{\lambda_0}(\lambda) \equiv 1$ in a neighborhood of λ_0 and $h_{\lambda_0}(\lambda) \equiv 0$ in a neighborhood of $\sigma(A) \setminus \{\lambda_0\}$. Define

$$E(\lambda_0) = h_{\lambda_0}(A). \tag{A.5}$$

Then Proposition A.2 implies the next theorem.

Theorem A.9. *Let $E(\lambda)$ be defined by (A.5) for $\lambda \in \mathbb{C}$. Then*

(i) $E(\lambda) \neq 0$ *if and only if* $\lambda \in \sigma(A)$;

(ii) $E(\lambda)^2 = E(\lambda)$ *and* $E(\lambda)E(\gamma) = 0$ *for* $\lambda \neq \gamma$;

(iii) $\sum_{\lambda \in \sigma(A)} E(\lambda) = I$ *(partition of unity)*.

Let $\sigma(A) = \{\lambda_1, \cdots, \lambda_k\}$ and let $M_i = R(E(\lambda_i))$ for $i = 1, \cdots, k$. It follows from Theorem A.9(ii) and (iii) that

$$\mathbb{C}^n = M_1 \oplus \cdots \oplus M_k.$$

Since $AE(\lambda_i) = E(\lambda_i)A$, we have $AM_i \subset M_i$ for each i. That is, each M_i of the above direct sum decomposition of $\mathbb{C}^n$ is an *invariant subspace* of A. Thus the task of studying the action of A on $\mathbb{C}^n$ can be reduced to that of studying the action of A on each of the subspaces M_i.

Since the analytic function $(\lambda - \lambda_i)^{\nu(\lambda_i)} h_{\lambda_i}(\lambda)$ has a zero of order at least $\nu(\lambda_i)$ at $\lambda_i \in \sigma(A)$, the matrix

$$(A - \lambda_i I)^{\nu(\lambda_i)} E(\lambda_i) = 0. \tag{A.6}$$

Therefore, $A - \lambda_i I : M_i \to M_i$ is a *nilpotent operator* for each i. The next result gives an exact relation between the index $\nu(\lambda_i)$ and the corresponding projection $E(\lambda_i)$ for each eigenvalue λ_i.

Theorem A.10. *Let* $\lambda \in \sigma(A)$. *Then*

$$R(E(\lambda)) = N((A - \lambda I)^{\nu(\lambda)}).$$

Proof Let $\lambda \in \sigma(A)$. Equality (A.6) shows that $R(E(\lambda)) \subset N((A - \lambda I)^{\nu(\lambda)})$. By the partition of unity $I = \sum_{\lambda \in \sigma(A)} E(\lambda)$, to show the reverse inclusion it is sufficient to prove that

$$N((A - \lambda I)^{\nu(\lambda)}) \cap N((A - \gamma I)^{\nu(\gamma)}) = \{0\}, \ \forall \, \gamma \neq \lambda, \ \gamma \in \sigma(A).$$

Suppose that $x \in N((A - \lambda I)^{\nu(\lambda)}) \cap N((A - \gamma I)^{\nu(\gamma)})$ is nonzero. Let r be the greatest integer such that $y = (A - \lambda I)^r x \neq 0$. Then $Ay = \lambda y$, so $(A - \gamma I)^{\nu(\gamma)} y = (\lambda - \gamma)^{\nu(\gamma)} \neq 0$. On the other hand,

$$\begin{aligned} (A - \gamma I)^{\nu(\gamma)} y &= (A - \gamma I)^{\nu(\gamma)} (A - \lambda I)^r x \\ &= (A - \lambda I)^r (A - \gamma I)^{\nu(\gamma)} x = 0, \end{aligned}$$

since $x \in N((A - \gamma I)^{\nu(\gamma)})$. This leads to a contradiction. $\square$

The projections $E(\lambda)$ for all the eigenvalues $\lambda \in \sigma(A)$ not only give the direct sum decomposition of $\mathbb{C}^n$ into the generalized eigenspaces of A, but also express $f(A)$ analytically in terms of the Taylor-like polynomials.

Theorem A.11. *If* $f \in \mathcal{H}(A)$, *then*

$$f(A) = \sum_{\lambda \in \sigma(A)} \sum_{r=0}^{\nu(\lambda)-1} \frac{f^{(r)}(\lambda)}{r!} (A - \lambda I)^r E(\lambda).$$

Proof Let $g \in \mathcal{H}(A)$ be defined by

$$g(\gamma) = \sum_{\lambda \in \sigma(A)} \sum_{r=0}^{\nu(\lambda)-1} \frac{f^{(r)}(\lambda)}{r!} (\gamma - \lambda)^r h_\lambda(\gamma).$$

Then $f^{(r)}(\lambda) = g^{(r)}(\lambda)$, $\forall\, r = 0, 1, \cdots, \nu(\lambda) - 1$, $\lambda \in \sigma(A)$. The result is from Proposition A.2 and the definition of $f(A)$. $\qquad\qquad\square$

Corollary A.3. *Let a sequence of functions* f_n, $n = 1, 2, \cdots$, *and* f *be in* $\mathcal{H}(A)$. *Then the matrix sequence* $\{f_n(A)\}$ *converges to the matrix* $f(A)$ *if and only if the number sequences* $\{f_n^{(r)}(\lambda)\}$ *converge to the numbers* $f^{(r)}(\lambda)$ *for all the eigenvalues* $\lambda \in \sigma(A)$ *and the nonnegative integers* $r < \nu(\lambda)$.

Proof The sufficiency part follows from Theorem A.11. Conversely, let the sequence $\{f_n(A)\}$ converge to $f(A)$ and let $\lambda \in \sigma(A)$ be given. Since $(A - \lambda I)^{\nu(\lambda)-1}E(\lambda) \neq 0$, there is x such that

$$(A - \lambda I)^{\nu(\lambda)-1}E(\lambda)x \neq 0.$$

Denote $y = E(\lambda)x$, and $y_r = (A - \lambda I)^r y$ for $0 \leq r < \nu(\lambda)$. Put $\nu(\lambda) - 1 = s$. Then $f_n(A)y_s = f_n(\lambda)y_s$, so the convergence of $\{f_n(A)\}$ to $f(A)$ implies the convergence of $\{f_n(\lambda)\}$ to $f(\lambda)$. Similarly, $f_n(A)y_{s-1} = f_n(\lambda)y_{s-1} + f_n'(\lambda)y_s$, so the sequence $\{f_n'(\lambda)\}$ converges to $f'(\lambda)$. By induction, we see that $\{f_n^{(r)}(\lambda)\}$ converges to $f^{(r)}(\lambda)$ for each $r = 0, 1, \cdots, \nu(\lambda) - 1$. $\quad\square$

Corollary A.4. *Let* $A \in \mathbb{C}^{n \times n}$. *The following statements are equivalent:*
  (i) *The sequence* $\{n^{-1}\sum_{j=0}^{n-1} A^j\}$ *converges.*
  (ii) $\lim_{n \to \infty} A^n/n = 0$.
  (iii) $|\sigma(A)| \leq 1$ *and* $\nu(\lambda) = 1$ *for those* $\lambda \in \sigma(A)$ *with* $|\lambda| = 1$.

Proof Let $f_n(\lambda) = n^{-1}\sum_{j=0}^{n-1} \lambda^j$ and $g_n(\lambda) = \lambda^n/n$. The sequence $\{f_n(\lambda)\}$ converges if and only if $|\lambda| \leq 1$, and $\{f_n^{(r)}(\lambda)\}$ converges for each fixed $r > 0$ if and only if $|\lambda| < 1$. Similarly, $\{g_n(\lambda)\}$ converges if and only if $|\lambda| \leq 1$, and $\{g_n^{(r)}(\lambda)\}$ converges for $r > 0$ if and only if $|\lambda| < 1$. The corollary is therefore from Theorem A.11. $\qquad\qquad\square$

The following representation of $f(A)$ in terms of the *Dunford integral* of a matrix-valued analytic function will be taken as the definition of $f(T)$ for bounded linear operators T in Section 6.1.

Theorem A.12. *Let* $f \in \mathcal{H}(A)$ *be analytic in a neighborhood* Ω *of* $\sigma(A)$, *and let* $U \subset \Omega$ *be open such that its boundary* ∂U *is piecewise smooth and its closure* $\overline{U} = U \cup \partial U \subset \Omega$. *Then*

$$f(A) = \frac{1}{2\pi i}\int_{\partial U} f(\lambda)R(\lambda, A)d\lambda.$$

Proof Let $\lambda \notin \sigma(A) = \{\lambda_1, \cdots, \lambda_k\}$, and let $g(\gamma) = (\lambda - \gamma)^{-1}$. Then Proposition A.2 and Theorem A.11 ensure that

$$R(\lambda, A) = g(A) = \sum_{j=1}^{k} \sum_{r=0}^{\nu(\lambda_j)-1} \frac{(A - \lambda_j I)^r}{(\lambda - \lambda_j)^{r+1}} E(\lambda_j).$$

Thus, for the given function $f \in \mathcal{H}(A)$, we have

$$\frac{1}{2\pi i} \int_{\partial U} f(\lambda) R(\lambda I, A) d\lambda$$

$$= \sum_{j=1}^{k} \sum_{r=0}^{\nu(\lambda_j)-1} (A - \lambda_j I)^r \frac{f^{(r)}(\lambda_j)}{r!} E(\lambda_j) = f(A). \qquad \square$$

A.4 Norms and Convergence

Many convergence theorems in matrix theory are related to the concepts of norms for vectors and corresponding norms for matrices, and the spectral analysis of a square matrix plays a key role in proving such results. We first review various norms for vectors and matrices and then establish some relations between the spectral radius of a matrix and its norms. Finally, several matrix convergence results will be presented.

A *vector norm* on $\mathbb{R}^n$ is a mapping $\| \ \| : \mathbb{R}^n \to \mathbb{R}$ which satisfies the following three conditions:

(i) $\|x\| \geq 0$, $\forall x \in \mathbb{R}^n$, and $\|x\| = 0$ if and only if $x = 0$;

(ii) $\|ax\| = |a| \|x\|$, $\forall x \in \mathbb{R}^n$, $a \in \mathbb{R}$;

(iii) $\|x + y\| \leq \|x\| + \|y\|$, $\forall x, y \in \mathbb{R}^n$ (*triangle inequality*).

Given a vector norm $\| \ \|$ on $\mathbb{R}^n$ and a vector norm $\| \ \|$ on $\mathbb{R}^m$, the *induced matrix norm* of a matrix $A \in \mathbb{R}^{m \times n}$ is defined as

$$\|A\| = \max_{\|x\|=1} \|Ax\|,$$

which is exactly the *operator norm* of A as a linear operator from the normed space $(\mathbb{R}^n, \| \ \|)$ into the normed space $(\mathbb{R}^m, \| \ \|)$. It is easy to see that the induced matrix norm satisfies the inequality

$$\|Ax\| \leq \|A\| \|x\|, \quad \forall x \in \mathbb{R}^n,$$

the properties (i) to (iii) above for a vector norm, and an additional property that for any $A \in \mathbb{R}^{m \times n}$ and $B \in \mathbb{R}^{n \times k}$,

$$\|AB\| \leq \|A\| \|B\|.$$

Thus, any vector norms on $\mathbb{R}^n$ and $\mathbb{R}^m$ induce a matrix norm on $\mathbb{R}^{m \times n}$ that satisfies $\|I\| = 1$. Norms on $\mathbb{C}^n$ and the induced matrix norms of complex matrices can be defined similarly.

There are several standard and useful norms on $\mathbb{R}^n$. Let $p \in [1, \infty)$. Then the *p-norm* of a vector $x \in \mathbb{R}^n$ is defined as

$$\|x\|_p = \left(\sum_{i=1}^{n} |x_i|^p \right)^{\frac{1}{p}}.$$

When $p = \infty$, the ∞-*norm* on $\mathbb{R}^n$ is defined by

$$\|x\|_\infty = \max\{|x_1|, \cdots, |x_n|\}.$$

It is not difficult to show that $\lim_{p \to \infty} \|x\|_p = \|x\|_\infty$, $\forall\, x \in \mathbb{R}^n$.

The induced matrix p-norms of $A \in \mathbb{R}^{m \times n}$ for $p = 1, 2, \infty$ can respectively be calculated by the formulas

$$\|A\|_1 = \max_{1 \leq j \leq n} \sum_{i=1}^{m} |a_{ij}|,$$

$$\|A\|_2 = \sqrt{r(A^T A)},$$

$$\|A\|_\infty = \max_{1 \leq i \leq m} \sum_{j=1}^{n} |a_{ij}|.$$

Suppose that λ is an eigenvalue of A and x is a corresponding eigenvector. Let $\|\ \|$ be any norm on $\mathbb{C}^n$. Then $|\lambda| \|x\| = \|\lambda x\| = \|Ax\| \leq \|A\| \|x\|$, so $|\lambda| \leq \|A\|$, which implies the inequality

$$r(A) \leq \|A\|.$$

On the other hand, the next result indicates that the spectral radius of A is just the greatest lower bound of all the norms of A.

Theorem A.13. *Let $A \in \mathbb{C}^{n \times n}$. Then, given any $\epsilon > 0$, there is a norm on $\mathbb{C}^n$ such that the induced matrix norm satisfies*

$$\|A\| \leq r(A) + \epsilon.$$

The above theorem implies the following important result.

Theorem A.14. *Suppose that $A \in \mathbb{C}^{n \times n}$. Then $\lim_{k \to \infty} A^k = 0$ if and only if $r(A) < 1$.*

The previous theorem implies a useful formula for $r(A)$.

Corollary A.5. *Let* $\| \ \|$ *be any matrix norm on* $\mathbb{C}^{n \times n}$. *Then*

$$r(A) = \lim_{k \to \infty} \|A^k\|^{\frac{1}{k}}.$$

The same condition for the spectral radius as in the previous theorem is used in the following fundamental perturbation result for square matrices, which can be extended to infinite dimensional bounded linear operators.

Theorem A.15. (Banach's lemma) *Let* $A \in \mathbb{C}^{n \times n}$ *be such that* $r(A) < 1$. *Then* $(I - A)^{-1}$ *exists and*

$$(I - A)^{-1} = \sum_{k=0}^{\infty} A^k.$$

Moreover, whenever $\|A\| < 1$ *for a matrix norm* $\| \ \|$,

$$\|(I - A)^{-1}\| \le \sum_{k=0}^{\infty} \|A\|^k = \frac{1}{1 - \|A\|}.$$

Banach's lemma implies the following general perturbation result for invertible matrices.

Corollary A.6. *Let* $A \in \mathbb{C}^{n \times n}$ *be nonsingular, and let* $E \in \mathbb{C}^{n \times n}$ *be such that* $\|EA^{-1}\| < 1$. *Then* $A + E$ *is also nonsingular, and furthermore,*

$$\|(A + E)^{-1}\| \le \frac{\|A^{-1}\|}{1 - \|EA^{-1}\|}.$$

Another consequence of Banach's lemma is the following Laurent series expansion of the resolvent $R(\lambda, A)$ of A at infinity.

Corollary A.7. *Let* $A \in \mathbb{C}^{n \times n}$. *Then*

$$R(\lambda, A) = \sum_{k=0}^{\infty} \frac{A^k}{\lambda^{k+1}}, \quad \forall\, \lambda \in \mathbb{C}, \ |\lambda| > r(A). \tag{A.7}$$

Appendix B

Basic Operator Theory

A natural extension of matrix theory is the theory for infinite dimensional linear operators defined on vector spaces. Of course, because of fundamental differences between finite dimensional vector spaces and infinite dimensional ones, for example, the closed unit ball of an infinite dimensional normed vector space is not compact, the theory of linear operators is more complicated. As a prerequisite for the study of the general theory of positive operators in the book, we review the basic operator theory. Most theorems presented here are usually covered in standard textbooks of functional analysis and so they are only stated without proof. More detailed coverage of linear operators can be found in the books [Dunford and Schwartz (1957); Goldberg (1966); Conway (1990)]. The reader with a good knowledge of functional analysis may skip this appendix and come back to review the concepts and theorems when needed.

In Section B.1 we introduce the concept of Banach spaces and their fundamental properties. Basic facts of Hilbert spaces will be briefly touched in Section B.2. Section B.3 is devoted to the general properties of bounded linear operators between normed spaces. In Section B.4 we study bounded linear functionals and dual operators. The spectral theory and ergodic theory of bounded linear operators based on the Dunford integral of linear operators are more advanced and specialized, so they will be treated in Sections 6.1 and 6.3 respectively as preliminaries for studying those topics for positive operators.

B.1   Banach Spaces

Let X be a vector space over $\mathbb{R}$ or $\mathbb{C}$. A *norm* on X is a nonnegative function $\| \ \| : X \to [0, \infty)$ such that

(i) $\|x\| = 0$ if and only if $x = 0$;

(ii) $\|ax\| = |a| \|x\|$ for all $x \in X$ and scalars a;

(iii) $\|x + y\| \leq \|x\| + \|y\|$ for all $x, y \in X$ (*triangle inequality*).

When a norm $\| \ \|$ is defined on a vector space X, then $(X, \| \ \|)$ is called a *normed space*. The norm defines a *distance* on X by

$$d(x, y) = \|x - y\|, \ \forall \ x, \ y \in X,$$

which makes (X, d) a *metric space* whose *topology* is generated by all the *open balls* $B_\epsilon(x) \equiv \{y \in X : d(y, x) < \epsilon\}$ centered at $x \in X$ with radii $\epsilon > 0$. The *closed ball* $\{y \in X : d(y, x) \leq \epsilon\}$ centered at x with radius ϵ is denoted by $\overline{B_\epsilon(x)}$. In particular, the set $\{x \in X : \|x\| \leq 1\}$ is called the *closed unit ball* of X and is denoted by $\overline{B(0)}$. The *distance* of a vector $x \in X$ to a subset A of X is defined to be

$$d(x, A) = \inf_{y \in A} \|x - y\|.$$

If the induced metric space (X, d) is *complete*, namely, if for any *Cauchy sequence* $\{x_n\}$ in X, that is, $\lim_{m,n \to \infty} \|x_m - x_n\| = 0$, there is a vector $x \in X$ such that $\lim_{n \to \infty} \|x_n - x\| = 0$, then the normed space $(X, \| \ \|)$ is called a *Banach space*.

Example B.1. The Euclidean n-space $\mathbb{R}^n$ and the unitary n-space $\mathbb{C}^n$ are Banach spaces with any vector norm.

Example B.2. For any set S, let $B(S)$ be the normed space of all bounded functions with the sup-*norm*

$$\|f\|_\infty = \sup_{s \in S} |f(s)|, \ \forall \ f \in B(S).$$

Then $B(S)$ is a Banach space.

Example B.3. For a compact metric space K, the closed subspace $C(K)$ of $B(K)$ consisting of all the continuous functions defined on K is a Banach space with the max-*norm*

$$\|f\|_\infty = \max_{x \in K} |f(x)|, \ \forall \ f \in C(K).$$

Let X be a normed space. A set $A \subset X$ is said to be *bounded* if there is a constant b such that $\|x\| \leq b$ for all $x \in A$. $A \subset X$ is called *open* in X if for any $x \in A$, there is $\epsilon > 0$ such that $B_\epsilon(x) \subset A$ and *closed* in X if its *complement* $A^c \equiv \{x \in X : x \notin A\}$ in X is open. Closed balls are closed sets of X while open balls are open sets. A point $x \in X$ is in the *boundary*

∂A of $A \subset X$ if any open ball $B_\epsilon(x)$ contains points of both A and A^c. The set $\overline{A} \equiv A \cup \partial A$ is called the *closure* of A in X.

A subset A of X is said to be *pre-compact* in X if each sequence in A contains a convergent subsequence. If in addition A is also closed, then A is said to be *compact* in X. A pre-compact subset of X must be bounded. A subset of a finite dimensional normed space is pre-compact if and only if it is bounded. A subset $A \subset C(K)$ is said to be *equi-continuous* if to every $\epsilon > 0$ and every $x \in K$ there corresponds an open set G_x containing x such that for all $y \in G_x$,

$$|f(y) - f(x)| < \epsilon, \ \forall f \in A.$$

The Ascoli-Arzelá theorem [Dunford and Schwartz (1957)] says that for a compact space K, a set $A \subset C(K)$ is compact in $C(K)$ if and only if it is closed, bounded, and equi-continuous.

Let M be a closed subspace of a normed space X. Two elements x and y of X are said to be *equivalent* if $x - y \in M$. The set X/M of all *equivalent classes* $[x] = x + M$ is a vector space under the following operations of vector additions and scalar multiplications:

$$[x] + [y] = [x + y], \quad a[x] = [ax].$$

X/M is called the *quotient space* with respect to M, and it is a normed space under the *quotient norm* defined by

$$\|[x]\| = d(x, M), \ \forall [x] \in X/M.$$

If X is a Banach space, then so is X/M.

Let $\{x_n\}$ be a sequence of vectors in a normed space X. The series $\sum_{n=1}^\infty x_n$ is said to *converge* to $x \in X$ if $\lim_{n\to\infty} \sum_{k=1}^n x_k = x$, and is *absolutely convergent* if $\sum_{n=1}^\infty \|x_n\| < \infty$.

Proposition B.1. *Let X be a Banach space. If the series $\sum_{n=1}^\infty x_n$ is absolutely convergent, then it is convergent.*

Since most concrete vector spaces in the book are function spaces and related bounded linear operators involve Lebesgue integrals, we briefly introduce the concept of Lebesgue integration and list the three fundamental convergence theorems and some other important results for Lebesgue integrals that will be used later.

Definition B.1. Let X be a set. A collection Σ of subsets of X is called a *σ-algebra* if $\emptyset \in \Sigma$ and
  (i) $A \in \Sigma$ implies $A^c \in \Sigma$;

(ii) $A_n \in \Sigma$, $n = 1, 2, \cdots$, imply $\bigcup_{n=1}^{\infty} A_n \in \Sigma$.

Σ is called an *algebra* if countable additivity (ii) is replaced by finite additivity, and a *σ-ring* if only (i) and (ii) are satisfied.

Every set A in a σ-algebra Σ is called a *measurable set*. A *measure* μ on (X, Σ) is a *countably additive* set function $\mu : \Sigma \to [0, \infty]$, namely $\mu(A)$ is a nonnegative number or infinity for all $A \in \Sigma$, and if $A_n \in \Sigma$, $n \geq 1$ are mutually disjoint, then

$$\mu \left(\bigcup_{n=1}^{\infty} A_n \right) = \sum_{n=1}^{\infty} \mu(A_n).$$

Here we have used the convention that $a + \infty = \infty$ for any real number a. If a set function μ is real- (or complex-) valued and is countably additive, then it is called a *real* (or *complex*) measure. Real or complex measures are sometimes called *signed measures*.

If Σ is a σ-algebra, then (X, Σ) is called a *measurable space*. (X, Σ, μ) is called a *measure space* if μ is a measure on (X, Σ). A measure space (X, Σ, μ) is called *finite* if $\mu(X) < \infty$. In particular, if $\mu(X) = 1$, then (X, Σ, μ) is called a *probability measure space*. (X, Σ, μ) is said to be *σ-finite* if there is a sequence $\{A_n\}$ in Σ such that $X = \bigcup_{n=1}^{\infty} A_n$ and $\mu(A_n) < \infty$ for each n. If a property is satisfied by all points of a measure space outside a subset of measure zero, then we say that this property is satisfied *almost everywhere* and it is denoted as a.e.

Example B.4. A simple example of measure spaces is the *counting measure space* (X, Σ, μ) in which X is any set, the σ-algebra Σ is the family of all subsets of X, and $\mu(A)$ is the number (including ∞) of elements in A. Another example is a finite *sample space* $X = \{x_1, \cdots, x_n\}$ in which the measure is defined by ascribing to each element $x_k \in X$ a nonnegative number p_k such that $\sum_{k=1}^{n} p_k = 1$.

Example B.5. Let X be a locally compact Hausdorff space, let $\mathcal{B} \equiv \mathcal{B}(X)$ be the *Borel σ-algebra* which is the smallest σ-algebra containing all the open subsets of X, and let μ be a measure defined on $\mathcal{B}$. Then the triple $(X, \mathcal{B}, \mu)$ is called a *Borel measure space*. In particular, if $X = \mathbb{R}^N$, then there exists a unique Borel measure m on the Borel σ-algebra $\mathcal{B}(\mathbb{R}^N)$ such that $\mu \left(\prod_{k=1}^{N} [a_k, b_k] \right) = \prod_{k=1}^{N} (b_k - a_k)$. This σ-finite measure m is called the *Lebesgue measure* on $\mathbb{R}^N$.

Definition B.2. Let (X, Σ) be a measurable space. A real-valued function on X is called *measurable* if the inverse image $f^{-1}([a, \infty)) \in \Sigma$, $\forall a \in \mathbb{R}$. A complex-valued function on X is *measurable* if its real part and imaginary part are measurable.

We define the Lebesgue integral of measurable functions. Let (X, Σ, μ) be a measure space. A measurable function s on X is called *simple* if it has only finitely many distinct values.

Let s be a nonnegative simple function of the form

$$s = \sum_{k=1}^{n} a_k \chi_{A_k},$$

where $a_1, \cdots, a_k$ are the distinct values of s and χ_{A_k} are the characteristic functions of A_k. Then the *Lebesgue integral* of s over X is defined to be

$$\int_X s d\mu = \sum_{k=1}^{n} a_k \mu(A_k),$$

where $0 \cdot \infty = 0$ is used. If f is a nonnegative measurable function, then the *Lebesgue integral* of f over X is defined as

$$\int_X f d\mu = \sup \left\{ \int_X s d\mu : \ 0 \le s \le f, \ s \text{ are simple functions} \right\}.$$

For a real-valued function f on X and $x \in X$, let

$$f^+(x) = \max\{f(x), 0\} \quad \text{and} \quad f^-(x) = \max\{-f(x), 0\}.$$

f^+ and f^- are called the *positive part* and *negative part* of f, respectively. Obviously, $|f| = f^+ + f^-$ and $f = f^+ - f^-$.

Definition B.3. Let f be a real-valued measurable function on X. Then the *Lebesgue integral* of f over X is defined to be

$$\int_X f d\mu = \int_X f^+ d\mu - \int_X f^- d\mu,$$

provided that at least one of $\int_X f^+ d\mu$ and $\int_X f^- d\mu$ is finite. If $\int_X f d\mu$ is finite, f is said to be *Lebesgue integrable*.

Definition B.4. Let $f = u + iv$ be a complex-valued measurable function on X. If $\int_X |f| d\mu < \infty$, then the function f is said to be *Lebesgue integrable* with respect to μ, and the *Lebesgue integral* of f over X is defined to be

$$\int_X f d\mu = \int_X u d\mu + i \int_X v d\mu.$$

The *Lebesgue integral* of f over a set $A \in \Sigma$ is defined to be

$$\int_A f d\mu = \int_X f \chi_A d\mu$$

provided the latter is well-defined.

We list three fundamental theorems of Lebesgue integration.

Theorem B.1. (Lebesgue's dominated convergence theorem) *Let $\{f_n\}$ be a sequence of measurable functions on X such that*

$$\lim_{n \to \infty} f_n(x) = f(x), \quad \forall\, x \in X \text{ a.e.}$$

If there is an integrable function g which satisfies the inequalities

$$|f_n(x)| \le g(x) \ \forall\, x \in X \text{ a.e.}, \quad n = 1, 2, \cdots,$$

then f and all f_n are integrable. Furthermore,

$$\lim_{n \to \infty} \int_X |f_n - f| d\mu = 0 \quad and \quad \lim_{n \to \infty} \int_X f_n d\mu = \int_X f d\mu.$$

Theorem B.2. (Lebesgue's monotone convergence theorem) *Let $\{f_n\}$ be a sequence of measurable functions on X such that*
  (i) $0 \le f_1(x) \le f_2(x) \le \cdots$ *for $x \in X$ a.e.;*
  (ii) $\lim_{n \to \infty} f_n(x) = f(x)$ *for $x \in X$ a.e.*
Then f is a measurable function and

$$\lim_{n \to \infty} \int_X f_n d\mu = \int_X f d\mu.$$

Theorem B.3. (Fatou's lemma) *Let $\{f_n\}$ be a sequence of nonnegative measurable functions defined on X. Then*

$$\int_X (\liminf_{n \to \infty} f_n) d\mu \le \liminf_{n \to \infty} \int_X f_n d\mu.$$

Definition B.5. Let μ be a measure on a σ-algebra Σ, and let ν be a measure or a signed measure on Σ. We say that ν is *absolutely continuous* with respect to μ, and write

$$\nu \ll \mu,$$

if $\nu(A) = 0$ whenever $A \in \Sigma$ with $\mu(A) = 0$. If μ and ν are both measures which are absolutely continuous with respect to each other, then they are said to be *equivalent*, written $\mu \cong \nu$.

The following important result of Radon-Nikodym characterizes absolutely continuous real or complex measures.

Theorem B.4. (The Radon-Nikodym theorem) *Let (X, Σ, μ) be a σ-finite measure space and let ν be a real (or complex) measure which is absolutely continuous with respect to the measure μ, then there exists a μ-integrable function $f : X \to \mathbb{R}$ (or $\mathbb{C}$) such that*

$$\nu(A) = \int_A f d\mu, \quad \forall\, A \in \Sigma.$$

The function f in the above theorem is called the *Radon-Nikodym derivative* of ν with respect to μ, written as $f = d\nu/d\mu$.

We now introduce L^p-spaces, the most important function spaces in analysis and applications.

Definition B.6. Let p be a real number with $1 \leq p < \infty$. The family of all measurable functions $f : X \rightarrow \mathbb{R}$ (or $\mathbb{C}$) satisfying

$$\int_X |f|^p d\mu < \infty$$

is denoted by $L^p(X, \Sigma, \mu)$. The space $L^\infty(X, \Sigma, \mu)$ is defined as the family of all bounded μ-a.e. real (or complex)-valued measurable functions on X. Here two functions f_1, $f_2 \in L^p(X, \Sigma, \mu)$ are considered the same if $f_1(x) = f_2(x)$, $x \in X$ μ-a.e. for $p \in [1, \infty]$.

We shall frequently write $L^p(X), L^p(\mu), L^1(\Sigma)$, or simply L^p instead of $L^p(X, \Sigma, \mu)$ if the measure space is understood from the context. L^1 is exactly the space of all μ-integrable functions, and for all real-valued functions f, $g \in L^1$ such that $f \geq g$,

$$\|f\|_1 = \|f - g\|_1 + \|g\|_1.$$

The number $\|f\|_p = \left(\int_X |f|^p d\mu\right)^{1/p}$ is called the L^p-*norm* of $f \in L^p$ for $p < \infty$ and the number $\|g\|_\infty = \text{ess sup}_{x \in X} |g(x)|$ is referred to as the L^∞-*norm* of $g \in L^\infty$. The L^1-norm $\|f\|_1$ of $f \in L^1$ is sometimes written as $\|f\|_\mu$ to emphasize the involved measure μ. These norms satisfy the three axioms for the definition of a norm:

   (i) $\|f\|_p = 0$ if and only if $f = 0$;

   (ii) $\|af\|_p = |a|\|f\|_p$ for all $f \in L^p$ and scalars a;

   (iii) $\|f + g\|_p \leq \|f\|_p + \|g\|_p$, $\forall f, g \in L^p$ (*Minkowski's inequality*).

Under the usual algebraic operations and the above norm, $L^p(X, \Sigma, \mu)$ is a Banach space for any $1 \leq p \leq \infty$.

B.2   Hilbert Spaces

A special class of Banach spaces are Hilbert spaces which generalize the notions of Euclidean spaces and unitary spaces. Let X be a complex vector space. An *inner product* on X is a mapping $(\cdot, \cdot) : X \times X \rightarrow \mathbb{C}$ such that

   (i) $(x, y) = \overline{(y, x)}$ for all x, $y \in X$;

   (ii) $(x + y, z) = (x, z) + (y, z)$ for all x, y, $z \in X$;

   (iii) $(ax, y) = a(x, y)$ for all x, $y \in X$ and $a \in \mathbb{C}$;

(iv) $(x, x) \geq 0$, $\forall\, x \in X$, and $(x, x) = 0$ if and only if $x = 0$.

An inner product on a real vector space can be defined similarly, in which (x, y) is a real number and the equality $(x, y) = (y, x)$ is valid for all vectors x, $y \in X$. When an inner product $(\cdot, \cdot)$ is defined on a vector space X, it is called an *inner product space*.

A fundamental inequality for an inner product space is

Theorem B.5. (The Cauchy-Schwarz inequality) *Let X be an inner product space. Then for any $x, y \in X$,*

$$|(x, y)|^2 \leq (x, x)(y, y).$$

Equality holds if and only if x and y are linearly dependent.

Corollary B.1. *Let X be an inner product space. Then*

$$\|x\| = \sqrt{(x, x)}, \ \ \forall\, x \in X$$

defines a norm on X.

The norm defined in Corollary B.1 is called the norm *induced* by the inner product. An inner product space will always be considered as a normed space with the induced norm. If $(X, \|\ \|)$ is a Banach space, then X is called a *Hilbert space*.

Corollary B.2. *The inner product $(\cdot, \cdot)$ on an inner product space X is a continuous function on the product space $X \times X$.*

Example B.6. $\mathbb{R}^n$ and $\mathbb{C}^n$ are Hilbert spaces with the inner product $(x, y) = y^T x$ on $\mathbb{R}^n$ and $(z, w) = w^H z$ on $\mathbb{C}^n$. Every finite dimensional inner product space is a Hilbert space.

Example B.7. The space $L^2(X, \Sigma, \mu)$ with the inner product

$$(f, g) = \int_X f\bar{g}\,d\mu, \ \ \forall\, f, g \in L^2(X)$$

is a Hilbert space. Its induced norm is exactly the L^2-norm.

Two vectors x and y in an inner product space X are called *orthogonal*, written $x \perp y$, if $(x, y) = 0$. A vector $x \in X$ is *orthogonal to* a set $A \subset X$, denoted $x \perp A$, if $x \perp y$ for all $y \in A$. The *orthogonal complement* $A^\perp$ of A is the set of all vectors in X orthogonal to A. Clearly $A^\perp$ is a closed subspace of X. A family of vectors in X is called *orthogonal* if distinct elements in the family are orthogonal, and *orthonormal* if in addition each member of the family has norm 1. For example, the sequence $\{\exp(i2\pi nt)\}_{n=-\infty}^{\infty}$ is orthonormal in the complex Hilbert space $L^2(0,1)$.

One of the most important properties of a Hilbert space is its *orthogonal decomposition* into a closed subspace and its orthogonal complement, as the following theorem shows.

Theorem B.6. *Let X be a Hilbert space and let M be a closed subspace of X. Then any $x \in X$ can be written uniquely as $u + v$, where $u \in M$ and $v \in M^{\perp}$. Furthermore,*

$$\|x - u\| = d(x, M) < \|x - z\|, \quad \forall\, z \in M, \ z \neq u.$$

The correspondence $x \to u$ in the above theorem defines an *orthogonal projection* $P_M : X \to X$ from X onto M. A class of vectors $\{v_s\}_{s \in \Lambda}$ is called a *basis* of a normed space X if each $x \in X$ is the limit of a sequence of linear combinations of vectors in $\{v_s\}_{s \in \Lambda}$. Given an orthonormal basis for M, there is a *Fourier series representation* of $P_M x$ for any $x \in X$.

Theorem B.7. *Let X be a Hilbert space and let $\{u_1, u_2, \cdots\}$ be an orthonormal basis of a closed subspace M. Then for each $x \in X$, the orthogonal projection $P_M x$ of x onto M is given by*

$$P_M x = \sum_k (x, u_k) u_k.$$

Furthermore, we have

$$\sum_k |(x, u_k)|^2 = \|P_M x\|^2 \quad \text{(Parseval's equality)}$$

and

$$\sum_k |(x, u_k)|^2 \leq \|x\|^2 \quad \text{(Bessel's inequality)}.$$

B.3 Bounded Linear Operators

Let X and Y be vector spaces over the same real or complex field. A transformation $T : X \to Y$ is called a *linear operator* from X into Y if the following two conditions are satisfied:

(i) $T(x + y) = Tx + Ty$ for all $x,\, y \in X$.

(ii) $T(ax) = aTx$ for all $x \in X$ and scalars a.

The vector subspace

$$N(T) = \{x \in X : Tx = 0\}$$

of X is called the *null space* of T, and the vector subspace

$$R(T) = \{Tx : x \in X\}$$

of Y is called the *range* of T.

A linear operator $T : X \to Y$ from a normed space X into a normed space Y over the same scalar field as X is said to be *bounded* if there exists a constant b such that

$$\|Tx\| \le b\|x\|, \ \ \forall \, x \in X.$$

For a bounded linear operator $T : X \to Y$, the number

$$\inf_{\|x\|\le 1} \|Tx\| = \inf_{\|x\|=1} \|Tx\|$$

is called the *norm* of T and is denoted by $\|T\|$. The identity operator $I : X \to X$ is a bounded linear operator with norm 1.

Theorem B.8. *Let X and Y be normed spaces. A linear operator $T : X \to Y$ is continuous if and only if T is bounded.*

Remark B.1. A linear operator from a normed space into a normed space is continuous everywhere if it is continuous at 0.

Example B.8. Define a linear operator $T : C[0,1] \to C[0,1]$ by

$$(Tf)(x) = \int_0^1 w(x,y)f(y)dy, \ \ \forall \, f \in C[0,1],$$

where w is a real-valued function which is continuous on the square $[0,1] \times [0,1]$. Then T is bounded. In fact

$$\|T\| = \max_{x\in[0,1]} \int_0^1 |w(x,y)|dy.$$

Let $B(X,Y)$ denote the set of all bounded linear operators from a normed space X into a normed space Y. When $X = Y$, we simply write $B(X,X)$ as $B(X)$. For any $T, S \in B(X,Y)$ and any scalar a, let $T+S$ and aT be defined naturally as

$$(T+S)x = Tx + Sx, \ (aT)x = aTx, \ \forall \, x \in X.$$

Then $B(X,Y)$ is a normed space under the operator norm.

Theorem B.9. *Let X be a normed space. Then $B(X,Y)$ is a Banach space if and only if Y is a Banach space.*

In the special situation that Y is the field $\mathbb{R}$ or $\mathbb{C}$ of scalars, which is itself a one dimensional Banach space, linear operators from a normed space X into Y are called *linear functionals*.

Definition B.7. The *dual space* X' of a normed space X is the Banach space of all bounded linear functionals on X.

When a transformation $T : X \to Y$ is one-to-one, then its *inverse* $T^{-1} :$ $R(T) \subset Y \to X$ is well-defined and satisfies the two equalities $T^{-1}Tx = x$ for all $x \in X$ and $TT^{-1}y = y$ for all $y \in R(T)$. If T is linear, so is T^{-1}. If T is one-to-one and onto, then T is said to be *invertible*.

Theorem B.10. *Suppose that $T : X \to Y$ is a linear operator from a normed space X into a normed space Y. Then the inverse operator T^{-1} of T exists and is continuous on its domain $R(T)$ if and only if there exists a number $\delta > 0$ such that*

$$\|Tx\| \geq \delta\|x\|, \ \ \forall \, x \in X.$$

There are several fundamental theorems in the theory of linear operators. They are *Banach's inverse mapping theorem, Banach's lemma* which generalizes Theorem A.15, the *closed graph theorem*, and the *principle of uniform boundedness*.

Theorem B.11. (Banach's inverse mapping theorem) *Let X and Y be Banach spaces. If $T \in B(X,Y)$ is invertible, then $T^{-1} : Y \to X$ is bounded, in other words, $T^{-1} \in B(Y,X)$.*

Theorem B.12. (Banach's lemma) *Let X be a Banach space and let $T \in B(X)$. If $\|T\| < 1$, then $I - T$ is invertible and $(I - T)^{-1} \in B(X)$ has the power series expression*

$$(I - T)^{-1} = \sum_{n=0}^{\infty} T^n$$

which converges in the operator norm of $B(X)$. Furthermore,

$$\|(I - T)^{-1}\| \leq \frac{1}{1 - \|T\|}.$$

Definition B.8. Let X and Y be normed spaces, let D be a subspace of X, and let $T : D \to Y$ be a linear operator. The *graph* $G(T)$ of T is the set

$$\{(x, Tx) : x \in D\} \subset X \times Y.$$

If $G(T)$ is closed in the normed space $X \times Y$ with the norm $\|(x, y)\| = \max\{\|x\|, \|y\|\}$, then T is called a *closed operator*. If T has a *linear extension* $\hat{T} : \hat{D} \subset X \to Y$, that is, $\hat{D} \supset D$ is a subspace of X and $\hat{T}x = Tx$ for all $x \in D$, which is a closed operator, then T is said to be *closable*.

Proposition B.2. *Let X and Y be normed spaces, let D be a subspace of X, and let $T : D \to Y$ be a linear operator. Then*

(i) *T is closed if and only if for any sequence $\{x_n\}$ in $D, x_n \to x$ and $Tx_n \to y$ imply that $x \in D$ and $Tx = y$;*

(ii) *if T is one-to-one and closed, then so is T^{-1};*

(iii) *if T is a closed operator, then $N(T)$ is a closed subspace;*

(iv) *if D is closed and T is continuous, then T is closed.*

The continuity of T does not necessarily imply the closedness of T when $D \neq X$, and the closedness of T also does not necessarily imply the continuity of T. However, we have

Theorem B.13. (The closed graph theorem) *Suppose that X and Y are Banach spaces. Let $T : X \to Y$ be a closed linear operator. Then T is a continuous linear operator.*

The following result is a consequence of the closed graph theorem, which is called the *principle of uniform boundedness*.

Theorem B.14. (Banach-Steinhaus) *Let X and Y be Banach spaces and let $\mathcal{T}$ be a family of bounded linear operators from X into Y. If $\sup_{T \in \mathcal{T}} \|Tx\| < \infty$ for every $x \in X$, then*

$$\sup_{T \in \mathcal{T}} \|T\| < \infty.$$

Corollary B.3. *Let X be a Banach space and let $\mathcal{F}$ be a subset of its dual space X' with the condition*

$$\sup_{f \in \mathcal{F}} |f(x)| < \infty, \quad \forall x \in X.$$

Then $\mathcal{F}$ is a bounded set in X'.

Definition B.9. Suppose that X is a normed space and M is a subspace of X. A bounded linear operator $P : X \to X$ is called a *projection* from X onto M if $P^2 = P$ with $R(P) = M$.

Again the closed graph theorem can be applied to show that when M is a closed subspace of a Banach space X, there is a one-to-one correspondence between the set of projections from X onto M and the set of certain closed subspaces of X. If M and N are subspaces of X such that $X = M + N$ and $M \cap N = \{0\}$, then X is said to be a *direct sum* of M and N, written $X = M \oplus N$, and N is an *algebraic complement* of M in X. If in addition M and N are both closed subspaces of X, we say that M and N are *topological complements* to each other in X.

Theorem B.15. *Let M be a closed subspace of a Banach space X. There exists a projection from X onto M if and only if there can be found a closed subspace N of X such that $X = M \oplus N$. In this case, there exist two numbers $c_1, c_2 > 0$ such that*

$$\|y + z\| \geq c_1 \|y\| \text{ and } \|y + z\| \geq c_2 \|z\|, \ \forall \, y \in M, \, z \in N.$$

B.4 Bounded Linear Functionals and Dual Operators

The dual space X' of X consists of all bounded linear functionals on X. In this section we present more properties of X'.

The value $\xi(x)$ of a functional $\xi \in X'$ at $x \in X$ is sometimes denoted by $\langle x, \xi \rangle$, and it is easily seen that

$$|\langle x, \xi \rangle| \leq \|x\| \|\xi\|, \ \forall \, x \in X, \, \xi \in X'.$$

Definition B.10. Suppose that X and Y are two normed spaces over the same field. An operator $T : X \to Y$ is called an *isometry* if $\|Tx\| = \|x\|$ for all $x \in X$. An isometry which is a linear operator is called a *linear isometry*. The normed spaces X and Y are said to be *equivalent* if there exists a linear isometry from X *onto* Y.

Since every linear functional on $\mathbb{C}^n$ can be written as

$$\xi^T z = \sum_{k=1}^{n} \xi_k z_k, \ \forall \, z \in \mathbb{C}^n$$

for a unique vector $\xi \in \mathbb{C}^n$, it is easy to see that the dual space of $\mathbb{C}^n$ is equivalent to itself. In general, the dual space of a finite dimensional normed space X is equivalent to itself. The following theorem characterizes the dual space of an L^p space.

Theorem B.16. *Let (X, Σ, μ) be a measure space and let $1 \leq p < \infty$. The dual space of the space $L^p(X, \Sigma, \mu)$ is equivalent to the space $L^q(X, \Sigma, \mu)$,* *where* $1/p + 1/q = 1$ *for* $p > 1$ *and* $q = \infty$ *for* $p = 1$. *The dual operation for* $f \in L^p(X)$ *and* $g \in L^q(X)$ *is*

$$\langle f, g \rangle = \int_X f g d\mu,$$

which satisfies the Cauchy-Hölder inequality

$$|\langle f, g \rangle| \leq \|f\|_p \|g\|_q, \quad \forall f \in L^p, \, g \in L^q.$$

Remark B.2. It should be noted that the dual space of $L^\infty(X)$ is not equivalent to $L^1(X)$ in general.

One of the most fundamental theorems in functional analysis is the following *Hahn-Banach extension theorem*. A real-valued functional p on a vector space X is called a *semi-norm* if

$$p(x + y) \leq p(x) + p(y), \quad p(ax) = |a|p(x)$$

for all $x, y \in X$ and scalars a. A semi-norm must be nonnegative.

Theorem B.17. (Hahn-Banach) *Suppose that* X *is a vector space. Let* M *be a subspace of* X *and let* p *be a semi-norm on* X. *If* ξ *is a linear functional on* M *such that*

$$|\xi(x)| \leq p(x), \quad \forall x \in M,$$

then ξ *can be extended to a linear functional* $\hat{\xi}$ *on* X *such that*

$$|\hat{\xi}(x)| \leq p(x), \quad \forall x \in X.$$

Theorem B.17 has several useful consequences.

Corollary B.4. *Let* M *be a subspace of a normed space* X *and let* $\xi \in M'$. *Then there exists* $\hat{\xi} \in X'$ *such that* $\hat{\xi}(x) = \xi(x)$ *for all* $x \in M$ *and furthermore,* $\|\hat{\xi}\| = \|\xi\|$.

Corollary B.5. *Let* M *be a subspace of a normed space* X *and let* $x \in X$ *with* $d(x, M) > 0$. *Then there exists* $\xi \in X'$ *such that*

$$\|\xi\| = 1, \quad \xi(M) = \{0\}, \quad \xi(x) = d(x, M).$$

Corollary B.6. *Let* X *be a normed space and let* $x \in X$. *There exists* $\xi \in X'$ *such that* $\|\xi\| = 1$ *and* $\xi(x) = \|x\|$.

Corollary B.7. *Let* X *be a normed space and let* $x \in X$. *Then*

$$\|x\| = \sup_{\xi \in X', \|\xi\| = 1} |\xi(x)|.$$

Another consequence of the above Hahn-Banach extension theorem is that every finite dimensional subspace of any Banach space must have a topological complement.

Corollary B.8. *There is always a projection from a normed space X onto any one of its finite dimensional subspaces.*

$x \in X$ and $\xi \in X'$ are said to be *orthogonal* to one another if $\langle x, \xi \rangle = 0$. A set $A \subset X$ is said to be *orthogonal* to a set $F \subset X'$ if $\langle x, \xi \rangle = 0$ for all $x \in A$ and $\xi \in F$. The *orthogonal complement* of A in X', denoted by $A^\perp$, is the set of all elements in X' which are orthogonal to A. The *orthogonal complement* of F in X, denoted by $^\perp F$, is the set of all elements in X which are orthogonal to F. $A^\perp$ and $^\perp F$ are closed subspaces of X' and X, respectively. Also $A^\perp = \overline{A}^\perp$ and $^\perp F = {}^\perp \overline{F}$.

Theorem B.18. *Let M be a subspace of a normed space X.*

(i) *The quotient space $X'/M^\perp$ is equivalent to the dual space M' under the linear isometry*

$$U : X'/M^\perp \to M'$$

defined by

$$U[\xi] = \xi|_M, \ \ \forall \, [\xi] \in X'/M^\perp.$$

(ii) *If M is closed, then the dual space $(X/M)'$ is equivalent to the orthogonal complement $M^\perp$ under the linear isometry*

$$V : (X/M)' \to M^\perp$$

defined by

$$(V\xi)(x) = \xi([x]), \ \ \forall \, \xi \in (X/M)'.$$

Every $x \in X$ defines a bounded linear functional on X' via

$$x(\xi) = \xi(x), \ \ \forall \, \xi \in X',$$

since $|x(\xi)| = |\xi(x)| \le \|\xi\| \|x\| = \|x\| \|\xi\|$. Corollary B.7 ensures that the above correspondence $J_X : X \to X'' \equiv (X')'$ is a linear isometry. If $R(J_X) = X''$, then X is called *reflexive*.

Proposition B.3. *Suppose that M is a subspace of X. Then $^\perp(M^\perp) = \overline{M}$. If N is a subspace of X', then $(^\perp N)^\perp \subset \overline{N}$, and the inclusion becomes an equality if X is a reflexive space.*

A reflexive normed space is necessarily a Banach space. Finite dimensional normed spaces are reflexive, as well as Hilbert spaces from the following Riesz representation theorem.

Theorem B.19. (The Riesz representation theorem) *Let X be a Hilbert space. For $\xi \in X'$ there is a unique $y \in H$ such that*

$$\xi(x) = (x, y), \ \forall\, x \in X.$$

Moreover, we have the equality $\|\xi\| = \|y\|$.

No matter whether a Banach space is reflexive or not, the principle of uniform boundedness implies the following result, which can be obtained by applying Corollary B.3 to X'':

Proposition B.4. *If A is a subset of a normed space X with*

$$\sup_{x \in A} |\xi(x)| < \infty, \ \forall\, \xi \in X',$$

then A is bounded.

Theorem B.16 implies that the Banach space $L^p(X)$ is reflexive for $1 < p < \infty$. However, $L^1(X)$, the most important function space in this textbook, is a non-reflexive Banach space as Remark B.2 indicates. Proposition B.5 that follows implies that $L^\infty(X)$ is also non-reflexive.

Proposition B.5. *Let X be a Banach space. Then*
(i) *X is reflexive if and only if X' is reflexive;*
(ii) *if X is reflexive, so is any closed subspace A of X.*

We now introduce the concept of weak convergence, which is a key mathematical tool in the study of positive operators.

Definition B.11. A sequence $\{x_n\}$ in a normed space X is said to *converge weakly* to $x \in X$ if $\lim_{n \to \infty} \xi(x_n) = \xi(x)$, $\forall\, \xi \in X'$, written as $x_n \overset{w}{\to} x$. A subset A of X is called *weakly pre-compact* if every sequence in A contains a weakly convergent subsequence, and *weakly compact* if in addition all such limits belong to A.

In particular, for $1 \le p < \infty$, a sequence of functions $\{f_n\} \subset L^p$ is weakly convergent to $f \in L^p$ if

$$\lim_{n \to \infty} \langle f_n, g \rangle = \langle f, g \rangle, \ \forall\, g \in L^q,$$

where $p^{-1} + q^{-1} = 1$, and *weekly Cesáro convergent* to $f \in L^p$ if

$$\lim_{n \to \infty} \frac{1}{n} \sum_{k=0}^{n-1} \langle f_k, g \rangle = \langle f, g \rangle, \quad \forall \, g \in L^q.$$

A weakly convergent sequence cannot have two weak limits. Since $|\xi(x) - \xi(y)| \leq \|\xi\| \|x - y\|$ for $\xi \in X'$, if $x_n \to x$ in norm, then $x_n \xrightarrow{w} x$. Thus, if a set is pre-compact, then it is weakly pre-compact. The converse is not true unless the space is finite dimensional, as the example $\{\sin nx\}_{n=1}^{\infty} \subset L^2(0, 1)$ shows.

A subset A of a normed space X is called *fundamental* if the closure $\overline{\text{span}} A$ of span A is the same as X.

Proposition B.6. *Let X be a Banach space. Then a sequence $\{x_n\}$ in X is weakly convergent if and only if*

(i) *$\{x_n\}$ is bounded in X;*

(ii) *there is a vector $x \in X$ such that $\lim_{n \to \infty} \xi(x_n) = \xi(x)$ for all ξ in a fundamental set A of X'.*

Proposition B.7. *Let X be a Banach space and suppose that $\{x_n\}$ is a sequence in X such that $x_n \xrightarrow{w} x \in X$. Then*

(i) *$x \in \overline{\text{span}} \{x_n\}$;*

(ii) *$\|x\| \leq \liminf_{n \to \infty} \|x_n\|$.*

A normed space X is *separable* if it has a countable fundamental set. One of the most attractive properties of reflexive spaces is given by the next theorem [Dunford and Schwartz (1957)].

Theorem B.20. (Banach-Alaoglu) *The closed unit ball of a separable reflexive Banach space is weakly compact.*

Since a Hilbert space is reflexive due to the Riesz representation theorem, a consequence of Theorem B.20 is

Corollary B.9. *Every closed and bounded set in a separable Hilbert space is weakly compact.*

$L^p(\Omega)$ is separable and reflexive when Ω is a bounded region of $\mathbb{R}^N$ and $1 < p < \infty$, so Theorem B.20 also implies

Corollary B.10. *Let $1 < p < \infty$. If Ω is a bounded region of $\mathbb{R}^N$, then closed bounded subsets of $L^p(\Omega)$ are weakly compact.*

The concept of weak compactness for L^1 spaces plays a key role in studying the asymptotic behavior of the iterates of Markov operators and Frobenius-Perron operators in the book. We give several useful criteria [Dunford and Schwartz (1957); Lasota and Mackey (1994)] of weak compactness of subsets in $L^1(X)$ as follows.

Proposition B.8. *Suppose that (X, Σ, μ) is a measure space and let $g \in L^1(X)$. Then the set*

$$\mathcal{F} = \{f \in L^1(X) : |f(x)| \leq |g(x)|, \ x \in X \text{ a.e.}\}$$

is weakly compact in $L^1(X)$. In particular, if $\mu(X) < \infty$, then any bounded subset of $L^\infty(X)$ is weakly pre-compact in $L^1(X)$.

Proposition B.9. *Suppose that $\mu(X) < \infty$ and $p > 1$. Then every bounded subset of $L^p(X)$ is weakly pre-compact in $L^1(X)$.*

When Ω is a bounded region of $\mathbb{R}^N$, the following result (see Theorem IV.8.9 of [Dunford and Schwartz (1957)]) characterizes the weak compactness of any subset of $L^1(\Omega)$.

Theorem B.21. *Let Ω be a bounded region of $\mathbb{R}^N$. Then $\mathcal{F} \subset L^1(\Omega)$ is weakly pre-compact if and only if*
  (i) $\mathcal{F}$ is bounded;
  (ii) the integrals $\int_A f \, dm$ are uniformly countably additive in the sense that for any $\epsilon > 0$, there is $\delta > 0$ such that

$$\left| \int_A f \, dm \right| < \epsilon, \ \forall f \in \mathcal{F}, \ A \subset \Omega, \ m(A) < \delta.$$

Definition B.12. Let X be a normed space. A sequence $\{\xi_n\}$ in X' is said to *converge to $\xi \in X'$ under the w'-topology* if $\xi_n(x) \to \xi(x)$ for every $x \in X$. This is written as $\xi_n \overset{w'}{\to} \xi$.

By Alaoglu's theorem [Dunford and Schwartz (1957)], the closed unit ball of X' is compact under the w'-topology. Thus, every bounded sequence $\{\xi_n\}$ of X' has a subsequence $\{\xi_{n_k}\}$ such that $\xi_{n_k} \overset{w'}{\to} \xi$ for some $\xi \in X'$.

Definition B.13. Let X and Y be two normed spaces and let $T \in B(X, Y)$. The *dual operator* of T is the linear operator $T' : Y' \to X'$ defined by

$$(T'\xi)(x) = \xi(Tx), \ \forall \xi \in Y', \ x \in X.$$

The dual operator is well-defined since if $\xi \in Y'$, as the composition of ξ and $T \in B(X, Y)$, we have $\xi \circ T \in X'$.

Theorem B.22. $T' : Y' \to X'$ *is continuous and* $\|T'\| = \|T\|$.

Theorem B.19 shows how a Hilbert space can be identified with its dual space, which motivates the definition of the adjoint operator of a bounded linear operator from a Hilbert space to another Hilbert space. Its following definition is meaningful because of the Riesz representation theorem.

Definition B.14. Let X and Y be Hilbert spaces and let $T \in B(X, Y)$. The operator $T^* : Y \to X$ defined by

$$(Tx, y) = (x, T^*y), \quad \forall \, x \in X, \, y \in Y$$

is called the *adjoint operator* of T.

Similar to Theorem B.22, $T^* \in B(Y, X)$ and $\|T^*\| = \|T\|$.

Like a matrix, a bounded linear operator T from a normed space X into a normed space Y gives rise to four important subspaces: the null space of T, the range of T, the null space of T', and the range of T'. The following gives their relations.

Theorem B.23. *Let* $T \in B(X, Y)$. *Then*
  (i) $R(T)^{\perp} = N(T')$;
  (ii) $\overline{R(T)} = {}^{\perp}N(T')$;
  (iii) ${}^{\perp}R(T') = N(T)$;
  (iv) $\overline{R(T')} \subset N(T)^{\perp}$.

Corollary B.11. *Let* X *and* Y *be Banach spaces. If* $T \in B(X, Y)$ *is invertible, then its dual operator* $T' \in B(Y', X')$ *is also invertible. Furthermore,*

$$(T')^{-1} = (T^{-1})'.$$

Theorem B.24. *Let* $T \in B(X, Y)$. *Then* $R(T') = X'$ *if and only if* T *has a bounded inverse operator defined on* $R(T)$.

Corollary B.12. *Let* X *be a normed space and let* Y *be a Banach space. Suppose that* $T \in B(X, Y)$. *If* $R(T) = Y$, *then* T' *has a bounded inverse.*

Bibliography

Albert, R. and Barabasi, A.-L. (2002). Statistical mechanics of complex networks, *Rev. Modern Phys.*, **74**, pp. 47–97.

Albeverio, S. and Hoegh-Krohn, R. (1978). Frobenius-theory for positive maps of von Neumann algebras, *Comm. Math. Phys.*, **64**, pp. 83–94.

Aliprantis, C. D. and Burkinshaw, O. (1985). *Positive Operators* (Academic Press).

Baladi, V. (2000). *Positive Transfer Operators and Decay of Correlations* (World Scientific).

Bapat, R. B. and Raghavan, T. E. S. (1997). *Nonnegative Matrices and Applications* (Cambridge University Press).

Barnsley, M. (1988). *Fractals Everywhere* (Academic Press).

Beck, C. and Schlögl, F. (1993). *Thermodynamics of Chaotic Systems* (Cambridge University Press).

Berman, A., Neumann, M. and Stern, R. J. (1989). *Nonnegative Matrices in Dynamical Systems* (John Wiley & Sons).

Berman, A. and Plemmons, R. J. (1979). *Nonnegative Matrices in the Mathematical Sciences* (Academic Press).

Bernstein, D. S. (2005). *Matrix Mathematics* (Princeton University Press).

Billingsley, P. (1968). *Convergence of Probability Measures* (Wiley).

Birkhoff, G. (1938). Dependent probabilities and the space (l), *Proc. Nat. Acad. Sci. U.S.A.*, **24**, pp. 154–159.

Birkhoff, G. (1967). *Lattice Theory* (Amer. Math. Soc., third edition).

Bose, C. and Murray, R. (2001). The exact rate of approximation in Ulam's method, *Disc. Cont. Dynam. Sys.*, **7**, 1, pp. 219–235.

Bowen, R. (1975). *Equilibrium States and the Ergodic Theory of Anosov Diffeomorphisms* (Springer-Verlag).

Boyarsky, A. and Góra, P. (1997). *Laws of Chaos: Invariant Measures and Dynamical Systems in One Dimension* (Birkhäuser).

Bryan, K. and Leise, T. (2006). The $25,000,000,000 eigenvector: the linear algebra behind Google, *SIAM Review*, **48**, pp. 569–581.

Campbell, S. L. and Meyer, C. D. (1979). *Generalized Inverses of Linear Transformations* (Pitman).

Chiu, C., Du, Q. and Li, T.-Y. (1992). Error estimates of the Markov finite approximation of the Frobenius-Perron operator, *Nonlinear Anal., TMA*, **19**, 4, pp. 291–308.

Choe, G. H. (2005). *Computational Ergodic Theory* (Springer-Verlag).

Chung, K. L. (1967). *Markov Chains with Stationary Transition Probabilities* (Springer-Verlag, second edition).

Ciarlet, P. G. (1978). *The Finite Element Method for Elliptic Problems* (North-Holland).

Conway, J. (1990). *A Course in Functional Analysis* (Springer-Verlag, second edition).

Cornfeld, I. P., Fomin, S. V. and Sinai, Y. G. (1982). *Ergodic Theory* (Springer-Verlag).

Dellnitz, M. and Junge, O. (1999). On the approximation of complicated dynamical behavior, *SIAM J. Numer. Anal.*, **36**, 2, pp. 491–515.

Deuflhard, P. and Schütte, C. (2004). Molecular conformation dynamics and computational drug design, *Applied Mathematics Entering the 21st Century*, pp. 91–119.

Deuflhard, P. and Weber, M. (2003). Robust Perron cluster analysis in conformation dynamics, *ZIB-Report 03-19*, pp. 91–119.

DeVore, R. A. (1972). *The Approximation of Continuous Functions by Positive Linear Operators* (Springer-Verlag).

Ding, J. (1990). *Finite Approximations of a Class of Frobenius-Perron Operators*, Ph.D. thesis, Michigan State University, East Lansing, MI, USA.

Ding, J. (1996). Computing invariant measures of piecewise convex transformations, *J. Stat. Phys.*, **83**, 3/4, pp. 623–635.

Ding, J. (1997). Decomposition theorems for Koopman operators, *Nonlin. Anal. Th. Meth. Appl.*, **28**, 6, pp. 1011–1018.

Ding, J. (1998). The point spectrum of Frobenius-Perron and Koopman operators, *Proc. Amer. Math. Sco.*, **126**, 5, pp. 1355–1361.

Ding, J., Eifler, L. and Rhee, N. (2007). Integral and nonnegativity preserving approximations of functions, *J. Math. Analy. Appl.*, **325**, 2, pp. 889–897.

Ding, J. and Fay, T. (2005). The Perron-Frobenius theorem and limits in geometry, *Amer. Math. Monthly*, **112**, 2, pp. 171–175.

Ding, J., Hitt, R. and Zhang, X.-M. (2003a). Markov chains and dynamical geometry of polygons, *Linear Alg. Its Appl.*, **367**, pp. 255–270.

Ding, J. and Hornor, W. E. (1994). A new approach to the Frobenius-Perron operator, *J. Math. Analy. Appl.*, **187**, 3, pp. 1047–1058.

Ding, J. and Li, T.-Y. (1991). Markov finite approximation of Frobenius-Perron operator, *Nonlinear Anal., Th. Meth. Appl.*, **17**, 8, pp. 759–772.

Ding, J. and Li, T.-Y. (1998). A convergence rate analysis for Markov finite approximations to a class of Frobenius-Perron operators, *Nonlinear Anal., Th. Meth. Appl.*, **31**, 5/6, pp. 765–777.

Ding, J. and Li, Z. (2009). On the dimension of Sierpiński pedal triangles, *Fractals*, **17**, 1, pp. 39–43.

Ding, J., Mao, D. and Zhou, A. (2003b). A Monte Carlo approach to piecewise linear Markov approximations of Markov operators, *Monte Carlo Methods*

Appl., **9**, 4, pp. 295–306.

Ding, J., Pye, W. and Zhao, L. (2006). Some results on structured *M*-matrices with an application to wireless communications, *Linear Alg. Appl.*, **416**, 2/3, pp. 608–614.

Ding, J. and Rhee, N. (2004). Approximations of Frobenius-Perron operators via interpolation, *Nonlinear Anal., Th. Meth. Appl.*, **57**, 5/6, pp. 831–842.

Ding, J. and Rhee, N. (2006). A modified piecewise lienar Markov approximation of Markov operators, *Appied Math. Comput.*, **174**, 1, pp. 235–251.

Ding, J. and Ye, N. (2006). Piecewise linear integral-preserving approximations of functions, *Inter. J. Math. Edu. Sci. Tech.*, **37**, 3, pp. 369–375.

Ding, J. and Zhou, A. (1995). Piecewise linear Markov approximations of Frobenius-Perron operators associated with multi-dimensional transformations, *Nonlinear Anal., Th. Meth. Appl.*, **25**, 4, pp. 399–408.

Ding, J. and Zhou, A. (1996). Finite approximations of Frobenius-Perron operators – a solution of Ulam's conjecture to multi-dimensional transformations, *Physica D*, **92**, pp. 61–68.

Ding, J. and Zhou, A. (2001). Constructive approximations of Markov operators on $L^1([0,1]^N)$, *J. Stat. Phys.*, **105**, 5/6, pp. 863–878.

Ding, J. and Zhou, A. (2002). Structure preserving finite element approximations of Markov operators, *Nonlinearity*, **15**, pp. 923–936.

Ding, J. and Zhou, A. (2008). Characteristic polynomials of some perturbed matrices, *Applied Math. Comput.*, **199**, 2, pp. 631–636.

Ding, J. and Zhou, A. (2009). *Statistical Properties of Deterministic Systems* (Tsinghua University Press and Springer-Verlag).

Dorogovstev, S. N. and Mendes, J. F. F. (2002). Evolution of networks, *Adv. in Phys.*, **51**, pp. 1079–1187.

Dunford, N. and Schwartz, L. (1957). *Linear Operators, Part I* (Interscience).

Evans, D. E. and Hoegh-Krohn, R. (1978). Spectral properties of positive maps on C^*-algebras, *J. London Math. Soc.*, **17**, 2, pp. 345–355.

Fannes, M., Nachtergaele, B. and Werner, R. F. (1992). Finitely correlated states on quantum spin chains, *Comm. Math. Phys.*, **144**, pp. 443–490.

Foguel, S. R. (1969). *The Ergodic Theory of Markov Processes* (Van Nostrand Reinhold).

Fréchet, M. (1933). Comportement asymptotique des solutions d'un systéme d'équations linéaires et homogénes aux différences finies du premier ordre á coefficients constants, *Publ. Fac. Sci. Univ. Masaryk*, **178**, pp. 1–24.

Frobenius, G. (1912). Über matrizen aus nicht negativen elementen, *S.-B. K. Preuss. Akad. Wiss. Berlin*, pp. 456–477.

Froyland, G. (1995). Finite approximations of Sinai-Bowen-Ruelle measures for Anosov systems in two dimensions, *Random Comp. Dynam.*, **3**, 4, pp. 251–264.

Giusti, E. (1984). *Minimal Surfaces and Functions of Bounded Variation* (Birkhäuser).

Goldberg, S. (1966). *Unbounded Linear Operators* (McGraw-Hill).

Góra, P. and Boyarsky, A. (1989). Absolutely continuous invariant measures for piecewise expanding C^2 transformations in $\mathbb{R}^N$, *Israel J. Math.*, **67**, 3, pp.

272–286.

Hitt, L. and Zhang, X. (2001). Dynamic geometry of polygons, *Elemente der Math.*, **56**, 1, pp. 21–37.

Horn, R. A. and Johnson, C. R. (1985). *Matrix Analysis* (Cambridge University Press).

Jabłoński, M. (1983). On invariant measures for piecewise C^2-transformations of the n-dimensional cube, *Ann. Polon. Math.*, **XLIII**, pp. 185–195.

Jenkinson, O. and Pollicott, M. (2004). Entropy, exponents and invariant measures for hyperbolic systems: dependence and computation, *Modern Dynamical Systems and Application*, , pp. 365–384.

Kakutani, S. (1941a). Concrete representation of abstract (L)-spaces and the mean ergodic theorem, *Ann. Math.*, **42**, pp. 523–537.

Kakutani, S. (1941b). Concrete representation of abstract (M)-spaces, *Ann. Math.*, **42**, pp. 994–1024.

Kantorovich, L. V. (1937). Linear operators in semi-ordered spaces, *Dokl. Akad. Nauk SSSR*, **14**, pp. 531–537.

Kantorovich, L. V., Vulikh, B. Z. and Pinsker, A. G. (1950). *Functional Analysis in Partially Ordered Spaces* (Gosudarstv. Izdat. Tehn. - Teor. Lit.).

Keane, M., Murrary, R. and Young, L.-S. (1998). Computing invariant measures for expanding circle maps, *Nonlinearity*, **11**, 1, pp. 27–46.

Keller, G. (1982). Stochastic stability in some chaotic dynamical systems, *Mona. Math.*, **94**, pp. 313–333.

Kingston, J. and Synge, J. (1988). The sequence of pedal triangles, *Amer. Math. Monthly*, **95**, 7, pp. 609–620.

Komornik, J. and Lasota, A. (1987). Asymptotic decomposition of Markov operators, *Bull. Polon. Acad. Sci. Math.*, **35**, pp. 321–327.

Koopman, B. O. (1931). Hamiltonian systems and transformations in Hilbert spaces, *Proc. Nat. Acad. Sci. USA*, **17**, pp. 315–318.

Krengel, U. (1985). *Ergodic Theorems* (Walter de Gruyter).

Langville, A. N. and Meyer, C. D. (2006). *Google's PageRank and Beyond: the Science of Search Engine Rankings* (Princeton University Press).

Lasota, A., Li, T.-Y. and Yorke, J. A. (1984). Asymptotic periodicity of the iterates of Markov operators, *Trans. Amer. Math. Soc.*, **286**, pp. 751–764.

Lasota, A. and Mackey, M. C. (1994). *Chaos, Fractals, and Noise: Stochastic Aspects of Dynamics* (Springer-Verlag).

Lasota, A. and Yorke, J. A. (1973). On the existence of invariant measures for piecewise monotonic transformations, *Trans. Amer. Math. Soc.*, **186**, pp. 481–488.

Lasota, A. and Yorke, J. A. (1982). Exact dynamical systems and the Frobenius-Perron operator, *Trans. Amer. Math. Soc.*, **273**, pp. 375–384.

Lax, P. (1990). The ergodic character of sequence of pedal triangles, *Amer. Math. Monthly*, **97**, pp. 377–381.

Lewin, M. (1971/72). On exponents of primitive matrices, *Numer. Math.*, **18**, pp. 154–161.

Li, T.-Y. (1976). Finite approximation for the Frobenius-Perron operator, a solution to Ulam's conjecture, *J. Approx. Th.*, **17**, pp. 177–186.

Li, T.-Y. and Yorke, J. A. (1978). Ergodic transformations from an interval into itself, *Trans. Amer. Math. Soc.*, **235**, pp. 183–192.

Lin, M. (1971). Mixing for Markov operators, *Z. Wahrscheinlichkeitstheorie Verw. Gebiete*, **19**, pp. 231–242.

Liverani, C. (2001). Rigorous numerical investigation of the statistical properties of piecewise expanding maps. a feasibility study, *Nonlinearity*, **14**, pp. 463–490.

MacCluer, C. R. (2000). The many proofs and applications of Perron's theorem, *SIAM Review*, **42**, 3, pp. 487–498.

Mandelbrot, B. (1982). *The Fractal Geometry of Nature* (Freeman).

Mann, W. R. (1953). Mean value methods in iteration, *Proc. Amer. Math. Soc.*, **4**, pp. 506–510.

Marcus, M. and Minc, H. (1962). Some results on doubly stochastic matrices, *Proc. Amer. Math. Soc.*, **76**, pp. 571–579.

Marcus, M., Minc, H. and Moyls, B. (1961). Some results on nonnegative matrices, *J. Res. Nat. Bur. Standards Sect. B*, **65**, pp. 205–209.

Marcus, M. and Newman, M. (1959). On the minimum of the permanent of a doubly stochastic matrix, *Duke Math. J.*, **26**, pp. 61–72.

Meyn, S. P. and Tweedie, R. L. (1993). *Markov Chains and Stochastic Stability* (Springer-Verlag).

Miller, W. M. (1994). Stability and approximation of invariant measures for a class of nonexpanding transformations, *Nonlin. Anal. Th. Meth. Appl.*, **23**, 8, pp. 1013–1025.

Miller, W. M. (1997). Discrete approximation of invariant measures for multidimensional maps, *DIMACS Ser. Disc. Math. Theoret. Compt. Sci.*, **34**, pp. 29–46.

Milnor, J. W. (1965). *Topology from the Differentiable Viewpoint* (The University of Virginia Press).

Minc, H. (1978). *Permanants* (Addison-Wesley).

Minc, H. (1988). *Nonnegative Matrices* (John Wiley & Sons).

Muirhead, R. F. (1903). Some methods applicable to identities and inequalities of symmetric algebraic functions of n letters, *Proc. Edinburgh Math. Soc.*, **21**, pp. 144–157.

Murray, R. (1998). Approximation error for invariant density calculations, *Disc. Cont. Dynam. Sys.*, **4**, 3, pp. 535–557.

Natanson, I. P. (1961). *Theory of Functions of a Real Variable* (F. Ungar).

Newman, M. E. J. (2003). The structure and function of complex networks, *SIAM Review*, **45**, 2, pp. 167–256.

Ortega, J. M. and Rheinboldt, W. C. (1970). *Iterative Solution of Nonlinear Equations in Several Variables* (Academic Press).

Perron, O. (1907). Zur theorie der matrizen, *Math. Ann.*, **64**, pp. 248–263.

Powell, M. J. D. (1981). *Approximation Theory and Methods* (Cambridge University Press).

Rohlin, V. A. (1964). Exact endomorphisms of Lebesgue spaces, *Amer. Math. Soc. Transl.*, **2**, pp. 1–36.

Roydon, H. L. (1968). *Real Analysis* (Macmillan).

Rudin, W. (1986). *Real and Complex Analysis* (McGraw-Hill).

Ruelle, D. (1968). Statistical mechanics of a one-dimensional lattice gas, *Commun. Math. Phys.*, **9**, pp. 267–278.

Ruelle, D. (1978). *Thermodynamic Formalism* (Addison-Wesley).

Schaefer, H. H. (1974). *Banach Lattices and Positive Operators* (Springer-Verlag).

Schaefer, H. H. (1980). On positive contractions in L^p spaces, *Trans. Amer. Math. Soc.*, **257**, pp. 261–268.

Schur, I. (1923). Über eine klasse von mittelbildungen mit anwendung auf die determinantentheorie, *Sber. Berliner Math. Ges.*, **22**, pp. 9–20.

Schütte, C. (1999). *Conformational Dynamics: Modeling, Theory, Algorithm, and Application to Biomoleculaes* (Freie Universität, Berlin, Germany), Habilitation Thesis.

Setti, G., Mazzini, G., Rovatti, R. and Callegari, S. (2002). Statistical modeling of discrete-time chaotic processes - basic finite-dimensional tools and applications, *Proc. IEEE*, **90**, pp. 662–690.

Terhesiu, D. and Froyland, G. (2008). Rigorous numerical approximation of Ruelle-Perron-Frobenius operators and topological pressure of expanding maps, *Nonlinearity*, **21**, 9, pp. 1953–1966.

Tolstov, G. P. (1976). *Fourier Series* (Dover).

Ulam, S. M. (1960). *A Collection of Mathematical Problems* (Interscience).

Ulam, S. M. and von Neumann, J. (1947). On combination of stochastic and deterministic processes, *Bull. Amer. Math. Soc.*, **53**, p. 1120.

Varga, R. (2000). *Matrix Iterative Analysis* (Springer-Verlag, secondn edition).

Walters, P. (1982). *An Introduction to Ergodic Theory* (Springer-Verlag).

Wielandt, H. (1950). Unzerlegbare nicht-negative matrizen, *Math. Z.*, **52**, pp. 642–648.

Wong, R. and Zigarovich, M. (2007). Tennis with Markov, *College Math. J.*, **38**, pp. 53–55.

Wong, S. (1978). Some metric properties of piecewise monotonic mappings of the unit interval, *Trans. Amer. Math. Soc.*, **246**, pp. 493–500.

Ye, N. and Ding, J. (2006). On the convergence of a linear recursive sequence with nonnegative coefficients, *Inter. J. Math. Edu. Sci. Tech.*, **37**, 8, pp. 998–1000.

Zaanen, A. C. (1997). *Introduction to Operator Theory in Riesz Spaces* (Springer-Verlag).

Zhang, X.-M. (1998). Optimization of Schur-convex functions, *Math. Inequ. Appl.*, **3**, pp. 319–330.

Zhang, X.-M., Hitt, R., Wang, B. and Ding, J. (2008). Sierpiński pedal triangles, *Fractals*, **2**, pp. 141–150.

Zhao, L., Mark, J. W. and Ding, J. (2006). Power distribution/allocation in multirate wideband CDMA systems, *IEEE Trans. Wireless Commun.*, **5**, 9, pp. 2458–2467.

Ziemer, W. P. (1989). *Weakly Differentiable Functions* (Springer-Verlag).

Index